AF352794

Lattice-Preferred Orientation and Microstructures of Minerals and Their Implications for Seismic Anisotropy

Lattice-Preferred Orientation and Microstructures of Minerals and Their Implications for Seismic Anisotropy

Editors

Haemyeong Jung
Munjae Park

MDPI • Basel • Beijing • Wuhan • Barcelona • Belgrade • Manchester • Tokyo • Cluj • Tianjin

Editors
Haemyeong Jung
Seoul National University
Korea

Munjae Park
Chungbuk National University
Korea

Editorial Office
MDPI
St. Alban-Anlage 66
4052 Basel, Switzerland

This is a reprint of articles from the Special Issue published online in the open access journal *Minerals* (ISSN 2075-163X) (available at: https://www.mdpi.com/journal/minerals/special_issues/LA).

For citation purposes, cite each article independently as indicated on the article page online and as indicated below:

LastName, A.A.; LastName, B.B.; LastName, C.C. Article Title. *Journal Name* **Year**, *Volume Number*, Page Range.

ISBN 978-3-0365-2642-3 (Hbk)
ISBN 978-3-0365-2643-0 (PDF)

Cover image courtesy of Haemyeong Jung

Contents

About the Editors

Haemyeong Jung is professor of Tectonophysics Laboratory at the School of Earth and Environmental Sciences, Seoul National University, South Korea. He is one of the leading scientists in the world in the field of study on lattice-preferred orientation and microstructures of minerals, and their effects on seismic anisotropy. He received his Ph.D. from the University of Minnesota, U.S.A. After working at Yale University for one year, he then served as a post-doc at the University of California for three years. His main research areas are rock deformation, deformation microstructures, and lattice-preferred orientation of minerals, the brittle/ductile behavior of minerals in the crust and mantle, and the resultant seismic anisotropy and seismic properties.

Munjae Park is assistant professor at the Department of Earth and Environmental Sciences, Chungbuk National University, South Korea. He received his Ph.D. in Structural Geology (specialization in microtectonics) at Seoul National University, South Korea. He has worked as a post-doc at Seoul National University and Korea University. Following this, he joined the Department of Earth and Environmental Sciences, Chungbuk National University. His research focuses on the evolution of deformation microstructures and fluid inclusions in the upper mantle.

Editorial

Editorial for Special Issue "Lattice-Preferred Orientation and Microstructures of Minerals and Their Implications for Seismic Anisotropy"

Haemyeong Jung [1],* and Munjae Park [2],*

[1] Tectonophysics Laboratory, School of Earth and Environmental Sciences, Seoul National University, Seoul 08826, Korea

[2] Department of Earth and Environmental Sciences, Chungbuk National University, Cheongju 28644, Korea

* Correspondence: hjung@snu.ac.kr (H.J.); mpark@chungbuk.ac.kr (M.P.)

Citation: Jung, H.; Park, M. Editorial for Special Issue "Lattice-Preferred Orientation and Microstructures of Minerals and Their Implications for Seismic Anisotropy". *Minerals* **2021**, *11*, 1152. https://doi.org/10.3390/min11111152

Received: 8 October 2021

Accepted: 15 October 2021

Published: 20 October 2021

Publisher's Note: MDPI stays neutral with regard to jurisdictional claims in published maps and institutional affiliations.

The lattice-preferred orientation (LPO) of minerals is important for interpreting seismic anisotropy [1–5], which occurs in the Earth's crust and mantle [6,7], and for understanding the internal structure of the deep interior of the Earth [6–8]. The characterization of microstructures, including LPO, grain size, grain shape, and misorientation, is important to determine the deformation conditions, deformation histories, kinematics, and seismic anisotropies in the crust and mantle [1,2,9]. LPOs develop in minerals in the crust, mantle, and subduction zones [1,2], where the rocks are deformed under high pressure and temperature [1–5].

This Special Issue [10–18] contains contributions pertaining to the LPOs of various minerals, including olivine, pyroxene, garnet, omphacite, muscovite, lawsonite, epidote, glaucophane, and natural ice in different settings in the crust, mantle, and subduction zones.

The articles published in this issue are divided according to the different rock types occurring in the mantle, subduction zones, and crust. The articles by Park et al. [10], Liu et al. [11], and S. Jung et al. [12] are case studies that investigate the microstructures, LPOs, and seismic properties of mantle peridotites.

Park et al. [10] have reported the deformation microstructures and LPOs of olivine and pyroxenes with respect to the petrogenesis of upper mantle xenoliths beneath the Baekdusan volcano, North Korea. Based on the petrographic features and deformation microstructures, they identified two textural categories for the peridotites: coarse and fine-grained harzburgites (CG and FG Hzb). They found that mineral composition, equilibrium temperature, LPO of olivine, stress, and extraction depth varied considerably with texture. Accordingly, they suggested that the A-type LPO of olivine in CG Hzb samples may be related to the pre-existing Archean cratonic mantle fabric (i.e., old frozen LPO) formed under anhydrous, high-temperature, and low-stress conditions. Conversely, they suggested that the D-type LPO of olivine in FG Hzb samples likely originated from later localized deformation events under low-temperature, high-stress, and anhydrous conditions after a high degree of partial melting.

Liu et al. [11] have carried out a comprehensive petro-structural and geochemical study to better elucidate the thermo-structural evolution of the Val Malenco peridotites in Italy. Their results showed that the Val Malenco serpentinized peridotites recorded both pre-alpine extension and alpine convergence events. The pre-alpine extension was recorded by microstructural and geochemical features preserved in clinopyroxene and olivine porphyroblasts, including partial melting and re-fertilization and high-temperature (900–1000 °C) deformation, followed by cooling and fluid–rock interactions. The following alpine convergence that occurred in a supra-subduction zone setting was documented by subduction-related prograde metamorphism that was preserved in the coarse-grained antigorite and olivine in the less-strained olivine-rich layers, with later low-temperature

(<350 °C) serpentinization of the fine-grained antigorite in the more strained antigorite-rich layers.

Jung et al. [12] have studied the microstructure of amphibole peridotites from Åheim, Norway, to understand the evolution of the LPO of olivine throughout the Scandian orogeny and its implications for seismic anisotropy in the subduction zone. Detailed microstructural analysis of the Åheim peridotites revealed multiple stages of deformation. The coarse grains showed an A-type LPO of olivine, which can be interpreted as the initial stage of deformation. The spinel-bearing samples showed a combination of B-type and C-type LPOs of olivine, which is considered to represent deformation under hydrous conditions. The recrystallized fine-grained olivine displayed a B-type LPO, which can be interpreted as the final stage of deformation. The microstructure and water content of olivine indicated that dislocation creep under hydrous conditions was the dominant deformation mechanism in olivine with B-type LPO. They found that the occurrence of the B-type LPO of olivine to be vital for trench-parallel seismic anisotropy in the mantle wedge. Moreover, the calculated seismic anisotropy of tremolite showed that tremolite could contribute to trench-parallel seismic anisotropy in the mantle wedge.

Several studies have focused on the LPOs of minerals and microstructures of eclogites and blueschists in subduction zones. Two articles in this issue pertain to eclogites (Lee et al. [13], Cao et al. [14]), while two focus on blueschists (Park et al. [15], Choi et al. [16]).

Lee et al. [13] have investigated retrograde eclogites from Xitieshan, northwestern China, to understand the seismic velocity, anisotropy, and seismic reflectance in the upper part of the subducting slab. In their study, S-type LPO of omphacite was observed in three samples. The LPOs of amphibole and omphacite were found to be similar in most samples. The misorientation angles between amphiboles and their neighboring omphacites were small, with a lack of intracrystalline deformation features in the amphiboles. This indicates that the LPO of amphibole was formed by its topotactic growth during the retrogression of eclogites. The seismic properties of retrograde eclogites and amphibole were similar, indicating that they were strongly affected by the amphibole growth. The contact between serpentinized peridotites and retrograde eclogites showed a high reflection coefficient, indicating that a reflected seismic wave could be easily detected at this boundary.

Cao et al. [14] have investigated the seismic velocity and seismic anisotropies of a unique olivine-rich eclogite from northwestern Flemsøya in the Nordøyane ultra-high pressure domain of the Western Gneiss Region in Norway. Detailed analyses of the seismic properties suggest that seismic anisotropy patterns of the Flem eclogite were largely controlled by the strength of the crystal-preferred orientation (CPO) and characterized by significant destructive effects of the CPO interactions, which together resulted in very weak bulk rock seismic anisotropies (AVp = 1.0–2.5%, max. AVs = 0.6–2.0%). The magnitudes of the seismic anisotropies of the Flem eclogite were similar to those of dry eclogite but markedly lower than those of gabbro, peridotite, hydrous phase-bearing eclogite, and blueschist. The average seismic velocities of the Flem eclogite were significantly affected by the relative volume proportions of omphacite and amphibole. The Vp (8.00–8.33 km/s) and Vs values (4.55–4.72 km/s) of the Flem eclogite were markedly larger than those of the hydrous phase-bearing eclogite, blueschist, and gabbro, but they were lower than those of dry eclogite and peridotite. The seismic features of the Flem eclogite can thus be used to distinguish olivine-rich eclogite from other common rock types (especially gabbro) in the deep continental crust or subduction channel when high-resolution seismic data are available.

Park et al. [15] have conducted deformation experiments on epidote blueschists in simple shear under high pressure (0.9–1.5 GPa) and high temperature (400–500 °C) to understand the LPO and deformation microstructures of blueschists occurring at the top of a subducting slab in a warm subduction zone. They discovered that the LPO of epidote and glaucophane changes with increasing shear strain. At low shear strain ($\gamma \leq 1$), the [001] axes of glaucophane were subparallel to the shear direction, and the (010) poles were subnormally aligned with the shear plane. At high shear strain ($\gamma > 2$), the [001] axes of glaucophane were subparallel to the shear direction, while the [100] axes were aligned subnormally to the shear plane. At a shear strain of $\gamma > 4$, the alignment of the (010) epidote poles altered from being subparallel to becoming subnormal to the shear plane, while the [001] axes were subparallel to the shear direction. Their experiments indicate that the magnitude of shear strain and rheological contrast between component minerals play an important role in the formation of LPOs for glaucophane and epidote.

Choi et al. [16] have investigated the effect of lawsonite twinning on the strength of CPO and seismic anisotropy in the lawsonite blueschists from Alpine Corsica (France) and the Sivrihisar Massif (Turkey). Lawsonite is an important mineral for understanding the seismic anisotropy of subducting oceanic crust due to its large elastic anisotropy and occurrence in cold subduction zones. The study concluded that twinned lawsonite could induce substantial seismic anisotropy reduction, particularly for the maximum S-wave anisotropy in lawsonite and whole rocks by up to 3.67% and 1.46%, respectively. This article was selected as the cover issue of *Minerals* on 11 April 2021.

One article focuses on the studies of LPOs and microstructures in crustal rocks (Han et al. [17]). Muscovite is a major constituent mineral in the continental crust that exhibits markedly strong seismic anisotropy. In this article, deformation microstructures of muscovite-quartz phyllites from the Geumseongri Formation in Gunsan, South Korea, were studied to investigate the relationship between muscovite and chlorite fabrics in strongly deformed rocks and the seismic anisotropy observed in the continental crust. Their results indicate that the modal composition and alignment of muscovite and chlorite significantly affect the magnitude and symmetry of seismic anisotropy. It was found that when their [001] axes are aligned in the same direction, the coexistence of muscovite and chlorite constructively contributes to seismic anisotropy.

Kim et al. [18] have reported the microstructures and fabric transitions of natural ice. They studied five ice core samples from the Styx Glacier, northern Victoria Land, Antarctica, and reported CPO changes in ice with depth. They interpreted that the change in CPOs at <140 m was related to a combination of vertical compression and shear on a horizontal plane, while the girdle CPOs at depths >140 m were the result of horizontal extension. Their results imply that, during burial, stress regimes are subject to changes due to external kinematic controls, such as the appearance of a small peak in the bedrock.

The articles in this Special Issue prove that studies of LPO and microstructures of minerals and rocks are a major research area and provide a foundation for interpreting seismic anisotropy in the crust, mantle, and subduction zones. Therefore, the authors hope that this Special Issue encompassing recent advances in the measurement of LPOs of different minerals under various tectonic settings will be a fundamental and valuable resource for the readers and researchers interested in exploring the deformation conditions of minerals and rocks as well as the interpretation of seismic anisotropy in the crust, mantle, and subduction zones.

Funding: This research was funded by the Mid-career Research Program through National Research Foundation of Korea (NRF: 2020R1A2C2003765) to H.J. and the research grant of the Chungbuk National University in 2021 to M.P.

Acknowledgments: The Guest Editors would like to thank all authors, reviewers, and the editorial staff of *Minerals* for their timely efforts to successfully complete this Special Issue.

Conflicts of Interest: The author declares no conflict of interest.

References

1. Almqvist, B.S.G.; Mainprice, D. Seismic properties and anisotropy of the continental crust: Predictions based on mineral texture and rock microstructure. *Rev. Geophys.* **2017**, *55*, 367–433. [CrossRef]
2. Jung, H. Crystal preferred orientations of olivine, orthopyroxene, serpentine, chlorite, and amphibole, and implications for seismic anisotropy in subduction zones: A review. *Geosci. J.* **2017**, *21*, 985–1011. [CrossRef]
3. Karato, S.; Jung, H.; Katayama, I.; Skemer, P. Geodynamic significance of seismic anisotropy of the upper mantle: New insights from laboratory studies. *Annu. Rev. Earth Planet. Sci.* **2008**, *36*, 59–95. [CrossRef]

4.	Mainprice, D. Seismic anisotropy of the deep Earth from a mineral and rock physics perspective. *Treatese Geophys.* **2007**, *2*, 437–491.

5.	Nicolas, A.; Christensen, N.I. Formation of anisotropy in upper mantle Peridotites—A review. In *Composition, Structure and Dynamics of the Lithosphere-Asthenosphere System*; Fuchs, K., Froidevaux, C., Eds.; AGU: Washington, DC, USA, 1987; pp. 111–123.

6.	Long, M.D.; Silver, P.G. The Subduction Zone Flow Field from Seismic Anisotropy: A Global View. *Science* **2008**, *319*, 315. [CrossRef]

7.	Zhao, D.; Yu, S.; Liu, X. Seismic anisotropy tomography: New insight into subduction dynamics. *Gondwana Res.* **2016**, *33*, 24–43. [CrossRef]

8.	Long, M.D.; Becker, T.W. Mantle dynamics and seismic anisotropy. *Earth Planet. Sci. Lett.* **2010**, *297*, 341–354. [CrossRef]

9.	Skemer, P.; Hansen, L.N. Inferring upper-mantle flow from seismic anisotropy: An experimental perspective. *Tectonophysics* **2016**, *668–669*, 1–14. [CrossRef]

10.	Park, M.; Kil, Y.; Jung, H. Evolution of Deformation Fabrics Related to Petrogenesis of Upper Mantle Xenoliths Beneath the Baekdusan Volcano. *Minerals* **2020**, *10*, 831. [CrossRef]

11.	Liu, W.; Cao, Y.; Zhang, J.; Zhang, Y.; Zong, K.; Jin, Z. Thermo-Structural Evolution of the Val Malenco (Italy) Peridotite: A Petrological, Geochemical and Microstructural Study. *Minerals* **2020**, *10*, 962. [CrossRef]

12.	Jung, S.; Jung, H.; Austrheim, H. Microstructural Evolution of Amphibole Peridotites in Åheim, Norway, and the Implications for Seismic Anisotropy in the Mantle Wedge. *Minerals* **2020**, *10*, 345. [CrossRef]

13.	Lee, J.; Jung, H. Lattice-Preferred Orientation and Seismic Anisotropy of Minerals in Retrograded Eclogites from Xitieshan, Northwestern China, and Implications for Seismic Reflectance of Rocks in the Subduction Zone. *Minerals* **2021**, *11*, 380. [CrossRef]

14.	Cao, Y.; Jung, H.; Ma, J. Seismic Properties of a Unique Olivine-Rich Eclogite in the Western Gneiss Region, Norway. *Minerals* **2020**, *10*, 774. [CrossRef]

15.	Park, Y.; Jung, S.; Jung, H. Lattice Preferred Orientation and Deformation Microstructures of Glaucophane and Epidote in Experimentally Deformed Epidote Blueschist at High Pressure. *Minerals* **2020**, *10*, 803. [CrossRef]

16.	Choi, S.; Fabbri, O.; Topuz, G.; Okay, A.I.; Jung, H. Twin Induced Reduction of Seismic Anisotropy in Lawsonite Blueschist. *Minerals* **2021**, *11*, 399. [CrossRef]

17.	Han, S.; Jung, H. Deformation Microstructures of Phyllite in Gunsan, Korea, and Implications for Seismic Anisotropy in Continental Crust. *Minerals* **2021**, *11*, 294. [CrossRef]

18.	Kim, D.; Prior, D.J.; Han, Y.; Qi, C.; Han, H.; Ju, H.T. Microstructures and Fabric Transitions of Natural Ice from the Styx Glacier, Northern Victoria Land, Antarctica. *Minerals* **2020**, *10*, 892. [CrossRef]

Article

Microstructural Evolution of Amphibole Peridotites in Åheim, Norway, and the Implications for Seismic Anisotropy in the Mantle Wedge

Sejin Jung [1], Haemyeong Jung [1,*] and Håkon Austrheim [2]

[1] Tectonophysics Laboratory, School of Earth and Environmental Sciences, Seoul National University, Seoul 08826, Korea; shazabi7@snu.ac.kr

[2] The Njord Centre, Department of Geoscience, University of Oslo, Blindern, N-0316 Oslo, Norway; h.o.austrheim@geo.uio.no

* Correspondence: hjung@snu.ac.kr; Tel.: +82-2-880-6733

Received: 26 March 2020; Accepted: 10 April 2020; Published: 12 April 2020

Abstract: The microstructure of amphibole peridotites from Åheim, Norway were analyzed to understand the evolution of the lattice-preferred orientation (LPO) of olivine throughout the Scandian Orogeny and its implication for the seismic anisotropy of the subduction zone. The Åheim peridotites had a porphyroclastic texture and some samples contained an abundant amount of hydrous minerals such as tremolite. Detailed microstructural analysis on the Åheim peridotites revealed multiple stages of deformation. The coarse grains showed an A-type LPO of olivine, which can be interpreted as the initial stage of deformation. The spinel-bearing samples showed a mixture of B-type and C-type LPOs of olivine, which is considered to represent the deformation under water-rich conditions. The recrystallized fine-grained olivine displays a B-type LPO, which can be interpreted as the final stage of deformation. Microstructures and water content of olivine indicate that the dominant deformation mechanism of olivine showing a B-type LPO is a dislocation creep under water-rich condition. The observation of the B-type LPO of olivine is important for an interpretation of trench-parallel seismic anisotropy in the mantle wedge. The calculated seismic anisotropy of the tremolite showed that tremolite can contribute to the trench-parallel seismic anisotropy in the mantle wedge.

Keywords: microstructural evolution; lattice preferred orientation; olivine in Åheim; amphibole; seismic anisotropy

---

1. Introduction

The deformation behavior of olivine is key to understanding the mantle flow and seismic anisotropy in the upper mantle [1–5]. Many experimental studies concerning the deformation of olivine have reported that it has various types of lattice-preferred orientations (LPOs), depending on the physicochemical conditions during its deformation, and that the different LPOs of olivine may influence the seismic anisotropy of the upper mantle [3,6–11]. For example, the fabric transition of olivine in the mantle wedge from an A-type to a B-type LPO of olivine is proposed as a possible mechanism for the change in the shear wave splitting pattern observed in the subduction zone [3,7,8,12–14]. Many studies have proposed a hypothesis for this change in fabric: a deformation under water-rich conditions [7,8], a deformation under high pressure [10,15,16], an enhancement of grain boundary sliding [17–19], a diffusional creep [20], or the presence of partial melt [21,22]. There have been many reports of such fabric transitions recorded in naturally deformed peridotites from various localities: The Bergen Arc, Norway [23], the Ronda massif, Spain [19], Lien, Almklovdalen, Norway [24], the Navajo volcanic field, USA [25], and the Calatrava volcanic field, Spain [26]. However, the exact mechanism for the fabric transition of olivine in the mantle wedge is still under debate.

In subduction zones, various hydrous minerals such as serpentines, chlorites, and amphiboles can be produced via chemical reactions that take place in the presence of fluid released from the subducting slab [27–30]. Hydrous minerals in and near the subducting slab are elastically highly anisotropic [31–34], and thus very important for understanding seismic anisotropy at a subduction zone. Many studies have suggested that the trench-parallel shear wave anisotropy observed at various subduction zones [35–37] can be induced by the LPO of hydrous minerals such as serpentine [38–40], chlorite [24,33,41], and amphibole [41–44]. Amphibole can be present in the mantle wedge above the subducting slab as a product of the hydration reaction of pyroxene [45,46]. However, studies on the influence of the amphibole fabric on seismic anisotropy in the mantle wedge is still very limited.

In this study, a detailed microstructural analysis of several samples of amphibole peridotite from Åheim, Norway, was performed to understand the evolution of its microstructures during the orogenic event and the subsequent exhumation processes. The LPO of the olivine in the Åheim amphibole peridotites was analyzed to study the mechanism underlying the fabric transition of the olivine in nature and its implication for seismic anisotropy at a subduction zone. The water content was measured using Fourier transform infrared (FTIR) analysis, and the dislocation microstructures in the olivine were observed to identify a possible mechanism leading to its deformation. In addition, the LPO of the amphibole was analyzed to estimate the influence of amphibole fabric on seismic anisotropy in a subduction zone.

2. Geological Setting and Sample Description

The Western Gneiss Region (WGR) in Norway is located between Bergen and Trondheim, with 25,000 km^2 of dominantly gneissic rocks representing the crustal root zone of the Caledonian mountain belt [47,48]. The Caledonian mountain belt was originated during the collision between Laurentia and Baltica [49–51]. The Scandian Orogenic event resulted in a series of high-pressure to ultra-high pressure (HP to UHP) metamorphism events in the WGR. Many orogenic peridotite bodies were emplaced into the crustal rocks of the WGR, which experienced multiple stages of metamorphism and deformation over the course of the Scandian Orogeny and the subsequent uplift [52–59]. During exhumation, many peridotite bodies were infiltrated by fluid and retrograded to amphibole peridotite or chlorite peridotite.

The WGR is predominantly composed of orthogneisses and paragneisses with abundant emplaced peridotite or eclogite bodies (Figure 1). The Proterozoic protolith, dated to 1654 ± 1 Ma, underwent UHP metamorphism and subsequent retrogression associated with the Scandian Orogenic event [60,61]. A continent–continent collision between Baltica and Laurentia from 425–400 Ma resulted in HP–UHP metamorphism at P = 1.8–3.6 GPa and T = 600–800 °C [52,59,62]. Following this collision (400–380 Ma), the WGR was uplifted to a depth of 15–20 km, where it underwent amphibolite facies retrogression at P = 0.5–1.5 GPa and T = 650–850 °C [48,57,63]. The orogenic peridotite bodies within the Nordfjord–Stadlandet UHP domains include garnet lherzolite and dunite trapped during the uplift stage. During the uplifting, these peridotites experienced multiple stages of deformation and associated recrystallization in the granulite facies (ol + opx + cpx + sp) and amphibolite facies (ol + opx + amp + chl) conditions [53–56,64]. The amphibole peridotite samples (426, 429, 443, 445, 446, 447, and 448) were collected from Gusdal quarry in Åheim, Western Norway (Figure 1) for a detailed study of their microstructures.

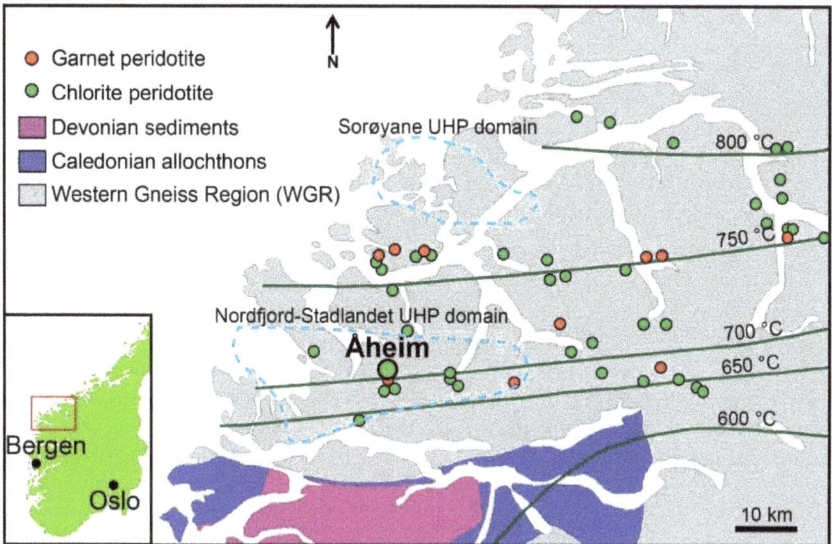

Figure 1. Simplified geological map and the distribution of peridotite bodies in the Western Gneiss Region, Norway (modified after Austrheim [47], Brueckner etl al. [53], Root et al. [63], and Wang et al. [16]). Samples are from Gusdal quarry in Åheim (larger green circle). The green lines indicate the approximate peak metamorphic temperature [57]. Locations of chlorite peridotite and a garnet peridotite body are marked with a green and red circle, respectively. Two UHP domains, Nordfjord-Stadlandet and Sorøyane, are marked by a dotted blue line.

3. Materials and Methods

3.1. The Chemical Composition of Minerals

The chemical compositions of the representative minerals were analyzed using a Shimadzu 1600 electron probe micro-analyzer (EPMA), with an accelerating voltage of 15 kV and a beam size of 1 μm, at the Korea Basic Science Institute in Jeonju. The chemical composition of the minerals was obtained from their cores. No chemical zoning was detected within the samples. The temperature of the spinel peridotite was estimated using an olivine–spinel geothermometer [65] and Al in an orthopyroxene geothermometer [66]. As there was an absence of garnet among the samples, and given the wide range in which amphibole peridotite is stable [27,33], it was difficult to precisely determine the pressure of these samples. However, it can be expected to have been lower than 1.6 GPa, the point at which the spinel–garnet peridotite boundary occurs at 800 °C [67].

3.2. Measurement of LPO and Seismic Anisotropy

The foliation of the samples was determined from the compositional layering of the olivine, amphibole, and chlorite. Lineation was determined by examining the shape-preferred orientation of the elongated olivines in the foliation plane using the projection-function method [68]. To study the LPO of the olivine and tremolite, a thin section was prepared in the x–z plane (x: lineation, z: normal to foliation). An electron backscattered diffraction (EBSD) detector attached to a scanning electron microscope (SEM) (JEOL JSM-6380) housed at the School of Earth and Environmental Sciences (SEES) at Seoul National University (SNU) was used to determine the LPO of each mineral. An HKL system with Channel 5 software was used for the EBSD analysis. The accelerating voltage and working distance in the SEM observation were 20 kV and 15 mm, respectively. To ensure an accurate solution, each EBSD pattern of the individual grains was analyzed manually. The fabric strength of the LPO of the olivine and tremolite was calculated using the M-index [69] and J-index [70].

To estimate the impact that the multiple stages of deformation and presence of hydrous minerals had on the seismic anisotropy, the seismic velocity and anisotropy of the Åheim amphibole peridotites were calculated using the software ANIS2k and VpG [71], on the basis of the LPO data. The ambient condition elastic constants for a single crystal of olivine [72], tremolite [73], and antigorite [74] were used for the crystallographic data. The thickness of the anisotropic layer (T_A) for a given delay time was estimated from the shear wave splitting via the following equation [75]: $T_A = (100 \times \delta \times < V_S >)/AV_S$, where δ is the delay time of the S-wave, $<V_S>$ is the average velocity of the fast and slow shear waves (V_{S1} and V_{S2}), and AV_S is the seismic anisotropy of the S-wave expressed as a percentage.

3.3. Measurement of Water Content in Olivine

The water content of the olivine was measured using FTIR spectroscopy. The FTIR specimens were thinned to a thickness of 100 μm and polished on both sides. Each sample was then heated at a temperature of T = 120 °C for 24 h to eliminate water from the surface and the grain boundary. The FTIR analysis was performed using a Nicolet 6700 FTIR spectrometer with a continuum IR microscope housed at the Tectonophysics laboratory in the SEES at SNU. Unpolarized transmitted light with an aperture size of 50 μm × 50 μm was used to obtain the FTIR spectra. For each sample, FTIR spectra were collected from 10 different olivine grains without any cracks or inclusions, to avoid interference, and averaged. To identify the constituents of the inclusions in the olivine, additional FTIR analyses were performed on the olivines with inclusions. A series of 128 scans were averaged for each spectrum to improve the quality of the spectra at a resolution of 4 cm^{-1}. The water content of the olivine was calculated at the wave numbers in the range 3400–3750 cm^{-1} using the calibration method described by Paterson [76].

3.4. Dislocation Microstructure

An oxygen decoration technique [77–79] was applied to allow observation of the dislocation microstructures in the olivine. The peridotite samples were polished on a single side and were then heated in the oven for 1 h at T = 800 °C. Each sample was polished with colloidal silica after oxidation in order to remove the thin layer of oxide from its surface. The polished samples were then coated with carbon to prevent charging during the SEM observation (JEOL JSM-6380). To observe the dislocation microstructures in the olivine, backscattered electron images (BEI) were taken with an accelerating voltage of 15 kV and at a working distance of 10 mm [77].

4. Results

4.1. Microstructures

The majority of the samples had a porphyroclastic texture and contained primarily olivine (>90%) with minor amounts of amphibole, orthopyroxene, chlorite, biotite, and chromite (Figure 2A–D). However, some samples showed different mineral assemblages and modal compositions. For example, some samples (429 and 445) included layers of Cr-rich spinel (Figure 2B), and one sample (443) included a tremolite-rich layer with an approximately 50% modal composition of tremolite (Figure 2D).

The average grain size of each sample was measured using the linear intercept method [80] and the range was found to be 0.32–1.2 mm (Table 1) with an average of 0.51 mm. The average size of the grains in sample 448, with its clear porphyroclastic texture, was 0.35 mm for the recrystallized fine grains and 1.2 mm for the coarse grains including the porphyroclasts (Table 1). The olivine porphyroclasts had curvy grain boundaries, indicating a recrystallization process that occurred via grain boundary migration (Figure 2A,C). Straight grain boundaries and triple junctions were often observed in the recrystallized small grains, which indicated that annealing occurred during exhumation (Figure 2C). A few four-grain junctions were also observed in the area of the recrystallized olivine grains (Figure 2E,F). Undulose extinction and subgrain boundaries were frequently observed in all samples (Figure 2A,C). Microstructures such as porphyroclastic texture, undulose extinction, subgrain boundaries, and abundant inclusions or fractures coincided with those of the grey peridotite described by Kostenko et al. [64].

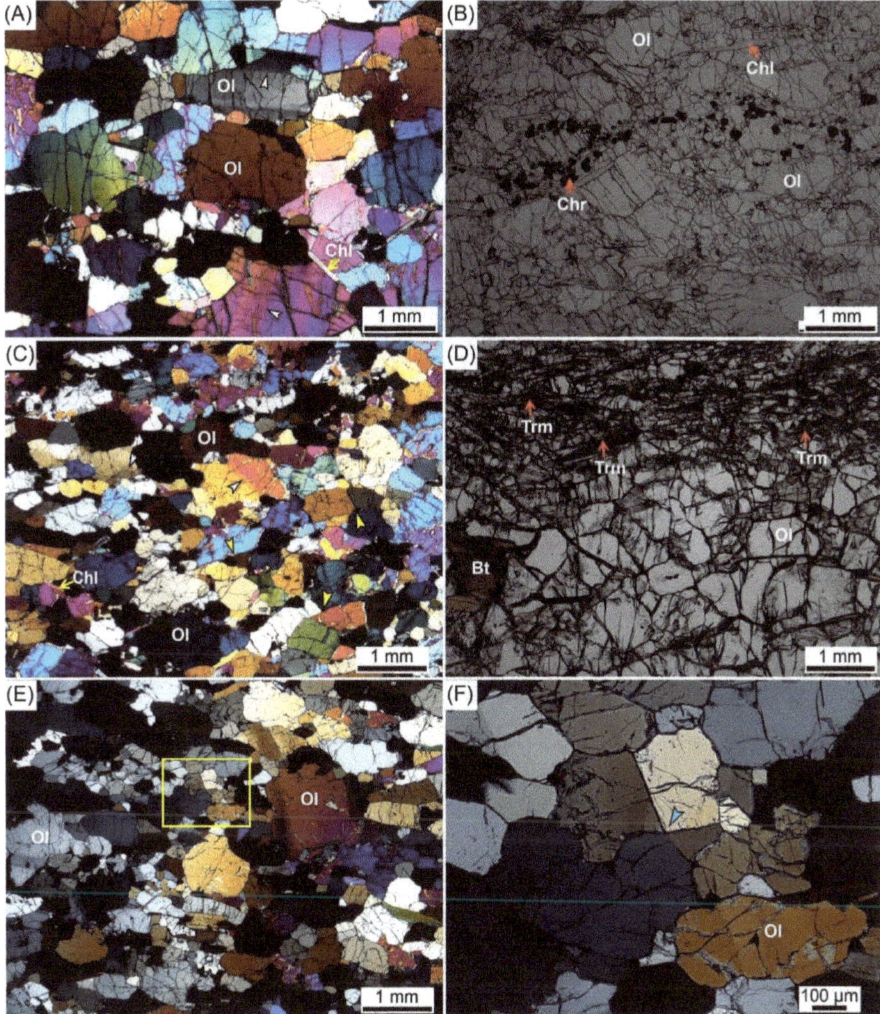

Figure 2. Optical photomicrographs of the samples in transmitted light. (**A**) Representative wide-view images of sample 448. White arrows mark the olivine crystals that show clear subgrain boundaries. Olivine and chlorite grain is indicated as Ol and Chl, respectively. (**B**) Spinel-rich layer observed in sample 429. Chromite and chlorite grains are indicated as Chr and Chl, respectively. (**C**) Representative wide-view images of sample 447. White arrows mark the olivine crystals showing clear subgrain boundaries. The yellow arrows mark the location of the tripple junctions. Olivine and chlorite grains are indicated as Ol and Chl, respectively. (**D**) Tremolite-rich layer observed from sample 443. Tremolite and biotite grains are indicated as Trm and Bt, respectively. (**E**) Wide view image of sample 447. The yellow rectangle indicates the location of Figure 2F. (**F**) Four-grain junction is observed in sample 447. The blue arrow marks the location of the four-grain junction.

Table 1. The lattice-preferred orientations (LPO), water content, and fabric strength of olivine.

Sample		LPO Type of Olivine	Water Content of Olivine [1] (ppm H/Si)	Average Grain Size [2] (mm)	M-Index [3]	J-Index [4]
426		A + B-type	500 ± 50	0.48	0.09	2.98
429	Spinel-rich	B + C-type	320 ± 50	0.36	0.05	2.00
	Spinel-poor	B-type		0.32	0.08	2.86
443	Tremolite-rich	B-type	230 ± 50	0.35	0.05 / 0.13 *	2.34 / 5.89 *
	Tremolite-poor	A-type		0.7	0.08	2.66
445		B + C-type	500 ± 50	0.51	0.05	1.89
446		B-type	310 ± 50	0.46	0.09	3.61
447		B-type	430 ± 50	0.38	0.19	4.55
448	Large grain	A-type	300 ± 50	1.2	0.1	3.25
	Small grain	B-type		0.35	0.08	4.46

[1] Water content of olivine was measured from the inclusion-free area. Paterson calibration was used to calculate water content [76]. [2] Average grain size was measured using the linear intercept method [80]. [3] Fabric strength of the LPO of olivine calculated using the M-index [69]. [4] Fabric strength of the LPO of olivine calculated using the J-index [70]. * Fabric strength of the LPO of tremolite.

4.2. Chemical Compositions of Minerals

The representative chemical compositions of the minerals obtained via the EPMA analysis are presented in Table 2. A high content of magnesium was found in the olivine and orthopyroxene, with an Mg# of 94 for the olivine and 93–95 for the orthopyroxene. No significant chemical difference was found between the coarse porphyroclasts and fine recrystallized grains. Spinel had a high concentration of chromium (Cr_2O_3 = 58.98 wt. %; Table 2) and can be classified as chromite. Amphiboles can be classified as tremolite with a very low aluminum concentration (Al_2O_3 = 1.1 wt. %; Table 2). The temperature of the Åheim amphibole peridotite was estimated at 586 ± 50 °C using the Ol–Sp geothermometer [65] and at 640 ± 50 °C using the Al in the orthopyroxene geothermometer [66].

Table 2. The chemical compositions of the representative minerals in the specimen.

Sample	447					448				
Mineral	ol-1	ol-2	opx-1	amp-1	bt-1	ol-1	ol-2	opx-1	chl-1	sp-1
SiO_2	41.10	41.61	57.85	56.33	42.68	41.49	41.60	58.22	31.33	0.00
TiO_2	0.03	0.01	0.00	0.01	0.06	0.01	0.01	0.00	0.02	0.01
Al_2O_3	0.01	0.00	0.10	1.10	12.18	0.00	0.00	0.08	13.36	3.03
Cr_2O_3	0.02	0.01	0.02	0.25	0.68	0.01	0.00	0.01	3.79	58.98
FeO	6.17	6.17	4.47	1.46	1.77	6.30	6.46	4.14	1.93	25.70
MnO	0.09	0.08	0.13	0.06	0.00	0.07	0.12	0.14	0.02	0.49
MgO	51.39	51.43	36.11	23.69	27.06	51.79	51.93	36.84	34.53	5.85
CaO	0.01	0.02	0.12	12.74	0.24	0.00	0.00	0.08	0.01	0.01
Na_2O	0.00	0.00	0.00	0.53	1.09	0.00	0.00	0.01	0.00	0.00
K2O	0.00	0.00	0.00	0.09	6.93	0.00	0.00	0.01	0.00	0.00
NiO	0.40	0.41	0.08	0.00	0.18	0.32	0.40	0.11	0.00	0.03
Total	99.21	99.75	98.88	96.26	92.86	99.99	100.52	99.62	84.97	94.10

wt. % oxides, ol: olivine, opx: orthopyroxene, amp: amphibole, bt: bitotite, chl: chlorite, sp: spinel.

4.3. LPO of Minerals

The LPOs of the olivine in the amphibole peridotite samples are illustrated in Figure 3. In samples 443 and 448, it is apparent that the [100] axes of the olivine are aligned subparallel to the lineation and the [010] axes are aligned subnormal to the foliation, which is known as an A-type LPO [7]. In samples 446 and 447, the [001] axes of the olivine are aligned subparallel to the lineation and the [010] axes are aligned subnormal to the foliation, which is a B-type LPO [7,8]. Some samples showed a combination of the two different types of LPOs. In sample 426, both the [100] and [001] axes of the olivine are aligned subparallel to the lineation and the [010] axes are aligned subnormal to the foliation; this is an A + B-type LPO [19,23,81]. In samples 429 and 445, the [001] axes of the olivine are aligned subparallel to the lineation and both the [100] and [010] axes are aligned subnormal to the foliation; this is known as a mixed B- and C-type LPO (B + C-type LPO). This B + C-type LPO of olivine is quite similar to the type III LPO of olivine which is reported by Prelicz (2005) [82]. The fabric strength of the samples, which was calculated as both the M-index and J-index values from the LPO of the olivine and tremolite, is listed in Table 1. Both the M- and J-index values tend to be slightly lower in the samples with a mixed olivine LPO, such as A + B or B + C (samples 426, 429, and 445; Table 1). The fabric strength of the tremolite was much higher than that of the olivine (Table 1).

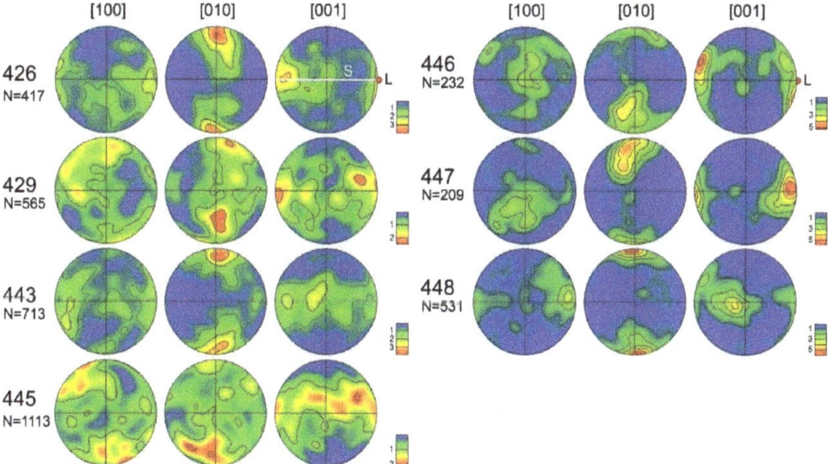

Figure 3. Pole figures of the olivine presented in the lower hemisphere using equal-area projection. The white line (S) represents foliation and the red dot (L) represents lineation. A half-scatter width of 20° was used. "N" represents the number of grains. The color-coding indicates the density of the data points. The numbers in the legend correspond to multiples of uniform distribution.

To understand the effects of grain size and mineral assemblage on the LPO of olivine, an additional EBSD analysis was performed for three samples: sample 429, 443, and 448. Sample 429 was divided into the spinel-rich layer (Figure 2B) and the spinel-poor layer. In the case of sample 443, the LPOs of the olivine in the tremolite-rich and tremolite-poor layers were determined separately (Figure 2D). Sample 448 was divided into the porphyroclasts and fine-grained recrystallized grains (Table 1). A detailed EBSD analysis of the samples (429, 443, and 448) revealed that the LPO of olivine has a tendency to be varied within the mineral assemblage of the layer and the grain size of the samples (Figure 4). In sample 429, the LPO of the olivine that was obtained from the spinel-rich layer (429 Sp-rich) was a B + C-type, but the LPO of the olivine obtained from the spinel-poor layer (429 Sp-poor) was a B-type (Figure 4A). In sample 443, the LPO of the olivine obtained from the tremolite-rich layer (443 Tr-rich) was a B-type, but the LPO of olivine obtained from the tremolite-poor layer (443 Tr-poor) was an A-type (Figure 4B). In sample 448, the LPO of the olivine observed from the coarse grains that included porphyroclasts (448 large grain) was an A-type, but the LPO of the olivine in the recrystallized fine grains (448 small grain) was a B-type (Figure 4C).

The LPO of the tremolite in the tremolite-rich layer from sample 443 is also shown in Figure 4B. The tremolite LPO exhibits the [001] axes and {010} poles forming a girdle distribution along the foliation with the [100] axes that are aligned subnormal to the foliation; this is known as a type-III LPO of amphibole [44].

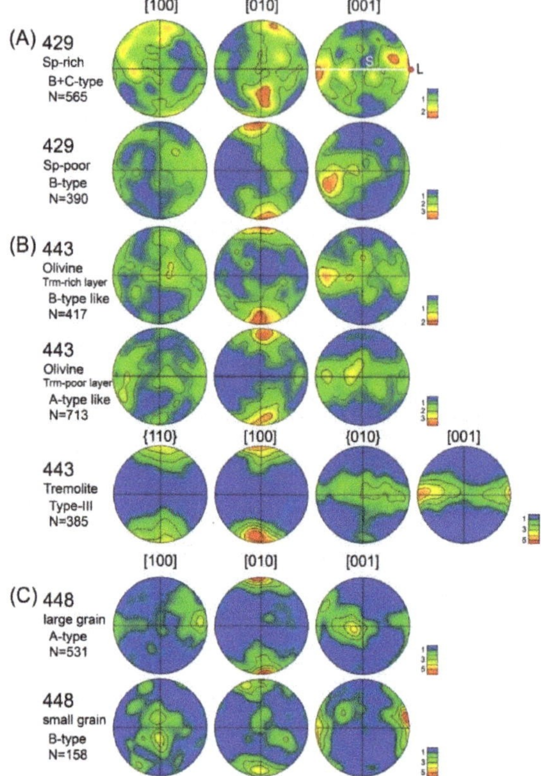

Figure 4. Pole figures of olivine presented in the lower hemisphere using an equal-area projection obtained from (**A**) the spinel-rich layer and a spinel-poor layer of sample 429, (**B**) the tremolite-poor layer and tremolite-rich layer of sample 443, and (**C**) large grains including the porphyroclasts and recrystallized small grains of sample 448. The pole figure of the tremolite from the tremolite-rich layer of sample 443 is also included in (**B**). The white line (S) represents foliation and the red dot (L) represents lineation. A half-scatter width of 20° was used. "N" represents the number of grains. The color-coding indicates the density of data points. The numbers in the legend correspond to multiples of uniform distribution.

4.4. Seismic Velocity and Anisotropy

To simulate the changes in the seismic properties from the degree of the recrystallization, the LPO data from the large porphyroclasts and fine-grained recrystallized grains were mixed with six different mixing ratios: 100:0, 80:20, 40:60, 60:40, 20:80, and 0:100. The LPO of the olivine in sample 448 was chosen as the original raw data because of its clear porphyroclastic texture (Figure 2A). Due to sample 448 mainly consisting of olivine (more than 90%), only the LPO of olivine was considered for the calculation. The seismic velocity and anisotropy of sample 448 are illustrated in Figure 5. In the large grains of sample 448 (100%), the P-wave anisotropy (AV$_P$) was found to be 5.7%, and the maximum S-wave anisotropy (AV$_S$) was 3.63%. The calculated AV$_P$ and maximum AV$_S$ values were similar to or slightly higher than those from the previous studies on the Almklovdalen chlorite peridotites; that is, 2.4–8.2% and 0.9–4.3% from Prelicz (2005) [82], 5.3–6.0% and 3.72–4.46% from Wang et al. (2013) [16], and 1.8–3.8% and 1.66–2.68% from Kim & Jung (2015) [24]. For large grains of sample 448, the polarization direction of the fast shear wave at the center of the stereonet was subparallel to the direction of flow (X-direction; Figure 5). With the increasing ratio of small grains, the polarization direction of the fast shear wave started tilting and became subnormal to the flow direction at the large grains (40%) and small grains (60%) (Figure 5). This result indicates that with the 60% of recrystallization rate, a trench-parallel shear wave anisotropy is expected at the mantle wedge, assuming a 2D corner flow of the upper mantle. The estimated thickness of the anisotropic layer for the given delay time is noted in Table 3.

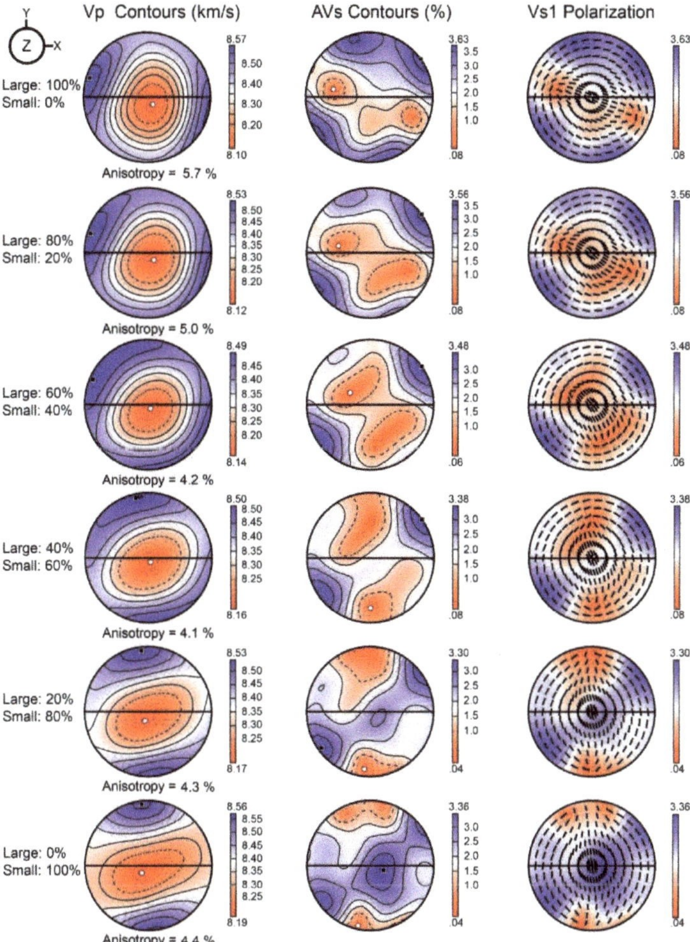

Figure 5. Effects of multiple stages of deformation on seismic velocity and anisotropy. LPO data of olivine from 448 large grains and 448 small grains (Figure 4C) was mixed with 6 different mixing ratios: 100:0, 80:20, 40:60, 60:40, 20:80, and 0:100. Seismic velocity and anisotropy are calculated from the mixed LPO data to estimate the effect that the secondary deformation event had on the seismic anisotropy. The P-wave velocity (V_P), the amplitude of the shear-wave anisotropy (AV_S), and the polarization direction of the faster shear wave (Vs1) are plotted in the lower hemisphere using an equal-area projection. The center of the stereonet corresponds to the direction normal to the foliation (Z), and the east-west direction corresponds to lineation (X).

Table 3. Estimated seismic anisotropy and the thickness of the anisotropic layer for the given delay time calculated from the LPOs of the olivine, amphibole, and serpentine.

The Mixing Ratio of 448 Large Grain and 448 Small Grain	Horizontal AV_S (%) [1]	$<V_S>$ (km/s) [1]	The Thickness of Anisotropic Layer for the Given Delay Time Calculated from the Seismic Anisotropy (km)		
			dt = 0.1 s	dt = 0.2 s	dt = 0.3 s
100:0	1.8	4.780	27	53	80
80:20	1.4	4.785	34	68	103
60:40	1.5	4.790	32	64	96
40:60	1.8	4.790	27	53	80
20:80	2.4	4.795	20	40	60
0:100	3.1	4.800	15	31	46
Mineral Assemblage					
Olivine (443) [2]	1.4	4.813	34	69	103
Olivine + tremolite [2] (70%) (30%)	1.7	4.665	27	55	82
Olivine + tremolite [2] (50%) (50%)	2.5	4.595	18	37	55
Tremolite (443) [2]	5.5	4.340	8	16	24
Antigorite (VM3) [3] [38]	1.6	3.735	23	47	70
Tremolite (443) [3]	6.0	3.975	7	13	20
Ol + Atg + Trm [3] (50%) (25%) (25%)	3.2	4.450	14	28	42

[1] Horizontal AV_S and $<V_S>$ were determined at the center of the stereonet (vertical S-wave propagation direction; Figure 5; Figure 6); [2] The LPO data was rotated with a dipping angle of 45° to simulate the effect of flow dipping along the subducting slab; [3] The LPO data was rotated with a dipping angle of 55° to simulate the effect of flow dipping along the subducting slab; AVs: anisotropy of S-wave velocity, <Vs>: the average velocity of the fast and slow shear waves (V_{S1} and V_{S2}), dt: delay time of S-wave, Ol; olivine, Atg: antigorite, Trm: tremolite.

The LPOs of the olivine and tremolite in the tremolite-rich layer of sample 443 (Figure 2D) were chosen to estimate the influence that the amphibole had on the seismic anisotropy of the mantle wedge. The LPO data of the olivine and tremolite in the tremolite-rich layer were mixed in 70:30 and 50:50 ratios and were rotated with a dipping angle of 45° to simulate the effect of flow dipping along the subducting slab in the mantle wedge. The seismic velocity and anisotropy of the tremolite-rich layer of sample 443 is shown in Figure 6. In the case of the olivine (443 Ol), the AV_P is 3.8%, the maximum AV_S is 2.68%, and the polarization direction of the fast shear wave at the center is oblique to the flow direction (Figure 6A). The tremolite of this sample (443 Trm) demonstrates an AV_P of 21.8%, a maximum AV_S of 14.19%, and the polarization direction of the fast shear wave at the center is aligned subnormal to the flow direction (Figure 6A). When the olivine and tremolite are mixed with the 70:30 ratio, the AV_P is 8.4%, the maximum AV_S is 5.79%, and the polarization direction of the fast shear wave at the center is aligned subnormal to the flow direction (Figure 6A). With the 50:50 mixing ratio, the AV_P is 12.0%, the maximum AV_S is 8.07%, and the polarization direction of the fast shear wave at the center is aligned subnormal to the flow direction (Figure 6A). This result suggests that the presence of amphibole may significantly contribute to the trench-parallel S-wave anisotropy, assuming that the flow dips along the subducting slab in the mantle wedge. In addition, the P- to S-wave velocity ratio (V_P/V_S) of the tremolite was smaller than that of the olivine (Figure 6A). The maximum V_P/V_S for the olivine, amphibole, and olivine–amphibole mixture (50:50) was 1.74, 1.70, and 1.71, respectively.

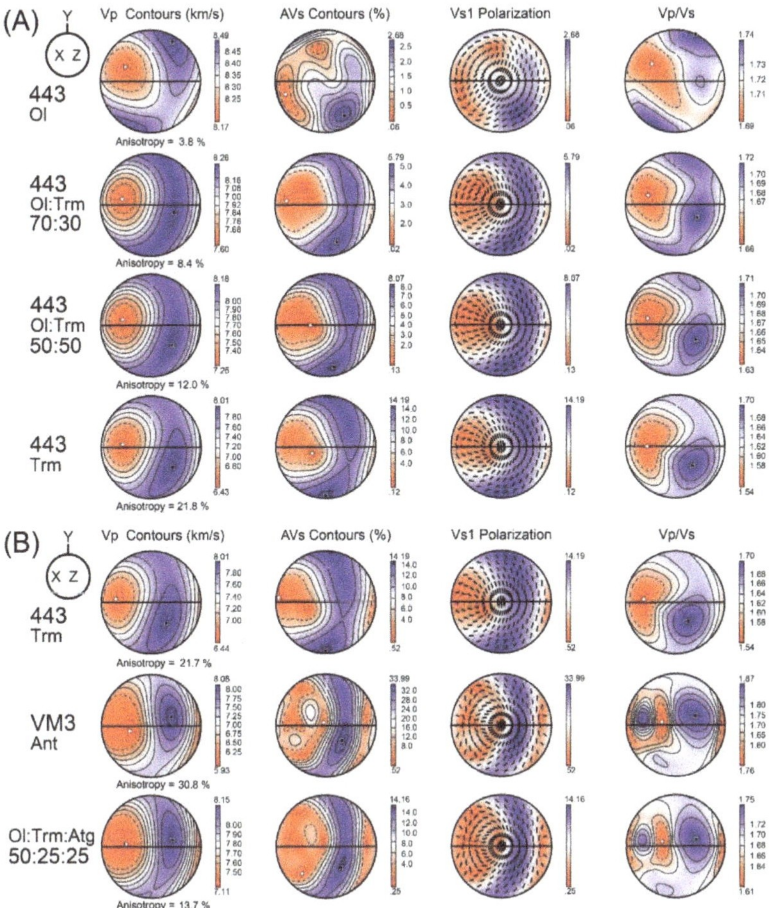

Figure 6. (**A**) Effect of amphibole on seismic velocity and anisotropy. LPO data of the olivine and tremolite in the tremolite-rich layer in sample 443 (Figure 4B) was used. (**B**) Effect of amphibole and serpentine on seismic velocity and anisotropy. LPO data of the olivine and tremolite in the tremolite-rich layer in sample 443 (Figure 4B) and LPO data of antigorite (VM3) reported by Jung [36] was used. The P-wave velocity (V_P), the amplitude of the shear-wave anisotropy (AV_S), the polarization direction of the faster shear wave (V_{S1}), and the P- to S-wave velocity ratio (V_P/V_S) are plotted in the lower hemisphere using an equal-area projection. The x direction and the z direction correspond to the lineation and the direction normal to the foliation, respectively. Ol: olivine, Trm: tremolite, Atg: antigorite.

In addition, the LPO data of the serpentine was mixed with the LPO of the olivine and tremolite in the tremolite-rich layer of sample 443 in order to compare the influence of amphibole and serpentine on the seismic anisotropy in subduction zones. The LPO data of the antigorite from Val Malenco, Italy (sample VM3), reported by Jung [38], was mixed with the LPO data of the olivine and tremolite in the tremolite-rich layer of sample 443 in a 50:25:25 (olivine–tremolite–antigorite) ratio. Because serpentine requires a high dipping angle to produce trench-parallel seismic anisotropy [36], the calculated data was rotated with a dipping angle of 55° in order to make a proper comparison. The calculated seismic velocity and anisotropy of the olivine, tremolite, and antigorite mixture are shown in Figure 6B. The mixing ratio of olivine, tremolite, and antigorite was 50:25:25. The AV_P is 13.7%, the maximum

AV_S is 14.16%, and the polarization direction of the fast shear wave at the center is aligned subnormal to the flow direction (Figure 6B). In addition, the pattern of the seismic velocity and anisotropy was similar to that of the antigorite (VM3; [38]), which indicates that the seismic anisotropy of the overall mixture is governed by the antigorite.

4.5. Water Content of Olivine

The representative FTIR spectra of the olivines are illustrated in Figure 7. In the case of the olivine grain without any inclusions or cracks, small absorption peaks were observed (Figure 7A). The average water content of the olivine calculated using the IR bands between 3400 and 3750 cm^{-1} was 370 ± 50 ppm H/Si (Table 1), which is above the boundary between the A-type and B-type LPO illustrated by the fabric diagram for the olivine [8]. The average water content of the olivine from this study was higher than that from previous studies on the Almklovdalen chlorite peridotites; that is, 7–115 ppm H/Si from Wang et al. (2013) [16] and 170–310 ppm H/Si from Kim & Jung (2015) [24]. No significant difference was apparent in the FTIR spectrum related to the grain size of the olivine or the mineral assemblage of the sample, and therefore all the FTIR spectra were averaged. Additional FTIR analyses were performed for the olivine grains with visible inclusions to study the components of these samples. The resulting FTIR spectra showed strong absorption bands in the range 3400–3750 cm^{-1} (Figure 7B,C). IR peaks were found at the wave numbers of 3689, 3684, 3649, and 3630 cm^{-1}, which indicates the presence of serpentine as a hydrous inclusion in the olivine [83–85].

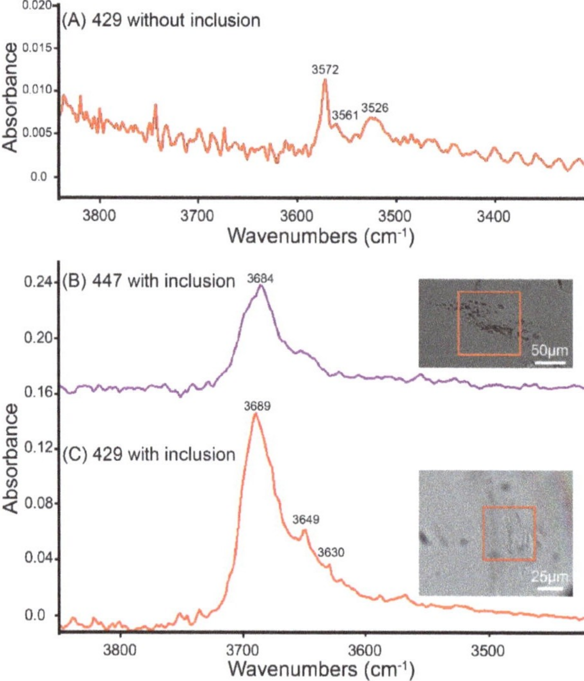

Figure 7. A representative unpolarized FTIR spectra of the olivine. (**A**) FTIR spectrum of olivine in sample 429 without any inclusions or grain boundaries. (**B**) FTIR spectrum of olivine in sample 447 where hydrous inclusions are found, and the optical micrograph image of the location from which the FTIR spectra are taken. (**C**) FTIR spectrum of olivine in sample 429 where a hydrous inclusion is found, and optical micrograph image of the location from which the FTIR spectra are taken.

4.6. Dislocation Microstructure of Olivine

Backscattered electron images of the dislocation microstructure of the olivine are shown in Figure 8. All samples showed a homogeneous distribution of the dislocations, and there was no significant difference between the samples with different LPOs of olivine. In many cases, the dislocations were observed to be curved (Figure 8A) or looped (Figure 8C), suggesting the strong influence that obstacles have on the dislocation glide. Abundant subgrain boundaries were observed in the olivine grains (Figure 8A–C), which were similar to those observed with the optical microscope (Figure 2A,C).

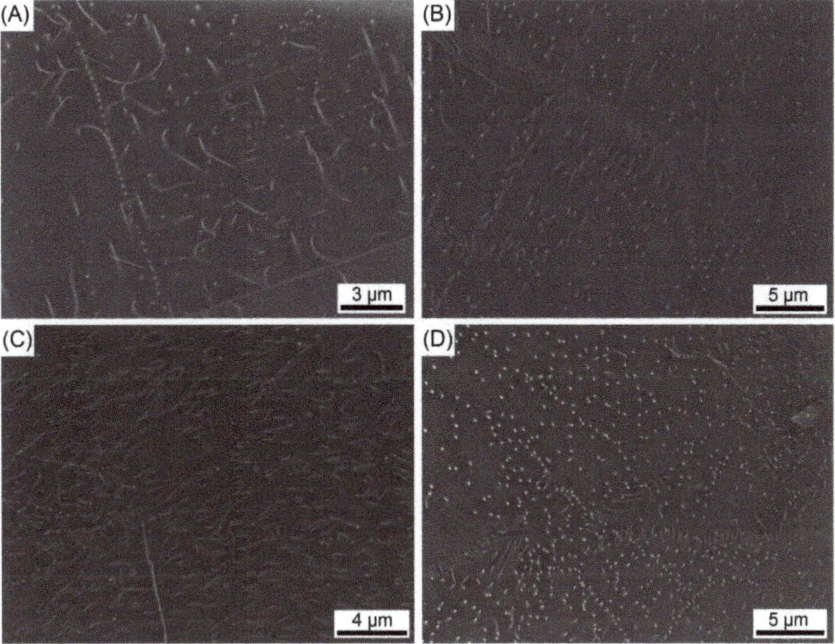

Figure 8. Backscattered electron images showing dislocation microstructures of olivine from (**A**) sample 426, (**B**) sample 429, (**C**) sample 447, and (**D**) sample 448. Dislocations are shown as white dots and lines.

5. Discussion

5.1. Development of LPO of Olivine

Previous studies on the Almklovdalen chlorite peridotites located at a distance of 3 km from the Åheim peridotites, reported both A- and B-type LPOs of olivine [16], A-, B-, C-, and E-type LPOs of olivine [24], and an A-, B + C-type, and axial [010] pattern LPOs of olivine [82]. Wang et al. [16] interpreted that the A-type and B-type LPOs of olivine were formed under dry conditions (7–115 ppm H/Si) and the B-type LPO of olivine was developed under conditions of high stress and high strain. On the other hand, Kim and Jung [24] reported that the A-type LPO was a result of deformation under dry conditions (170 ± 30 ppm H/Si), whereas the B-type LPO was developed due to the deformation of olivine under wet conditions (210–310 ± 30 ppm H/Si). Prelicz (2005) [82] reported that the B + C-type (type-III) LPO of olivine developed in a mantle peridotite under water-rich retrograde conditions possible during the emplacement in crustal rock.

In this study, we observed that the LPO of the olivine in the Åheim amphibole peridotites has a close relationship with the grain size of the olivine and the mineral assemblage of the peridotite (Table 1, Figure 4). The A-type LPO of olivine was observed in the coarse grains of sample 448 (with an average grain size of 1.2 mm, Table 1, Figure 2A) and the tremolite-poor layer of sample 443 (with an average grain size of 0.7 mm, Table 1, Figure 2D). On the other hand, the B-type LPO was observed from the recrystallized fine grains of the olivine in sample 448 (with an average grain size of 0.35 mm, Table 1, Figure 2A) and in the tremolite-rich layer of sample 443 (with an average grain size of 0.35 mm, Table 1, Figure 2D). One possible mechanism for this fabric transition from an A-type to B-type LPO observed in the olivine from the Åheim amphibole peridotites is an olivine deformation under water-rich conditions. The water content of the olivine measured via the FTIR spectroscopy was in the range of 230–500 ± 50 ppm H/Si, which indicates that the olivine was deformed under water-rich conditions [7,8]. The presence of hydrous minerals such as amphibole and chlorite (Figure 2D), and the abundant hydrous inclusions observed in the olivine (Figure 7B,C) suggest that the deformation of the recrystallized olivine grains occurred in the presence of a fluid. In addition, numerous dislocations were observed in the olivine samples (Figure 8). These results indicate that the dominant deformation mechanism of olivine showing a B-type LPO in the Åheim peridotites is a dislocation creep under water-rich conditions.

The other possible mechanism for a fabric change from an A-type to B-type LPO of olivine is an enhancement of dislocation-accommodated grain boundary sliding (DisGBS) [17–19,86]. Precigout and Hirth [19] proposed that a fabric transition from an A-type to B-type LPO in olivine that was observed in the Ronda massif, Spain could have resulted from the enhancement of GBS as the grain size decreased. The average grain size of the olivine in the Ronda peridotite was ~1 mm in the tectonites and 0.05–0.3 mm in the upper mylonite [19], which is similar to that of the Åheim amphibole peridotite (448 fine: 0.35 mm, and 448 coarse: 1.2 mm; Table 1). In addition, the few four-grain junctions (Figure 2E,F) observed from the recrystallization of the olivine grains with a B-type olivine fabric (Figure 3) could be further evidence for a minor contribution from deformation via GBS [87,88].

A B + C-type LPO of olivine was observed in the olivine of sample 445 and the spinel-rich layer of sample 429. A C-type LPO of olivine is known to be produced under conditions of low stress and water-rich conditions [7,8,89]. Water may have been lost from the olivine during exhumation. A high water content (up to 500 ± 50 ppm H/Si; Table 1) and the presence of hydrous inclusions such as serpentine in the olivine (Figure 7C) indicate that the B + C-type LPO of olivine observed in samples 429 and 445 can be related to deformation under water-rich conditions. In addition, numerous dislocations were observed in the olivine samples (Figure 8B). These results indicate that the dominant deformation mechanism of olivine demonstrating the C-type LPO in the Åheim peridotites is a dislocation creep under water-rich conditions. A C-type LPO of olivine was also reported in the other previous studies where peridotites were deformed in water-rich conditions at various localities such as Cima di Gagnone in the Central Alps [90], Otrøy Island in Western Norway [91], the North Quidam UHP belt, NW China [92], and the Rio Grande rift, USA [93].

5.2. The Deformation History of Åheim Peridotite

The LPO of the olivine in the Åheim amphibole peridotites showed four different types of olivine fabric: A-, B-, A + B-, and B + C-type LPOs of olivine (Figures 3 and 4). The Åheim peridotite bodies represent the mantle wedge where it was entrapped in the crust during the uplift process [53,54]. An A-type LPO of olivine was observed in the coarse olivine grains including the porphyroclasts (Figure 4B,C, Table 1), which can, therefore, be interpreted as the original mantle fabric prior to uplift. Deformation was localized in the fine recrystallized grain area of the samples (Figure 2B–D). Considering that a significant amount of strain is required to alter the LPO of olivine [6,94], the strain of the deformation during the exhumation process is considered insufficient to change the pre-existing LPO of olivine porphyroclasts. During the uplift process, fluid infiltrated the samples in the amphibole peridotite stability field [53,64], enhancing the recrystallization of the olivine under water-rich conditions [77], which is considered to have resulted in the fabric transition of the olivine from an A-type to a C- and B-type.

The secondary olivine fabric which was developed in the Åheim peridotites is considered to be the C-type LPO of olivine, which is preserved as the B + C-type LPO of olivine in sample 445 and the spinel-rich layer of sample 429 (Figure 4A). The C-type fabric in the Åheim amphibole peridotite can be correlated with the spinel bearing assemblage (Figures 2B and 4A), which is related to the granulite facies (ol + opx + cpx + sp) condition during the process of exhumation [53,54,58]. After the granulite facies condition, a localized deformation associated with the fluid infiltration may have resulted in the fabric transition to the C-type LPO of olivine (Figure 7C). As granulite facies metamorphism was recorded in the Åheim amphibole peridotite prior to amphibolite facies [53,54,58], this C-type LPO can be regarded as a secondary olivine fabric.

The last olivine fabric which was developed in the Åheim peridotites is considered to be as the B-type LPO of olivine observed in the small recrystallized olivine grains (Figure 4B,C). Samples with the B-type LPO of olivine were of the smallest grain size. The B-type LPO of olivine can be related to the deformation of samples during amphibolite facies (ol + opx + amp + chl) conditions, following the granulite facies [53,54,58]. Fluid infiltration and the enhanced recrystallization of olivine at the amphibolite facies condition could lead to the fine-grained olivine grains that were deformed under water-rich conditions, resulting in the B-type LPO of olivine in the small recrystallized olivine grains. Both the A + B-type or B + C-type LPO of the olivine observed in samples 426, 429, and 445 (Figure 3) can be interpreted as a B-type LPO of olivine overprinting a preexisting A- or C-type LPO of the olivine. These mixed LPOs of olivine are believed to be the last olivine fabrics preserved in the Åheim amphibole peridotites.

5.3. Implications for the Seismic Anisotropy

Trench-parallel seismic anisotropy has been observed at various subduction zones around the world [13,37,95–98]. After experiments concerning the deformation of olivine under wet conditions at high pressures, a water-induced fabric change of the olivine in the subduction zone was suggested as one of the possible mechanisms for this phenomenon [7,8]. As the fabric transition from an A-type to B-type LPO of the olivine in the Åheim amphibole peridotite was driven by the fluid infiltration associated with the exhumation process during/after the Scandian orogeny, the olivine fabric data from this study cannot be directly applied to the current mantle wedge above the subducting slab. However, the olivine fabric transition observed from the Åheim amphibole peridotites could be a good example of water-induced fabric change to olivine in naturally deformed peridotites.

Prelicz (2005) reported the ultrasound velocities of the P- and S-waves measured on the cores obtained from the Almklovdalen chlorite peridotites [82]. The mean V_P and V_{S1} at a confining pressure of 400 MPa were around 7.9–8.2 km/s and 4.76–5.1 km/s, respectively [82], The seismic velocities calculated from the LPO of olivine in sample 448 showed that the V_P was slightly higher (8.1–8.57 km/s; Figure 5) and the $<V_S>$ was similar to the ultrasound velocity data (4.78–4.8 km/s; Table 3). There are several explanations for these differences in seismic velocities: (1) Since sample 448 mainly consisted of olivine, only the LPO of olivine was considered for the calculation. Therefore, the effect of secondary minerals was excluded in this case, especially for the effect of hydrous minerals such as tremolite and chlrotite. (2) In this study, the elastic constant of a single crystal olivine at an ambient condition was used for calculation. Since the elastic constant of olivine is dependent on the pressure and temperature [72], there may be minor errors related to the elastic constant because of the low pressure and temperature conditions of the rocks. (3) As the calculated seismic anisotropy only consider the volume fraction, density, elastic constants, and the LPO of the mineral [71], the effects of thermal cracks or pores on the seismic velocities were neglected.

As S-wave anisotropies in the A- and B-type LPOs of olivine are negatively interfered with [7,8], the mixing of these two olivine fabrics may result in a significant weakening of the seismic anisotropy [99] or change in the S-wave anisotropy [14]. In the case of sample 448, large grains (i.e., porphyroclasts) showed an A-type LPO of olivine and the small recrystallized grains (448 small grain) showed a B-type LPO of olivine (Figure 4C). The seismic velocity and anisotropy calculated from the LPOs of the olivine in sample 448 showed that with more than 60% of fabric transition from an A-type to B-type LPO of olivine, the polarization direction of the fast shear wave at the center of the stereonet (vertical propagation of S-wave) changed, to become subnormal to the flow direction (trench-parallel; Figure 5). Comparing the seismic anisotropy of the 60% fabric transition (60:40) with that of the 100% fabric transition to a B-type LPO (0:100), the S-wave anisotropy (AV_S) at the center of the stereonet was decreased to ~42%, and the estimated thickness of the anisotropic layer for a given delay time was increased to ~80% (Table 3). Assuming that the olivine fabric is partially changed from an A- to B-type LPO of olivine in the forearc mantle wedge (Figure 9), the shear wave seismic anisotropy of the olivine in the forearc mantle wedge would be significantly decreased (Table 3). This mixed olivine fabric in the mantle wedge could be a possible explanation for the relatively small delay time (~0.2 s) observed at various subduction zones such as in NE Japan or Mexico [99–102].

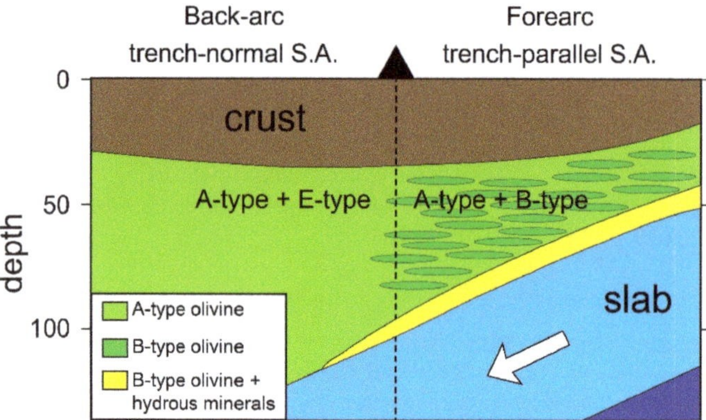

Figure 9. Schematic diagram of the proposed olivine fabrics in the mantle wedge of a cold subduction zone. Hydrous minerals: serpentine, chlorite, tremolite, etc. S.A.: Seismic anisotropy.

To understand trench-parallel seismic anisotropy in subduction zones and assuming a mantle flow that is dipping alongside the slab, it is important to consider the effect that hydrous minerals are expected to have on the seismic anisotropy of a subduction zone [2,33], such as the formation of serpentine [38,39] and chlorite [24]. Under a slab dipping angle of approximately 45°, the polarization direction of the fast shear wave was oblique to the lineation (flow) direction when only the LPO of the olivine of sample 443 was considered (Figure 6A). However, the polarization direction of the fast shear wave became subnormal to the flow direction when the tremolite LPO was mixed with the LPO of the olivine in sample 443 (Figure 6A). Comparing the seismic anisotropy of the Ol–Trm mixture (50:50) with that of the olivine, the AV_S was increased to ~79%, and the estimated thickness of the anisotropic layer for the given delay time was decreased to ~47% (Table 3). This result suggested that the amphibole can have a similar effect on the trench-parallel shear wave splitting as other hydrous minerals such as serpentine and chlorite in the mantle wedge (Figure 9).

Due to the strong seismic anisotropy of the antigorite, the overall seismic velocity and anisotropy of the olivine, antigorite, and tremolite mixture were governed by the LPO of antigorite (Figure 6B). The maximum AV_S of the antigorite was 2.4 times larger than that of the tremoloite, and the pattern of the seismic velocity and anisotropy of Ol–Atg–Trm mixture mostly followed that of the antigorite. However, the AV_S values of the tremolite (Tremolite 443) for the vertically propagating seismic wave (at the center of the stereonet) was 3.75 times higher than that of antigorite (VM3) with a slab dipping angle of 55° (Table 3). In addition, amphibole can produce trench-parallel seismic anisotropy (Figure 6A) in a relatively lower slab dipping angle (45°) than that of the serpentine [38].

The maximum V_P/V_S values of the olivine, tremolite, and olivine–tremolite mixture were 1.74, 1.70, and 1.71, respectively (Figure 6A). The calculated maximum V_P/V_S value of the tremolite (1.70) was close to the isotropic V_P/V_S value ($\beta_0 = 1.718$) in the amphibolite reported by Ji et al. [42]. The seismic velocity and anisotropy calculated from the LPOs of the olivine and tremolite in the tremolite-rich layer of sample 443 showed that the V_P/V_S ratio of the tremolite was smaller than that of the olivine (Figure 6A). The V_P/V_S ratio of the mixed olivine–tremolite layer was also smaller than that of the olivine. High V_P/V_S zones have been observed at various subduction zones such as those near Cascadia, Nankai, and in Central Mexico [103–107]. These high V_P/V_S zones are usually interpreted as being the result of a regionally high pore fluid pressure [103,105] or presence of talc [108]. Our result indicates that, unlike talc or serpentine, amphiboles do not contribute much to the formation of the high V_P/V_S zones.

6. Conclusions

The microstructures of the amphibole peridotites from the Gusdal quarry in Åheim, Norway were studied and evidence for the multiple stages of deformation during the Scandian Orogeny and subsequent exhumation was found. The Åheim amphibole peridotites showed a porphyroclastic texture with abundant undulose extinctions and subgrain boundaries in the olivine. The LPOs of the olivine in the Åheim amphibole peridotites were closely related to the grain size of the olivine and mineral assemblage of the samples. The coarse grains including porphyroclasts showed an A-type LPO of the olivine, which corresponds to the initial stage of deformation in the mantle. During the exhumation process, deformation was associated with the infiltration of fluid and enhanced dynamic recrystallization under water-rich conditions that resulted in the fabric transition of the olivine from an A-type to C- or B-type LPO. Olivines in the spinel-bearing assemblage showed a C-type LPO of olivine which can be interpreted as a result of deformation under water-rich conditions after granulite facies metamorphism. The small recrystallized olivine grains showed a B-type LPO of olivine, which represents the deformation in amphibolite facies. A high water content (370 ± 50 ppm H/Si) and abundant dislocations in the olivine suggest that the B-type LPOs of olivine that were observed in the recrystallized olivine grains were developed via deformation by a dislocation creep under water-rich conditions. In addition, the existence of small recrystallized grains and four-grain junctions indicate that the fabric transition from A-type to B-type was also influenced by the DisGBS process related to the grain size reduction. The microstructural evolution of the Åheim amphibole peridotites can be a good example of the fabric change of the olivine from an A- to B-type that is observed in naturally deformed peridotites. Seismic anisotropy of the olivine in the Åheim amphibole peridotite calculated with a different ratio of coarse grains (showing an A-type LPO) and the fine recrystallized grains (showing a B-type LPO) of olivine indicates that with the 60% recrystallization rate, a trench-parallel S-wave anisotropy is expected from the mantle wedge. Tremolites from the tremolite-rich layer showed a type-III LPO of amphibole. A stronger fabric strength was observed in the tremolite than that in the olivine, and the resultant seismic anisotropy of the tremolite implies that amphibole can perform a similar role to other hydrous minerals such as serpentine and chlorite on the formation of trench-parallel seismic anisotropy.

Author Contributions: Conceptualization, H.J.; methodology, H.J.; software, S.J.; validation, S.J., H.J. and H.A.; formal analysis, S.J.; investigation, S.J.; resources, S.J. and H.A.; data curation, S.J.; writing—original draft preparation, S.J.; writing—review and editing, S.J., H.J. and H.A.; visualization, S.J.; supervision, H.J.; project administration, H.J.; funding acquisition, H.J. All authors have read and agreed to the published version of the manuscript.

Funding: This research was funded by the Mid-career Research Program through National Research Foundation of Korea (NRF: 2017R1A2B2004688 and 2020R1A2C2003765) to Haemyeong Jung.

Acknowledgments: The authors are grateful to the anonymous reviewers for constructive comments and corrections.

Conflicts of Interest: The authors declare no conflict of interest.

References

1. Ben Ismail, W.; Mainprice, D. An olivine fabric database: An overview of upper mantle fabrics and seismic anisotropy. *Tectonophysics* **1998**, *296*, 145–157. [CrossRef]
2. Jung, H. Crystal preferred orientations of olivine, orthopyroxene, serpentine, chlorite, and amphibole, and implications for seismic anisotropy in subduction zones: A review. *Geosci. J.* **2017**, *21*, 985–1011. [CrossRef]
3. Karato, S.; Jung, H.; Katayama, I.; Skemer, P. Geodynamic significance of seismic anisotropy of the upper mantle: New insights from laboratory studies. *Annu. Rev. Earth Planet. Sci.* **2008**, *36*, 59–95. [CrossRef]
4. Nicolas, A.; Christensen, N.I. Formation of anisotropy in upper mantle peridotites: A review. *Geodyn. Ser.* **1987**, *16*, 111–123. [CrossRef]
5. Skemer, P.; Hansen, L.N. Inferring upper-mantle flow from seismic anisotropy: An experimental perspective. *Tectonophysics* **2016**, *668*, 1–14. [CrossRef]
6. Boneh, Y.; Skemer, P. The effect of deformation history on the evolution of olivine CPO. *Earth Planet. Sci. Lett.* **2014**, *406*, 213–222. [CrossRef]
7. Jung, H.; Karato, S. Water-induced fabric transitions in olivine. *Science* **2001**, *293*, 1460–1463. [CrossRef]
8. Jung, H.; Katayama, I.; Jiang, Z.; Hiraga, T.; Karato, S. Effect of water and stress on the lattice-preferred orientation of olivine. *Tectonophysics* **2006**, *421*, 1–22. [CrossRef]
9. Katayama, I.; Jung, H.; Karato, S. New type of olivine fabric from deformation experiments at modest water content and low stress. *Geology* **2004**, *32*, 1045–1048. [CrossRef]
10. Ohuchi, T.; Kawazoe, T.; Nishihara, Y.; Nishiyama, N.; Irifune, T. High pressure and temperature fabric transitions in olivine and variations in upper mantle seismic anisotropy. *Earth Planet. Sci. Lett.* **2011**, *304*, 55–63. [CrossRef]
11. Soustelle, V.; Manthilake, G. Deformation of olivine-orthopyroxene aggregates at high pressure and temperature: Implications for the seismic properties of the asthenosphere. *Tectonophysics* **2017**, *694*, 385–399. [CrossRef]
12. Long, M.D.; van der Hilst, R.D. Shear wave splitting from local events beneath the Ryukyu arc: Trench-parallel anisotropy in the mantle wedge. *Phys. Earth Planet. Inter.* **2006**, *155*, 300–312. [CrossRef]
13. Nakajima, J.; Hasegawa, A. Shear-wave polarization anisotropy and subduction-induced flow in the mantle wedge of northeastern Japan. *Earth Planet. Sci. Lett.* **2004**, *225*, 365–377. [CrossRef]
14. Precigout, J.; Almqvist, B.S.G. The Ronda peridotite (Spain): A natural template for seismic anisotropy in subduction wedges. *Geophys. Res. Lett.* **2014**, *41*, 8752–8758. [CrossRef]
15. Jung, H.; Mo, W.; Green, H.W. Upper mantle seismic anisotropy resulting from pressure-induced slip transition in olivine. *Nat. Geosci.* **2009**, *2*, 73–77. [CrossRef]
16. Wang, Q.; Xia, Q.K.; O'Reilly, S.Y.; Griffin, W.L.; Beyer, E.E.; Brueckner, H.K. Pressure- and stress-induced fabric transition in olivine from peridotites in the Western Gneiss Region (Norway): Implications for mantle seismic anisotropy. *J. Metamorph. Geol.* **2013**, *31*, 93–111. [CrossRef]
17. Cao, Y.; Jung, H.; Song, S.G. Olivine fabrics and tectonic evolution of fore-arc mantles: A natural perspective from the Songshugou dunite and harzburgite in the Qinling orogenic belt, central China. *Geochem. Geophys. Geosyst.* **2017**, *18*, 907–934. [CrossRef]
18. Hansen, L.N.; Zimmerman, M.E.; Kohlstedt, D.L. Grain boundary sliding in San Carlos olivine: Flow law parameters and crystallographic-preferred orientation. *J. Geophys. Res. Solid Earth* **2011**, *116*, 16. [CrossRef]
19. Precigout, J.; Hirth, G. B-type olivine fabric induced by grain boundary sliding. *Earth Planet. Sci. Lett.* **2014**, *395*, 231–240. [CrossRef]
20. Sundberg, M.; Cooper, R.F. Crystallographic preferred orientation produced by diffusional creep of harzburgite: Effects of chemical interactions among phases during plastic flow. *J. Geophys. Res. Solid Earth* **2008**, *113*, 16. [CrossRef]
21. Holtzman, B.K.; Kohlstedt, D.L.; Zimmerman, M.E.; Heidelbach, F.; Hiraga, T.; Hustoft, J. Melt segregation and strain partitioning: Implications for seismic anisotropy and mantle flow. *Science* **2003**, *301*, 1227–1230. [CrossRef]
22. Qi, C.; Hansen, L.N.; Wallis, D.; Holtzman, B.K.; Kohlstedt, D.L. Crystallographic preferred orientation of olivine in sheared partially molten rocks: The source of the" a-c Switch". *Geochem. Geophys. Geosyst.* **2018**, *19*, 316–336. [CrossRef]

23. Jung, S.; Jung, H.; Austrheim, H. Characterization of olivine fabrics and mylonite in the presence of fluid and implications for seismic anisotropy and shear localization. *Earth Planets Space* **2014**, *66*, 21. [CrossRef]

24. Kim, D.; Jung, H. Deformation microstructures of olivine and chlorite in chlorite peridotites from Almklovdalen in the Western Gneiss Region, southwest Norway, and implications for seismic anisotropy. *Int. Geol. Rev.* **2015**, *57*, 650–668. [CrossRef]

25. Behr, W.M.; Smith, D. Deformation in the mantle wedge associated with Laramide flat-slab subduction. *Geochem. Geophys. Geosyst.* **2016**, *17*, 2643–2660. [CrossRef]

26. Puelles, P.; Abalos, B.; Ibarguchi, J.I.G.; Sarrionandia, F.; Carracedo, M.; Fernandez-Armas, S. Petrofabric and seismic properties of lithospheric mantle xenoliths from the Calatrava volcanic field (Central Spain). *Tectonophysics* **2016**, *683*, 200–215. [CrossRef]

27. Fumagalli, P.; Poli, S. Experimentally determined phase relations in hydrous peridotites to 6.5 GPa and their consequences on the dynamics of subduction zones. *J. Petrol.* **2005**, *46*, 555–578. [CrossRef]

28. Hacker, B.R.; Abers, G.A.; Peacock, S.M. Subduction factory 1. Theoretical mineralogy, densities, seismic wave speeds, and H_2O contents. *J. Geophys. Res. Solid Earth* **2003**, *108*, 26. [CrossRef]

29. Ohtani, E. Water in the mantle. *Elements* **2005**, *1*, 25–30. [CrossRef]

30. Wada, I.; Behn, M.D.; Shaw, A.M. Effects of heterogeneous hydration in the incoming plate, slab rehydration, and mantle wedge hydration on slab-derived H_2O flux in subduction zones. *Earth Planet. Sci. Lett.* **2012**, *353*, 60–71. [CrossRef]

31. Almqvist, B.S.G.; Mainprice, D. Seismic properties and anisotropy of the continental crust: Predictions based on mineral texture and rock microstructure. *Rev. Geophys.* **2017**, *55*, 367–433. [CrossRef]

32. Lee, J.; Jung, H.; Klemd, R.; Tarling, M.S.; Konopelko, D. Lattice preferred orientation of talc and implications for seismic anisotropy in subduction zones. *Earth Planet. Sci. Lett.* **2020**, *537*, 1–11. [CrossRef]

33. Mainprice, D.; Ildefonse, B. Seismic anisotropy of subduction zone minerals–contribution of hydrous phases. In *Subduction Zone Geodynamics*; Springer: Berlin, Germany, 2009; pp. 63–84.

34. Mookherjee, M.; Mainprice, D. Unusually large shear wave anisotropy for chlorite in subduction zone settings. *Geophys. Res. Lett.* **2014**, *41*, 1506–1513. [CrossRef]

35. Walpole, J.; Wookey, J.; Kendall, J.M.; Masters, T.G. Seismic anisotropy and mantle flow below subducting slabs. *Earth Planet. Sci. Lett.* **2017**, *465*, 155–167. [CrossRef]

36. Liu, X.; Zhao, D.P. Depth-varying azimuthal anisotropy in the Tohoku subduction channel. *Earth Planet. Sci. Lett.* **2017**, *473*, 33–43. [CrossRef]

37. Long, M.D. Constraints on subduction geodynamics from seismic anisotropy. *Rev. Geophys.* **2013**, *51*, 76–112. [CrossRef]

38. Jung, H. Seismic anisotropy produced by serpentine in mantle wedge. *Earth Planet. Sci. Lett.* **2011**, *307*, 535–543. [CrossRef]

39. Katayama, I.; Hirauchi, H.; Michibayashi, K.; Ando, J. Trench-parallel anisotropy produced by serpentine deformation in the hydrated mantle wedge. *Nature* **2009**, *461*, 1114–1117. [CrossRef]

40. Nagaya, T.; Walker, A.M.; Wookey, J.; Wallis, S.R.; Ishii, K.; Kendall, J.M. Seismic evidence for flow in the hydrated mantle wedge of the Ryukyu subduction zone. *Sci. Rep.* **2016**, *6*, 13. [CrossRef]

41. Kang, H.; Jung, H. Lattice-preferred orientation of amphibole, chlorite, and olivine found in hydrated mantle peridotites from Bjørkedalen, southwestern Norway, and implications for seismic anisotropy. *Tectonophysics* **2019**, *750*, 137–152. [CrossRef]

42. Ji, S.C.; Shao, T.B.; Michibayashi, K.; Long, C.X.; Wang, Q.; Kondo, Y.; Zhao, W.H.; Wang, H.C.; Salisbury, M.H. A new calibration of seismic velocities, anisotropy, fabrics, and elastic moduli of amphibole-rich rocks. *J. Geophys. Res. Solid Earth* **2013**, *118*, 4699–4728. [CrossRef]

43. Kim, J.; Jung, H. New crystal preferred orientation of amphibole experimentally found in simple shear. *Geophys. Res. Lett.* **2019**, *46*, 12996–13005. [CrossRef]

44. Ko, B.; Jung, H. Crystal preferred orientation of an amphibole experimentally deformed by simple shear. *Nat. Commun.* **2015**, *6*, 10. [CrossRef]

45. Cao, Y.; Song, S.G.; Su, L.; Jung, H.; Niu, Y.L. Highly refractory peridotites in Songshugou, Qinling orogen: Insights into partial melting and melt/fluid-rock reactions in forearc mantle. *Lithos* **2016**, *252*, 234–254. [CrossRef]

46. Evans, B.W. Metamorphism of alpine metaperidotite and serpentinite. *Annu. Rev. Earth Planet. Sci.* **1977**, *5*, 397–447. [CrossRef]

47. Austrheim, H. Fluid and deformation induced metamorphic processes around Moho beneath continent collision zones: Examples from the exposed root zone of the Caledonian mountain belt, W-Norway. *Tectonophysics* **2013**, *609*, 620–635. [CrossRef]
48. Hacker, B.R.; Andersen, T.B.; Johnston, S.; Kylander-Clark, A.R.C.; Peterman, E.M.; Walsh, E.O.; Young, D. High-temperature deformation during continental-margin subduction & exhumation: The ultrahigh-pressure Western Gneiss Region of Norway. *Tectonophysics* **2010**, *480*, 149. [CrossRef]
49. Soper, N.J.; Strachan, R.A.; Holdsworth, R.E.; Gayer, R.A.; Greiling, R.O. Sinistral transpression and the closure of Iapetus. *J. Geol. Soc.* **1992**, *149*, 871–880. [CrossRef]
50. Gee, D.G. Tectonic model for central part of Scandinavian Caledonides. *Am. J. Sci.* **1975**, *A275*, 468–515.
51. Roberts, D. The Scandinavian Caledonides: Event chronology, palaeogeographic settings and likely, modern analogues. *Tectonophysics* **2003**, *365*, 283–299. [CrossRef]
52. Beyer, E.E.; Brueckner, H.K.; Griffin, W.L.; O'Reilly, S.Y. Laurentian provenance of archean mantle fragments in the proterozoic baltic crust of the Norwegian Caledonides. *J. Petrol.* **2012**, *53*, 1357–1383. [CrossRef]
53. Brueckner, H.K.; Carswell, D.A.; Griffin, W.L.; Medaris, L.G.; Van Roermund, H.L.M.; Cuthbert, S.J. The mantle and crustal evolution of two garnet peridotite suites from the Western Gneiss Region, Norwegian Caledonides: An isotopic investigation. *Lithos* **2010**, *117*, 1–19. [CrossRef]
54. Carswell, D.A. The metamorphic evolution of Mg-Cr type Norwegian garnet peridotites. *Lithos* **1986**, *19*, 279–297. [CrossRef]
55. Cordellier, F.; Boudier, F.; Boullier, A.M. Structural study of the Almklovdalen peridotite massif (southern-Norway). *Tectonophysics* **1981**, *77*, 257–281. [CrossRef]
56. Jamtveit, B.; Carswell, D.A.; Mearns, E.W. Chronology of the high-pressure metamorphism of Norwegian garnet peridotites pyroxenites. *J. Metamorph. Geol.* **1991**, *9*, 125–139. [CrossRef]
57. Kylander-Clark, A.R.C.; Hacker, B.R.; Mattinson, J.M. Slow exhumation of UHP terranes: Titanite and rutile ages of the Western Gneiss Region, Norway. *Earth Planet. Sci. Lett.* **2008**, *272*, 531–540. [CrossRef]
58. Lapen, T.J.; Medaris, L.G.; Beard, B.L.; Johnson, C.M. The Sandvik peridotite, Gurskoy, Norway: Three billion years of mantle evolution in the Baltica lithosphere. *Lithos* **2009**, *109*, 145–154. [CrossRef]
59. Vrijmoed, J.C.; Van Roermund, H.L.M.; Davies, G.R. Evidence for diamond-grade ultra-high pressure metamorphism and fluid interaction in the Svartberget Fe-Ti garnet peridotite-websterite body, Western Gneiss Region, Norway. *Mineral. Petrol.* **2006**, *88*, 381–405. [CrossRef]
60. Austrheim, H.; Corfu, F.; Bryhni, I.; Andersen, T.B. The Proterozoic Hustad igneous complex: A low strain enclave with a key to the history of the Western Gneiss Region of Norway. *Precambrian Res.* **2003**, *120*, 149–175. [CrossRef]
61. Hacker, B.R. Ascent of the ultrahigh-pressure Western Gneiss Region, Norway. *Geol. Soc. Am. Spec. Pap.* **2007**, *419*, 171–184. [CrossRef]
62. Dobrzhinetskaya, L.F.; Eide, E.A.; Larsen, R.B.; Sturt, B.A.; Tronnes, R.G.; Smith, D.C.; Taylor, W.R.; Posukhova, T.V. Microdiamond in high-grade metamorhic rocks of the Western Gneiss Region, Norway. *Geology* **1995**, *23*, 597–600. [CrossRef]
63. Root, D.B.; Hacker, B.R.; Gans, P.B.; Ducea, M.N.; Eide, E.A.; Mosenfelder, J.L. Discrete ultrahigh-pressure domains in the Western Gneiss Region, Norway: Implications for formation and exhumation. *J. Metamorph. Geol.* **2005**, *23*, 45–61. [CrossRef]
64. Kostenko, O.; Jamtveit, B.; Austrheim, H.; Pollok, K.; Putnis, C. The mechanism of fluid infiltration in peridotites at Almklovdalen, western Norway. *Geofluids* **2002**, *2*, 203–215. [CrossRef]
65. O'Neill, H.S.C.; Wall, V.J. The olivine-orthopyroxene-spinel oxygen geobarometer, the nickel precipitation curve, and the oxygen fugacity of the Earth's upper mantle. *J. Petrol.* **1987**, *28*, 1169–1191. [CrossRef]
66. Witteickschen, G.; Seck, H.A. Solubility of Ca and Al in orthopyroxene from spinel peridotite—An improved version of an empirical geothermometer. *Contrib. Mineral. Petrol.* **1991**, *106*, 431–439. [CrossRef]
67. O'Neill, H.S.C. The transition between spinel lherzolite and garnet lherzolite, and its use as a geobarometer. *Contrib. Mineral. Petrol.* **1981**, *77*, 185–194. [CrossRef]
68. Panozzo, R. Two-dimensional strain from the orientation of lines in a plane. *J. Struct. Geol.* **1984**, *6*, 215–221. [CrossRef]
69. Skemer, P.; Katayama, I.; Jiang, Z.; Karato, S. The misorientation index: Development of a new method for calculating the strength of lattice-preferred orientation. *Tectonophysics* **2005**, *411*, 157–167. [CrossRef]
70. Bunge, H.-J. *Texture Analysis in Materials Science: Mathematical Models*; Butterworths: London, UK, 1982.

71. Mainprice, D. A fortran program to calculate seismic anisotropy from the lattice preferred orientation of minerals. *Comput. Geosci.* **1990**, *16*, 385–393. [CrossRef]
72. Abramson, E.H.; Brown, J.M.; Slutsky, L.J.; Zaug, J. The elastic constants of San Carlos olivine to 17 GPa. *J. Geophys. Res.* **1997**, *102*, 12253–12263. [CrossRef]
73. Brown, J.M.; Abramson, E.H. Elasticity of calcium and calcium-sodium amphiboles. *Phys. Earth Planet. Inter.* **2016**, *261*, 161–171. [CrossRef]
74. Bezacier, L.; Reynard, B.; Bass, J.D.; Sanchez-Valle, C.; Van de Moortele, B.V. Elasticity of antigorite, seismic detection of serpentinites, and anisotropy in subduction zones. *Earth Planet. Sci. Lett.* **2010**, *289*, 198–208. [CrossRef]
75. Pera, E.; Mainprice, D.; Burlini, L. Anisotropic seismic properties of the upper mantle beneath the Torre Alfina area (Northern Appennines, Central Italy). *Tectonophysics* **2003**, *370*, 11–30. [CrossRef]
76. Paterson, M.S. The determination of hydroxyl by infrared-absorption in quartz, silicate-glasses and similar materials. *Bull. Mineral.* **1982**, *105*, 20–29. [CrossRef]
77. Jung, H.; Karato, S. Effects of water on dynamically recrystallized grain-size of olivine. *J. Struct. Geol.* **2001**, *23*, 1337–1344. [CrossRef]
78. Karato, S. Scanning electron microscope observation of dislocations in olivine. *Phys. Chem. Miner.* **1987**, *14*, 245–248. [CrossRef]
79. Kohlstedt, D.L.; Goetze, C.; Durham, W.B.; Vandersande, J. New technique for decorating dislocations in olivine. *Science* **1976**, *191*, 1045–1046. [CrossRef]
80. Gifkins, R.C. *Optical Microscopy of Metals*; Elsevier: New York, NY, USA, 1970.
81. Chatzaras, V.; Kruckenberg, S.C.; Cohen, S.M.; Medaris, L.G.; Withers, A.C.; Bagley, B. Axial-type olivine crystallographic preferred orientations: The effect of strain geometry on mantle texture. *J. Geophys. Res. Solid Earth* **2016**, *121*, 4895–4922. [CrossRef]
82. Prelicz, R.M. Seismic Anisotropy in Peridotites from the Western Gneiss Region (Norway) Laboratory Measurements at High PT Conditions and Fabric Based Model Predictions. Ph.D. Thesis, ETH Zürich, Zürich, Switzerland, 2005.
83. Jung, H. Deformation fabrics of olivine in Val Malenco peridotite found in Italy and implications for the seismic anisotropy in the upper mantle. *Lithos* **2009**, *109*, 341–349. [CrossRef]
84. Khisina, N.R.; Wirth, R.; Andrut, M.; Ukhanov, A.V. Extinction and intrinsic mode of hydrogen occurrence in natural olivine: FTIR and TEM investigation. *Phys. Chem. Miner.* **2001**, *28*, 291–301.
85. Miller, G.H.; Rossman, G.R.; Harlow, G.E. The natural occurrence of hydroxide in olivine. *Phys. Chem. Miner.* **1987**, *14*, 461–472. [CrossRef]
86. Hirth, G.; Kohlstedt, D. Rheology of the upper mantle and the mantle wedge: A view from the experimentalists. *Geophys. Monogr.* **2003**, *138*, 83–105. [CrossRef]
87. Ashby, M.F.; Verral, R.A. Diffusion accommodated flow and superplasticity. *Acta Metall.* **1973**, *21*, 149–163. [CrossRef]
88. Goldsby, D.L.; Kohlstedt, D.L. Superplastic deformation of ice: Experimental observations. *J. Geophys. Res. Solid Earth* **2001**, *106*, 11017–11030. [CrossRef]
89. Katayama, I.; Karato, S. Effect of temperature on the B- to C-type olivine fabric transition and implication for flow pattern in subduction zones. *Phys. Earth Planet. Inter.* **2006**, *157*, 33–45. [CrossRef]
90. Frese, K.; Trommsdorff, V.; Kunze, K. Olivine 100 normal to foliation: Lattice preferred orientation in prograde garnet peridotite formed at high H_2O activity, Cima di Gagnone (Central Alps). *Contrib. Mineral. Petrol.* **2003**, *145*, 75–86. [CrossRef]
91. Katayama, I.; Karato, S.; Brandon, M. Evidence of high water content in the deep upper mantle inferred from deformation microstructures. *Geology* **2005**, *33*, 613–616. [CrossRef]
92. Jung, H.; Lee, J.; Ko, B.; Jung, S.; Park, M.; Cao, Y.; Song, S.G. Natural type-C olivine fabrics in garnet peridotites in North Qaidam UHP collision belt, NW China. *Tectonophysics* **2013**, *594*, 91–102. [CrossRef]
93. Park, M.; Jung, H.; Kil, Y. Petrofabrics of olivine in a rift axis and rift shoulder and their implications for seismic anisotropy beneath the Rio Grande rift. *Isl. Arc.* **2014**, *23*, 299–311. [CrossRef]
94. Boneh, Y.; Morales, L.F.G.; Kaminski, E.; Skemer, P. Modeling olivine CPO evolution with complex deformation histories: Implications for the interpretation of seismic anisotropy in the mantle. *Geochem. Geophys. Geosyst.* **2015**, *16*, 3436–3455. [CrossRef]

95. Long, M.D.; Silver, P.G. The subduction zone flow field from seismic anisotropy: A global view. *Science* **2008**, *319*, 315–318. [CrossRef] [PubMed]

96. Mehl, L.; Hacker, B.R.; Hirth, G.; Kelemen, P.B. Arc-parallel flow within the mantle wedge: Evidence from the accreted Talkeetna arc, south central Alaska. *J. Geophys. Res. Solid Earth* **2003**, *108*, 18. [CrossRef]

97. Savage, M.K. Seismic anisotropy and mantle deformation: What have we learned from shear wave splitting? *Rev. Geophys.* **1999**, *37*, 65–106. [CrossRef]

98. Smith, G.P.; Wiens, D.A.; Fischer, K.M.; Dorman, L.M.; Webb, S.C.; Hildebrand, J.A. A complex pattern of mantle flow in the Lau backarc. *Science* **2001**, *292*, 713–716. [CrossRef]

99. Cao, Y.; Jung, H.; Song, S.G.; Park, M.; Jung, S.; Lee, J. Plastic deformation and seismic properties in fore-arc mantles: A petrofabric analysis of the Yushigou Harzburgites, North Qilian Suture Zone, NW China. *J. Petrol.* **2015**, *56*, 1897–1943. [CrossRef]

100. Huang, Z.C.; Zhao, D.P.; Wang, L.S. Shear wave anisotropy in the crust, mantle wedge, and subducting Pacific slab under northeast Japan. *Geochem. Geophys. Geosyst.* **2011**, *12*, 17. [CrossRef]

101. Long, M.D.; Wirth, E.A. Mantle flow in subduction systems: The mantle wedge flow field and implications for wedge processes. *J. Geophys. Res. Solid Earth* **2013**, *118*, 583–606. [CrossRef]

102. Soto, G.L.; Ni, J.F.; Grand, S.P.; Sandvol, E.; Valenzuela, R.W.; Speziale, M.G.; Gonzalez, J.M.G.; Reyes, T.D. Mantle flow in the Rivera-Cocos subduction zone. *Geophys. J. Int.* **2009**, *179*, 1004–1012. [CrossRef]

103. Audet, P.; Bostock, M.G.; Christensen, N.I.; Peacock, S.M. Seismic evidence for overpressured subducted oceanic crust and megathrust fault sealing. *Nature* **2009**, *457*, 76–78. [CrossRef]

104. Kim, Y.; Clayton, R.W.; Jackson, J.M. Geometry and seismic properties of the subducting Cocos plate in central Mexico. *J. Geophys. Res. Solid Earth* **2010**, *115*, 22. [CrossRef]

105. Kodaira, S.; Iidaka, T.; Kato, A.; Park, J.O.; Iwasaki, T.; Kaneda, Y. High pore fluid pressure may cause silent slip in the Nankai Trough. *Science* **2004**, *304*, 1295–1298. [CrossRef] [PubMed]

106. Peacock, S.M.; Christensen, N.I.; Bostock, M.G.; Audet, P. High pore pressures and porosity at 35 km depth in the Cascadia subduction zone. *Geology* **2011**, *39*, 471–474. [CrossRef]

107. Shelly, D.R.; Beroza, G.C.; Ide, S.; Nakamula, S. Low-frequency earthquakes in Shikoku, Japan, and their relationship to episodic tremor and slip. *Nature* **2006**, *442*, 188–191. [CrossRef] [PubMed]

108. Kim, Y.; Clayton, R.W.; Asimow, P.D.; Jackson, J.M. Generation of talc in the mantle wedge and its role in subduction dynamics in central Mexico. *Earth Planet. Sci. Lett.* **2013**, *384*, 81–87. [CrossRef]

Article

Seismic Properties of a Unique Olivine-Rich Eclogite in the Western Gneiss Region, Norway

Yi Cao [1,*], Haemyeong Jung [2] and Jian Ma [1]

[1] State Key Laboratory of Geological Processes and Mineral Resources, School of Earth Sciences, China University of Geosciences, Wuhan 430074, China; wjmajian@gmail.com

[2] Tectonophysics Laboratory, School of Earth and Environmental Sciences, Seoul National University, Seoul 08826, Korea; hjung@snu.ac.kr

* Correspondence: caoyi0701@126.com

Received: 24 July 2020; Accepted: 29 August 2020; Published: 31 August 2020

Abstract: Investigating the seismic properties of natural eclogite is crucial for identifying the composition, density, and mechanical structure of the Earth's deep crust and mantle. For this purpose, numerous studies have addressed the seismic properties of various types of eclogite, except for a rare eclogite type that contains abundant olivine and orthopyroxene. In this contribution, we calculated the ambient-condition seismic velocities and seismic anisotropies of this eclogite type using an olivine-rich eclogite from northwestern Flemsøya in the Nordøyane ultrahigh-pressure (UHP) domain of the Western Gneiss Region in Norway. Detailed analyses of the seismic properties data suggest that patterns of seismic anisotropy of the Flem eclogite were largely controlled by the strength of the crystal-preferred orientation (CPO) and characterized by significant destructive effects of the CPO interactions, which together, resulted in very weak bulk rock seismic anisotropies (AVp = 1.0–2.5%, max. AVs = 0.6–2.0%). The magnitudes of the seismic anisotropies of the Flem eclogite were similar to those of dry eclogite but much lower than those of gabbro, peridotite, hydrous-phase-bearing eclogite, and blueschist. Furthermore, we found that amphibole CPOs were the main contributors to the higher seismic anisotropies in some amphibole-rich samples. The average seismic velocities of Flem eclogite were greatly affected by the relative volume proportions of omphacite and amphibole. The Vp (8.00–8.33 km/s) and Vs (4.55–4.72 km/s) were remarkably larger than the hydrous-phase-bearing eclogite, blueschist, and gabbro, but lower than dry eclogite and peridotite. The Vp/Vs ratio was almost constant (avg. ≈ 1.765) among Flem eclogite, slightly larger than olivine-free dry eclogite, but similar to peridotite, indicating that an abundance of olivine is the source of their high Vp/Vs ratios. The Vp/Vs ratios of Flem eclogite were also higher than other (non-)retrograded eclogite and significantly lower than those of gabbro. The seismic features derived from the Flem eclogite can thus be used to distinguish olivine-rich eclogite from other common rock types (especially gabbro) in the deep continental crust or subduction channel when high-resolution seismic wave data are available.

Keywords: seismic anisotropy; seismic velocity; olivine-rich eclogite; Western Gneiss Region

---

1. Introduction

Eclogite is a unique high-pressure to ultrahigh-pressure (HP–UHP) metamorphic rock that commonly originates from the subduction of an oceanic or continental crust or derives from the thickening of an orogenic crust during a continental collision [1–4]. A typical eclogite consists of a mainly bi-mineralic assemblage (i.e., garnet and omphacite), which is metamorphosed from mafic components, such as gabbro, diabase, and basalt [4]. Hydrous minerals, such as amphibole (e.g., glaucophane and hornblende), epidote, lawsonite, and mica (e.g., phengite and biotite) are also common in the eclogite and indicate different equilibrated pressure–temperature (P–T) conditions and hydrous states of the eclogite facies' metamorphisms [5–7]. Magnesium-rich minerals, such as olivine and orthopyroxene, can also be sporadically abundant in the eclogite when its protolith is rich in olivine (i.e., olivine gabbro or troctolite) [8].

Investigating the seismic properties (i.e., P- and S-wave velocities and their anisotropies) of eclogite is crucial for constraining the presence of eclogite in the deep crust and upper mantle, which has profound implications for interpreting the composition, density, and thermal and mechanical structures of the subducted crust, continental lithosphere, and upper mantle, as well as for reconstructing the geodynamic evolutions of the subduction and collision zones [9–19]. For this purpose, numerous previous studies have addressed the seismic properties of dry eclogite (i.e., bi-mineralic eclogite), e.g., [20–27]; retrograded eclogite (i.e., amphibolized eclogite), e.g., [16,19,23,28–31]; epidote/glaucophane eclogite [12,14,28]; lawsonite eclogite [13,32]. However, because of the sample rarity, the seismic properties of the olivine and orthopyroxene rich eclogite have not been studied yet.

The eclogite in the Flem Gabbro from Flemsøya in the Western Gneiss Region (WGR), Norway, is exceptional due to its peculiar mineral assemblage of olivine and orthopyroxene, which were inherited from an olivine gabbro protolith [8,33]. This eclogite body may once have been situated in a continental subduction channel (the interface between the subducted slab and the wedge of an overlying crust and mantle) or a continental root zone (around Moho) owing to continental subduction and collision [2]. In this context, it can provide a valuable opportunity to examine the seismic properties of this rare type of eclogite and its implications for the detectability of eclogite bodies in the deep crust.

2. Geological Background

The Western Gneiss Region (WGR) in Norway is one of the largest, best-exposed, and most studied HP–UHP terranes in the world. It resulted from the deep subduction of the Baltica basement and subsequent collision with the Laurentia continent ≈425–400 Ma [34–38]. The peak metamorphic pressure and temperature (P–T) conditions increase from SE to NW with the highest P–T estimates preserved in three UHP domains (Nordfjord, Sørøyane, and Nordøyane) along the west Norwegian coast (Figure 1a). The Flemsøya is located in the Nordøyane UHP domain, which has the highest peak P–T condition (>800 °C and 3.0–4.0 GPa; see Terry et al. [39], Carswell et al. [40], and Butler et al. [41]) among the three UHP domains. The eclogite samples analyzed in this study were collected from northwestern Flemsøya in the Nordøyane UHP domain (Figure 1b). The Flem Gabbro (or Sandvikhaugane Gabbro) is the largest gabbro block that experienced heterogeneous eclogitization [33]. The studied eclogite (named as Flem eclogite hereafter) is located in the northernmost margin of the Flem gabbro. It is almost completely eclogitized and features variously deformed structures (massive and foliated structures) in the outcrop [42]. Our recent petrological work proposed that the Flem eclogite experienced UHP metamorphism at ≈2.7–3.7 GPa and ≈700–820 °C, in accordance with the previous P-T estimates in the Nordøyane UHP domain [42].

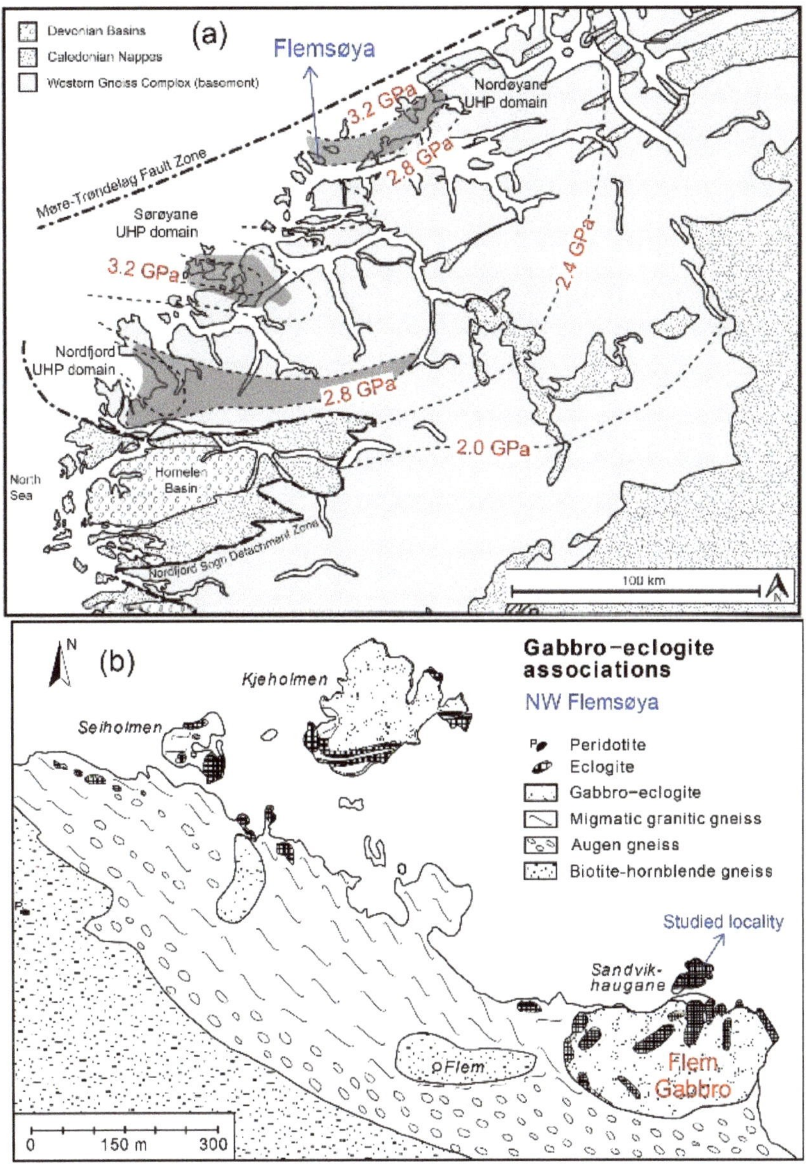

Figure 1. (**a**) Geological map showing the Western Gneiss Region (WGR) (modified from Renedo et al. [43]). The dark gray shaded area includes three ultrahigh-pressure (UHP) domains (Nordfjord, Sørøyane, and Nordøyane) along the western coast of the WGR. The dashed lines are isobars of peak metamorphic pressures. The blue arrow indicates the Flemsøya region where our eclogite samples were collected. (**b**) Geological map of the NW part of Flemsøya showing the distributions of eclogite (modified from Mørk [33] and Terry and Robinson [38]). The blue arrow indicates the locality where our eclogite samples were collected.

3. Samples and Methods

3.1. Sample Descriptions

Based on the field occurrences and deformation microstructures, the Flem eclogite could be divided into massive eclogite (MEC) and foliated eclogite (FEC). In the hand specimen, MEC hardly shows discernable foliation and lineation, whereas FEC developed these structural frameworks conspicuously. The MEC was mainly composed of olivine (Ol), garnet (Grt), omphacite (Omp), orthopyroxene (Opx), phlogopite (Phl), ilmenite (Ilm), amphibole (Amp), minor spinel (Sp), and plagioclase (Pl) (Table 1). The Ol crystals were mostly (sub)euhedral and often rimmed with small Opx and Omp grains (Figure 2a). The Ol crystals lacked a shape-preferred orientation (SPO) and presented as an irregularly shaped Grt matrix (Figure 2a). In contrast, the FEC often displayed a pronounced foliation characterized by mineral layering, Grt bands, and elongated Ol porphyroclasts (Figure 2b). The mineral assemblage of FEC was similar to but had relatively higher proportions of Omp, Grt, and Amp, but lower Ol content, than the MEC (Table 1).

Table 1. Normalized phase volume proportions (vol.%) based on electron backscatter diffraction (EBSD) mapping.

Sample	Rock Type	Grt	Omp	Ol	Opx	Amp	Phl	Ilm	Pl	Sp
NW1140	MEC	21.64	29.37	40.12	2.78	2.97	1.29	1.40	0.40	0.03
NW1153	MEC	24.41	21.10	40.68	6.39	2.73	1.19	2.11	1.03	0.34
NW1142	FEC (subgroup 1)	30.55	13.25	16.03	4.37	33.28	1.15	0.98	0.37	0.02
NW1143	FEC (subgroup 1)	22.95	30.03	30.00	2.63	12.05	1.00	0.82	0.50	0.02
NW1144	FEC (subgroup 1)	31.82	26.42	21.06	2.10	16.47	1.00	0.93	0.17	0.02
NW1147	FEC (subgroup 2)	36.56	38.29	18.45	0.36	4.71	0.88	0.55	0.19	0.03
NW1148	FEC (subgroup 2)	29.79	24.25	27.74	2.71	11.99	2.25	0.77	0.46	0.04

Mineral abbreviations: garnet (Grt), omphacite (Omp), olivine (Ol), orthopyroxene (Opx), amphibole (Amp), phlogopite (Phl), ilmenite (Ilm), plagioclase (Pl), and spinel (Sp).

Figure 2. Optical photomicrographs under crossed polarizers showing the microstructures of (**a**) massive eclogite (MEC) and (**b**) foliated eclogite (FEC). Mineral abbreviations: olivine (Ol), garnet (Grt), amphibole (Amp), and orthopyroxene (Opx).

3.2. Crystal-Preferred Orientation (CPO) Measurement and Analysis

The CPOs of the Flem eclogite were measured by investigating finely polished thin sections using an electron backscatter diffraction (EBSD) system. The thin sections, parallel to the lineation and perpendicular to the foliation (i.e., XZ plane), were used for the FEC, whereas the thin sections, parallel to the foliation of the neighboring FEC in the outcrop (named the apparent XY plane or X'Y' plane), were used for the MEC owing to its vague foliation and lineation. The EBSD system we used was installed in a Quanta 450 Field Emission Scanning Electron Microscope (FESEM) at the State Key Laboratory of Geological Processes and Mineral Resources in the China University of Geosciences (Wuhan), China. The operational settings of the EBSD system were: an acceleration voltage of 20 kV, a working distance of 25 mm, and a spot size of 60 in a low-vacuum mode. The CPOs were presented in the pole figures with marked foliation and/or lineation, and their strengths were quantified using the J-index (ranging from 1 to infinity; see [44]) and the M-index (ranging from 0 to 1; see [45]). The CPO analyses were completed using the MTEX toolbox (ver. 5.2.beta2) in MATLAB (ver. 2017b) (http://mtex-toolbox.github.io/) [46,47]. More details about the EBSD measurement and data treatment of studied samples are available in Cao et al. [42].

3.3. Seismic Property Calculation

To calculate the seismic property (velocity and anisotropy) of the MEC and FEC, the elastic stiffness, CPOs, densities, and volume proportions of their constituent minerals are required before inserting them into the Christoffel equation to solve for the seismic velocities (i.e., Vp, Vs1, and Vs2) along every direction in 3D space [48]. The volume proportions of the constituent minerals were obtained directly from the EBSD phase maps (Table 1). The densities of the bulk rock were averaged from the densities of each mineral weighted by their volume proportions. The elastic stiffness of the single-crystal (Cij) Amp (kataphorite), Opx (enstatite80), Phl, Pl (albite), and Sp were adopted from Brown and Abramson [49], Webb and Jackson [50], Alexandrov and Ryzhova [51], Brown et al. [52], and Duan et al. [53], respectively. Because Omp in MEC and FEC has the composition of Jd20-40Di50-75 [42], the (Cij) of the Omp that has a similar composition (Jd30Di70) recently reported by Hao et al. [54] is used. The elastic stiffness of single-crystal Ol (avg. Fo# $\approx$ 68) was estimated using the (Cij) of their forsterite [55] and fayalite [56] endmembers, which were linearly averaged by their molar proportions. Likewise, the elastic stiffness of single-crystal Grt (Grs10–15, Prp40–45, and Alm+Sps45) was calculated using the (Cij) of grossular [57], pyrope [58], and alamdine-spessartite [59] endmembers. Since the single-crystal elastic stiffness of ilmenite ($FeTiO_3$) was not reported, we constructed a fictive isotropic tensor using the bulk and shear modulus of ilmenite ($MgTiO_3$) [60]. A list of single-crystal elastic stiffness tensors for common minerals was recently provided by Almqvist and Mainprice [61]. The elastic stiffness tensors <Cij> of mono-mineralic polycrystal and bulk rock were calculated by integrating the elastic stiffness tensor (Cij) of all grains (and phases) over their individual orientations using the Vogit-Reuss-Hill (VRH) averaging scheme. In this study, the calculation was accomplished using the MTEX toolbox (ver. 5.2.beta2) in MATLAB (ver. 2017b) [47,62]. Since this calculation method yields seismic velocities along discrete orientations in 3D space, we could derive the average Vp and Vs by simply averaging their maximum and minimum values (Table 2).

To understand the contributions of different mineral phases and/or their CPOs to the bulk rock seismic properties, we also calculated the seismic properties of major mono-mineralic aggregates (e.g., olivine, orthopyroxene, garnet, omphacite, and amphibole). These results were compared with the seismic properties of bulk rock, providing a straightforward approach to characterizing the constructive and destructive effects among different mineral aggregates.

4. Results

4.1. Crystal-Preferred Orientation

Both MEC and FEC exhibited obvious but weak CPOs for most of their constituent minerals (Figures 3 and 4). Notably, in most FEC samples, Opx, Omp, and Amp tended to align the maximum of their [001] axes subparallel to the lineation and the maximum of the [010] or [100] axes sub-perpendicular to the foliation (Figures 3 and 4). These CPO patterns are common and are also observed in many other deformed Opx, Omp, and Amp bearing rocks. The weak CPO strengths agreed with their fairly low J- (<2.5) and M-indices (<0.06). For more details of the EBSD analytical method, descriptions of the microstructures and CPOs in the studied MEC and FEC samples, readers are referred to Cao et al. [42].

4.2. Seismic Properties

4.2.1. Olivine Polycrystals

The Ol polycrystals in the MEC showed the maxima of Vp at low angles to the apparent foliation, subparallel either to the X′ or Y′ direction, while the Vp was low when a seismic ray propagated at moderate to high angles to the apparent foliation (Figure S1a,b). The degree of shear wave splitting or S-wave polarization anisotropy and velocities of the fast shear wave (Vs1) displayed no regular patterns among MEC samples. In the FEC, the Vp, AVs, and Vs1 all lacked consistent distributions between the samples (Figure S1c–g). Regardless of the MEC and FEC samples, both the fast P-wave and the polarization direction of the fast shear wave (S1) tended to align parallel to the maxima of the [100] axes of the Ol CPOs (Figure 3). Notably, the P-wave anisotropy (AVp) and maximum polarization anisotropy of the S-wave (max. AVs) were higher in the MEC (AVp = 2.04–3.42%, max. AVs = 1.69–2.53%) than the FEC (AVp = 0.86–1.65%, max. AVs = 0.84–1.28%), which is in agreement with the stronger Ol CPO strength in the MEC than the FEC (Figure 3).

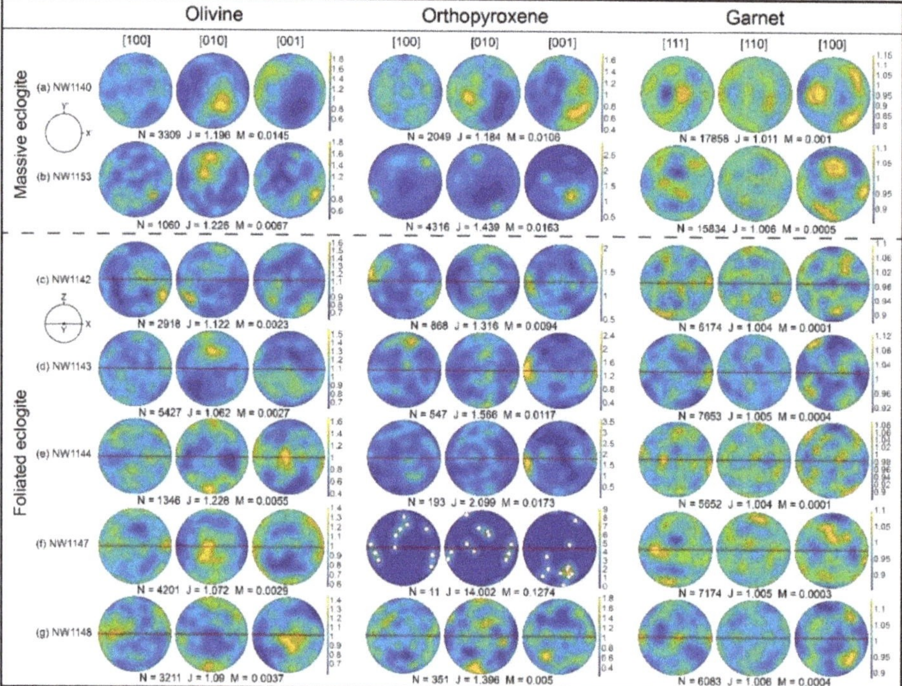

Figure 3. Crystal-preferred orientations (CPOs) of Ol (left column), Opx (middle column), and Grt (right column) in the MEC and FEC (modified from Cao et al. [42]). The pole figures are presented with an equal area, in the lower hemisphere, and contoured with a resolution of 1°. The superimposed circles in (**f**) are scattered orientations of Opx grains. X, Y, and Z denote the lineation, direction perpendicular to the lineation and parallel to the foliation, and foliation-normal direction, respectively. X′ and Y′ are the apparent lineation and the direction perpendicular to the lineation on the apparent foliation (X′Y′ plane). N: number of grains, M: M-index [45], and J: J-index [44].

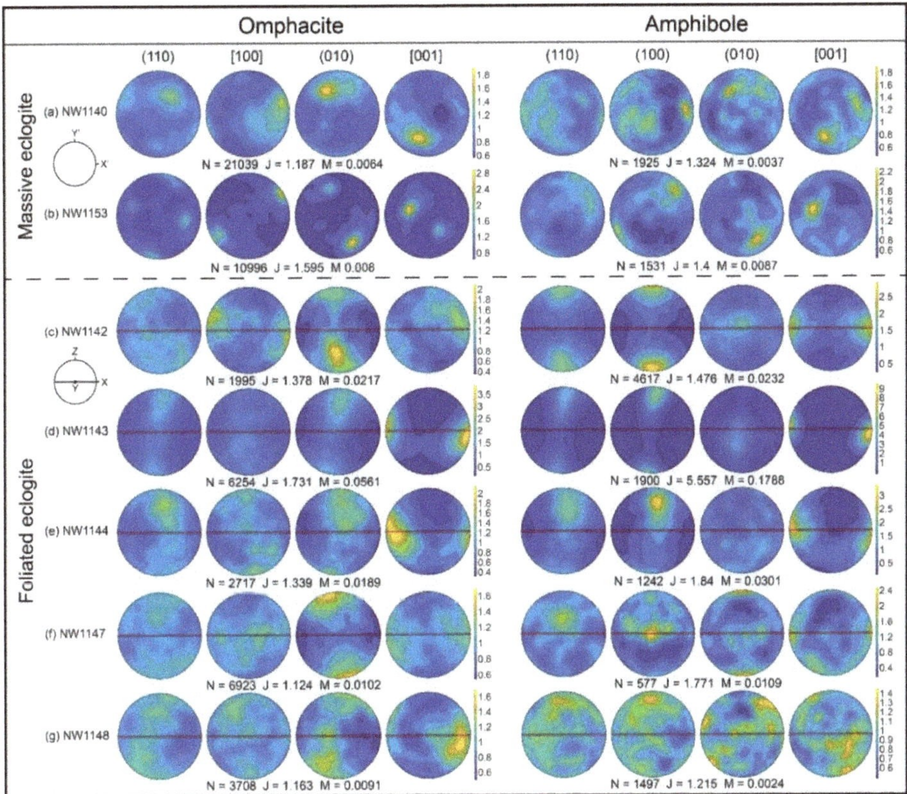

Figure 4. CPOs of the Omp (left column) and Amp (right column) in the MEC and FEC (modified from Cao et al. [42]). The method used to plot the pole figures and the associated parameters were the same as those used in Figure 3.

4.2.2. Orthopyroxene Polycrystals

The Vp, AVs, and Vs1 patterns of Opx polycrystals were irregular in the MEC, but they were more consistent in the FEC (Figure S2). Except for one FEC sample (Figure S2f), the max. Vp was aligned subparallel to the lineation and the low Vp tended to distribute in a girdle perpendicular to the lineation with their minima mostly subnormal to the foliation (Figure S2c–e,g). Despite the lack of consistency in the distributions of AVs and Vs1 in the FEC, the fast S-wave always tended to polarize parallel to the lineation just like the fast P-wave (Figure S2c–g). The magnitudes of the AVp and max. AVs were low and varied in similar ranges for both the MEC (AVp = 1.30%, max. AVs = 1.20–1.72%) and FEC (AVp = 0.84–1.52%, max. AVs = 0.82–1.70%).

4.2.3. Garnet Polycrystals

In both the MEC and FEC, the Grt polycrystals displayed orthogonal patterns for the max. and min. Vp (Figure S3). The max. Vp mainly corresponded to the <100> axes in the Grt CPO (Figure 3), which were the fastest P-wave directions for the Grt single crystal. The AVs and Vs1 also displayed some regular clustering and girdling without consistent spatial correlation with the foliation and lineation. The AVp and max. AVs were very low in both the MEC (AVp = 0.03–0.04%, max. AVs = 0.06–0.09%) and FEC (AVp = 0.01–0.03%, max. AVs = 0.03–0.05%).

4.2.4. Omphacite Polycrystals

The Omp polycrystals exhibited disparate and irregular Vp, AVs, and Vs1 patterns in the MEC, whereas they were more consistent and regular in the FEC (Figure S4). Specifically, the max. Vp was always subparallel to the lineation, complying with the maxima of the Omp [001] axes in the FEC (Figure 4). In contrast, the slow P-wave was perpendicular to the lineation with minima mainly at high angles to the foliation, matching the distribution of the Omp [010] pole (Figure 4). The fast S-wave tended to polarize subparallel to the lineation when the seismic ray traveled at high angles to the lineation. The intensities of the AVp and max. AVs were also larger in the MEC (AVp = 3.09–4.98%, max. AVs = 2.09–5.29%) than in the FEC (AVp = 1.87–2.69%, max. AVs = 0.83–2.12%).

4.2.5. Amphibole Polycrystals

No consistent Vp, AVs, and Vs1 patterns were observed in the Amp polycrystals in the MEC (Figure S5a,b). In contrast, these patterns were more regular in the FEC (Figure S5). Excluding two FEC samples that showed XZ-girdled (Figure S5f) and off-lineation clustering (Figure S5g) of the fast Vp, the max. Vp was aligned subparallel to the lineation in the other FEC samples (Figure S5c–e). Furthermore, the min. Vp was always found in the plane sub-perpendicular to the lineation. Notably, the max. and min. Vps were well correlated with the distributions of the [001] and [100] axes of Amp, respectively (Figure 4). The AVs was large when the seismic ray propagated along the moderate Vp directions and with S1 polarizing parallel to the max. Vp directions. Except for one FEC sample that had the lowest AVp (1.58%) and max. AVs (1.45%) (Figure S5g), the seismic anisotropies of the FEC samples (AVp = 3.29–16.95%, max. AVs = 2.56–12.00%) were much higher than those of the MEC (AVp = 2.29–2.80%, max. AVs = 2.00–2.44%).

4.2.6. Bulk Rocks

Overall, the bulk rock Vp distribution patterns of the MEC were similar to those of the Omp polycrystals (cf. Figures S4a,b and 5a,b). In contrast, three FEC samples (subgroup 1) presented Vp, AVs and Vs1 patterns resembling those of the Amp polycrystals (cf. Figures S5c–e and 5c–e); the other two FEC samples (subgroup 2) showed similar Vp, AVs, and Vs1 patterns to those of the Omp or Ol polycrystals (Figures S1f,g, S4f,g and 5f,g). Apart from one MEC sample (Figure 5a), the max. Vp and polarization direction of the fast S-wave were aligned subparallel (occasionally oblique) to the (apparent) lineation in other eclogite samples (Figure 5b–g). The two MEC samples showed an intermediate AVp (1.06–1.08%) and max. AVs (1.00–1.38%) between the subgroup 1 FECs (Figures 5c–e, 6 and 7), which display the highest AVp (1.78–2.46%) and max. AVs (1.30–1.99%), and the subgroup 2 FECs (Figures 5f,g, 6 and 7), which exhibit the lowest AVp (0.99–1.09%) and max. AVs (0.63–0.65%) (Table 2). Taking all samples together, the intensities of the AVp and max. AVs increased markedly with the volume proportion of Amp (Figure 8) and CPO strength (Figure 9).

Table 2. Calculated seismic properties of the studied MEC and FEC samples.

Sample	Rock Type	Density (g/cm^3)	Vp (km/s) [1]	Vs (km/s) [1]	Vp/Vs	AVp (%) [2]	max. AVs (%) [2]	AVs1 (%) [3]	AVs2 (%) [3]	A(Vp/Vs1) (%) [4]	A(Vp/Vs2) (%) [4]
NW1140	MEC	3.562	8.17	4.61	1.770	1.06	1.00	0.62	0.91	1.22	1.19
NW1153	MEC	3.579	8.18	4.62	1.769	1.08	1.38	1.12	0.95	1.56	1.60
NW1142	FEC (subgroup 1)	3.519	8.00	4.55	1.758	2.46	1.57	1.13	1.08	1.59	1.96
NW1143	FEC (subgroup 1)	3.517	8.13	4.61	1.765	2.42	1.99	1.60	1.26	1.95	2.06
NW1144	FEC (subgroup 1)	3.549	8.18	4.64	1.762	1.78	1.30	0.97	0.86	1.20	1.41
NW1147	FEC (subgroup 2)	3.575	8.33	4.72	1.763	1.09	0.63	0.30	0.49	1.11	0.91
NW1148	FEC (subgroup 2)	3.547	8.15	4.62	1.766	0.99	0.65	0.48	0.42	0.88	0.75

[1] $Vp = \left(Vp_{max} + Vp_{min}\right)/2$ and $Vs = (Vs1_{max} + Vs1_{min} + Vs2_{max} + Vs2_{min})/4$; [2] $AVp = 200 \times \left(Vp_{max} - Vp_{min}\right)/\left(Vp_{max} + Vp_{min}\right)$ and $AVs = 200 \times (Vs1 - Vs2)/(Vs1 + Vs2)$; [3] $AVs1 = 200 \times (Vs1_{max} - Vs1_{min})/(Vs1_{max} + Vs1_{min})$ and $AVs2 = 200 \times (Vs2_{max} - Vs2_{min})/(Vs2_{max} + Vs2_{min})$; [4] $A(Vp/Vs1) = 200 \times (Vp/Vs1_{max} - Vp/Vs1_{min})/(Vp/Vs1_{max} + Vp/Vs1_{min})$ and $A(Vp/Vs2) = 200 \times (Vp/Vs2_{max} - Vp/Vs2_{min})/(Vp/Vs2_{max} + Vp/Vs2_{min})$.

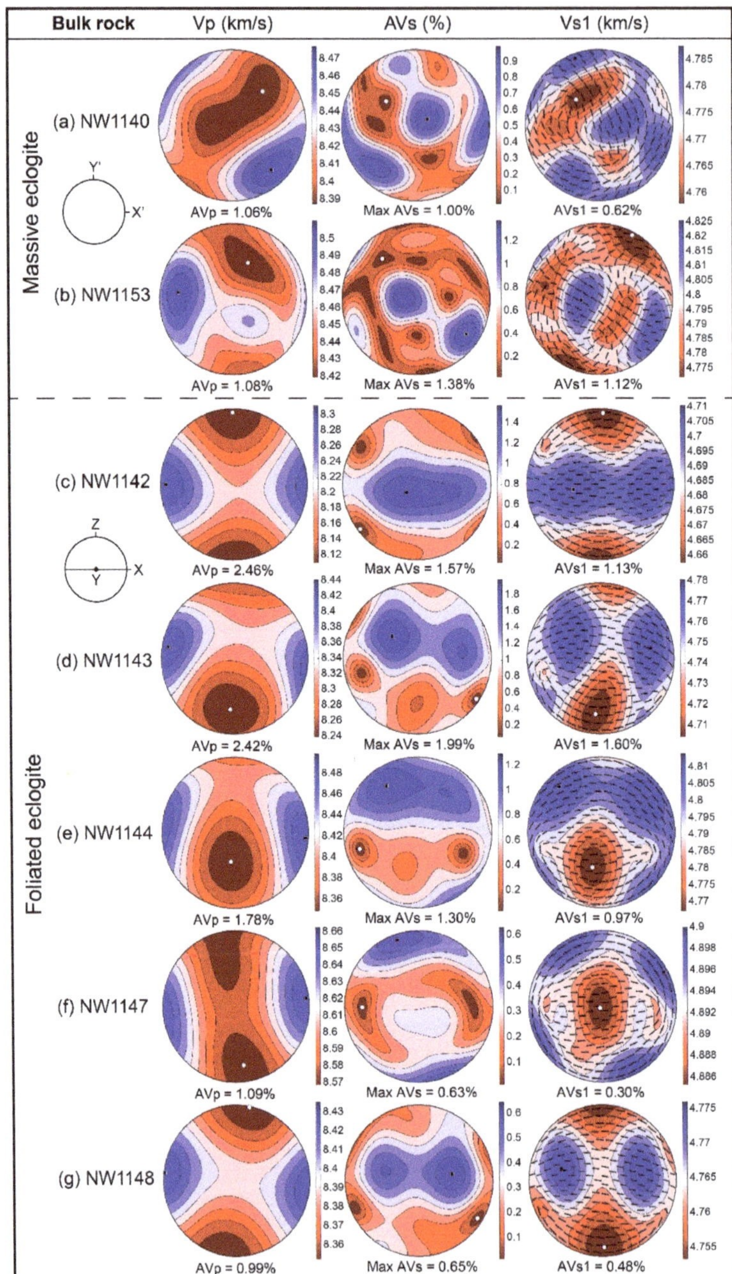

Figure 5. Bulk rock seismic properties of the MEC and FEC in ambient conditions. (**a,b**) MEC, (**c–g**) FEC. Data are presented using equal area and upper hemisphere stereonets. First column: P-wave velocity (Vp) and its anisotropy (AVp); second column: shear wave splitting or S-wave polarization anisotropy (AVs) and its maximum (max. AVs); third column: fast S-wave velocity (Vs1) and its anisotropy (AVs1). The black bars in the Vs1 stereonets indicate the polarization directions of the fast shear wave. The structural references (X, Y, Z, X′, and Y′) are the same as in Figure 3.

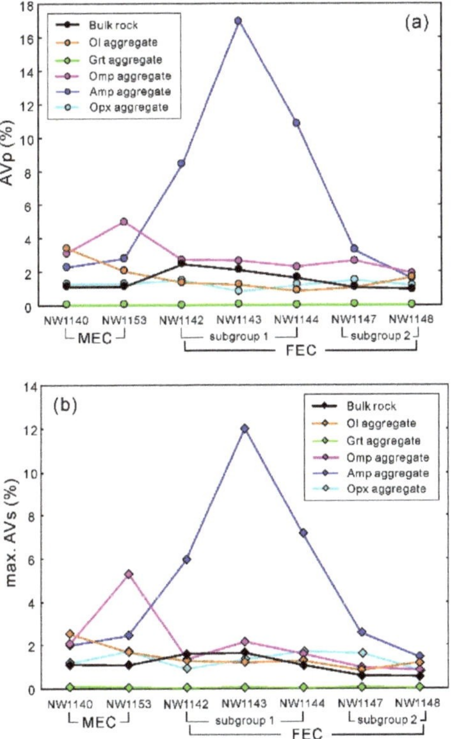

Figure 6. (a) AVp and (b) max. AVs of the bulk rock and mono-mineralic polycrystals in the MEC and FEC.

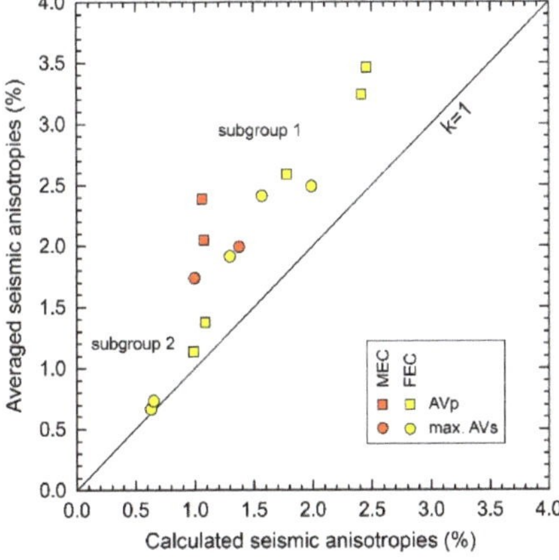

Figure 7. Comparisons of the bulk rock AVp and max. AVs between the directly calculated values (AVp$_{cal}$ and max. AVs$_{cal}$) and the indirectly averaged values (AVp$_{avg}$ and max. AVs$_{avg}$). The former were calculated conventionally using CPOs and volume proportions of all constituent minerals. The latter were estimated from the seismic anisotropies of all mono-mineralic polycrystals weighted by their volume proportions. The six and four yellow symbols in the upper and lower parts denote the FEC subgroups 1 and 2, respectively.

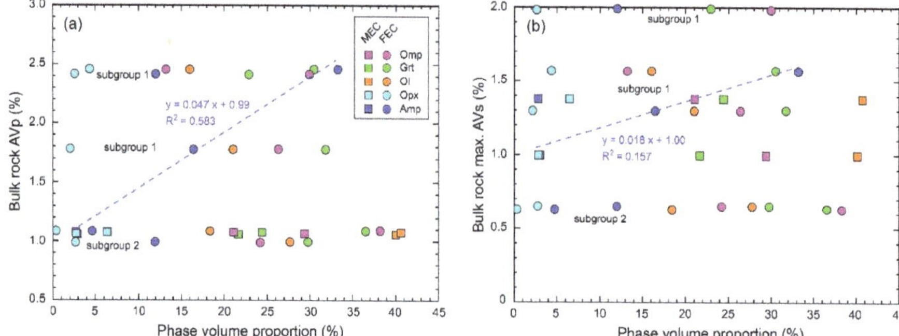

Figure 8. Relations between the phase volume proportions and the bulk rock (**a**) P-wave anisotropy (AVp) and (**b**) maximum S-wave polarization anisotropy (max. AVs) in the MEC and FEC. The blue dashed lines are the best-fitting curves along with their fitting functions for Amp.

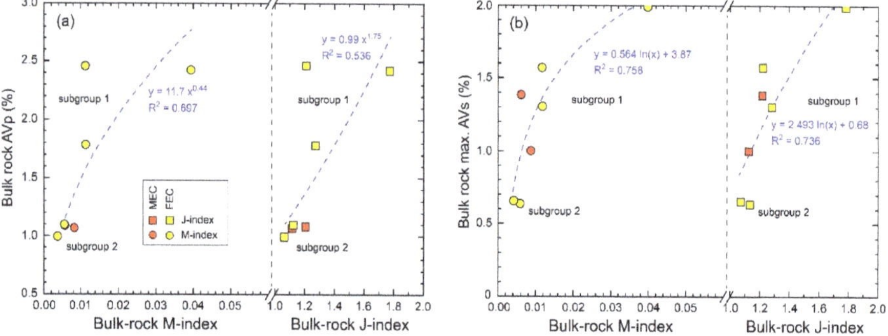

Figure 9. Relations between the bulk rock CPO strength (J- and M-index) and bulk rock (**a**) AVp and (**b**) max. AVs in the MEC and FEC. The bulk rock J- and M-index were averaged J- and M-index of the constituent minerals weighted by their volume proportions. The six and four yellow symbols in the upper and lower parts of each diagram are the FEC subgroups 1 and 2, respectively. The blue dashed lines are the best-fitting curves along with their fitting functions.

5. Discussion and Implications

5.1. Factors Controlling the Seismic Anisotropy of Flem Eclogite

Several factors can affect the seismic anisotropy of a natural rock, including microcracks, mineral assemblage, and the strength of CPOs and their interactions [13,19,63–66]. The role of microcracks is nontrivial in natural rocks; however, its contribution to seismic anisotropy decreases dramatically with depth, where the effect is negligible at pressures above ≈300 MPa (≈10 km in depth) owing to the fast closure of cracks under an increasing confining pressure [20,21,23], though a few studies suggest that some spherical micropores may still exist up to 1 GPa [67,68]. Nevertheless, these pressures are much lower than the case of our high-pressure MEC and FEC that were formed at ≈1.0–3.7 GPa or ≈30–110 km depths [42], where the effect of microcrack should be minimized. Because the single-crystal seismic anisotropies can be highly different between minerals, their volume proportions can tremendously influence the seismic anisotropies of the bulk rock that they constitute. This factor is generally the main contributor to the disparate seismic anisotropies between different rock types, such as blueschist and eclogite [12–14]. However, the bulk rock seismic anisotropies (AVp and max. AVs) of Flem eclogite show no obvious correlations with the volume proportions of major phases, except for Amp (Figure 8), indicating that bulk rock seismic anisotropies are not mainly controlled by the mineral assemblage.

The weak positive relations between the seismic anisotropies and the volume proportions of Amp, which is relatively stronger for AVp (R^2 = 0.583) than max. AVs (R^2 = 0.157), reveal that the high Amp content can partly account for the higher seismic anisotropies in the Amp-rich FEC subgroup 1 (Figure 8).

The stronger CPO strength (denoted by larger J- and M-indices) means that the crystal lattices were aligned more toward a similar orientation, resulting in greater seismic anisotropies. Despite a weak CPO strength, positive correlations between the bulk rock seismic anisotropies and the bulk rock CPO strength were still recognizable in the MEC and FEC (Figure 9). This result suggests that the variations of bulk rock seismic anisotropies in the Flem eclogite were largely affected by their CPO strength, which is also corroborated by previous studies, especially the variations of seismic anisotropies in the same rock type [13,14].

The interactions of CPO can either reinforce or weaken the bulk rock seismic anisotropies, depending on the distribution patterns of the constituent mono-mineralic polycrystals. If the fast velocities (Vp or Vs) of different mono-mineralic polycrystals are aligned in similar orientations (same for the slow velocities), the CPOs are thus thought to coherently contribute to the bulk rock seismic anisotropies (i.e., the constructive effect of CPO interactions). Otherwise, the seismic anisotropies of bulk rock are much weaker than those of mono-mineralic polycrystals, when the fast and slow velocities are subparallel and thus counteract with each other (i.e., destructive effect of CPO interactions) [13,69,70]. In the FEC subgroup 1, the Omp, Amp, and Opx polycrystals presented overall similar distribution patterns for Vp (constructive effect; Figures S2c–e, S4c–e and S5c–e), thus partly contributing to the highest bulk rock seismic anisotropies for the FEC subgroup 1. In contrast, the Vp distribution patterns were less consistent in the FEC subgroup 2 (destructive effect; Figures S2f,g, S4f,g and S5f,g), hence leading to the lowest bulk rock seismic anisotropies in collaboration with the lowest bulk rock CPO strength (Figure 9). The Vp patterns of Ol polycrystals mostly differed from those of Omp, Amp, and Opx (Figure S1), which is indicative of the destructive effect of Ol CPO. The Grt polycrystals showed almost negligible seismic anisotropies (Figure S3) and thus weakened the bulk rock seismic anisotropies. A comparison of the seismic anisotropies between bulk rock and constituent mono-mineralic polycrystals illustrated that the bulk rock seismic anisotropies were indeed the combined results of each of the mono-mineralic polycrystals (Figure 6). Especially for the FEC subgroup 1, their highest seismic anisotropies were mainly due to the Amp polycrystals. The calculated seismic anisotropies were all smaller than the arithmetically averaged seismic anisotropies that approximated the purely constructive contributions of all constituent mono-mineralic polycrystals (Figure 7). This result clearly reflects the existence of destructive effects of CPOs on weakening the bulk rock seismic anisotropies [14]. The degree of deviations (calculated using $100 \times \left(\text{AVp}_{avg} - \text{AVp}_{cal} \right) / \text{AVp}_{cal}$ and $100 \times \left(\text{max. AVs}_{avg} - \text{max. AVs}_{cal} \right) / \text{max. AVs}_{cal}$) from the dividing line (k = 1) were much larger for the MEC (90–126% for AVp and 45–74% for max. AVs) and FEC subgroup 1 (34–45% for AVp and 25–54% for max. AVs) than FEC subgroup 2 (15–26% for AVp and 6–13% for max. AVs), implying that the destructive effects of the CPOs were not constant in the Flem eclogite, i.e., they were more remarkable in the MEC and FEC subgroup 1 than in FEC subgroup 2 (Figure 7).

It is noteworthy that the three factors above are probably not mutually exclusive. The highest seismic anisotropies in FEC subgroup 1 were contributed jointly by their higher Amp contents, larger bulk rock CPO strength, and constructive CPO interactions. The enrichment of Amp in the FEC subgroup 1 may favor the development of a stronger Amp CPO (especially in the Amp-rich layers where strain is likely more localized) and result in a larger bulk rock CPO strength (combined consequences of stronger Amp CPO and higher Amp volume proportion). The stronger CPOs of Amp, Omp, and Opx in FEC subgroup 1 may also imply that more coherent CPO developments and resultant constructive CPO interactions reinforce the seismic anisotropies.

5.2. Seismic Properties and the Implications of Flem Eclogite

The Flem eclogite is a unique type of eclogite because of its uncommon mineral assemblage of abundant Ol and Opx. FEC subgroup 1 showed an AVp (1.78–2.46%) and max. AVs (1.30–1.99%) similar to previously reported dry eclogite that consisted of dominantly bi-mineralic Omp and Grt (AVp = 1.2–2.9%, max. AVs = 0.74–2.02%; Bascou et al. [22]), while the seismic anisotropies of the MEC (AVp = 1.06–1.08%, max. AVs = 1.00–1.38%) and FEC subgroup 2 (AVp = 0.99–1.09%, max. AVs = 0.63–0.65%) were even lower (Figure 10). The seismic anisotropies of the Flem eclogite were also strikingly lower than the seismic anisotropies of other types of eclogite that bear hydrous phases, such as epidote (Ep), glaucophane (Gln), and lawsonite (Lws) (AVp = 1.4–10.2%, max. AVs = 1.35–8.09%; Figure 10). Therefore, compared to the eclogite free of Ol and Opx, and despite the large strain and high Amp content (as reflected by the Amp-rich FEC), the presence of Ol and Opx actually weakened the bulk rock seismic anisotropies, mainly by reducing the mineral CPO strength (Figure 9) through non-dislocation creep mechanisms, such as diffusion creep, grain or phase boundary sliding, and rigid-body-like rotations [42]. The Ep- and Lws-bearing blueschist, which is the hydrous precursor of eclogite, displayed the highest seismic anisotropies, implying a gradual decrease of the elastic anisotropies in the subducted oceanic crust with increasing depth [12–14]. Peridotite is the rock that is typically enriched with Ol and Opx. However, peridotite had a higher AVp and max. AVs (Figure 10), indicating that it can be distinguished from Flem eclogite and dry eclogite, but not Ep-, Gln-, and Lws-bearing eclogite, in terms of the magnitudes of seismic anisotropies [71]. The Flem eclogite is surrounded by non- or weakly metamorphosed gabbro (Figure 1b; see Mørk [33]). Although the seismic anisotropies of the surrounding gabbro are unknown, an examination of the seismic anisotropies of the gabbro from other localities revealed similar intensities to the peridotite [72] (Figure 10). This fact thus raises the feasibility of using the intensities of seismic anisotropies to distinguish Flem eclogite and dry eclogite from their gabbroic country rocks in the deep continental crust.

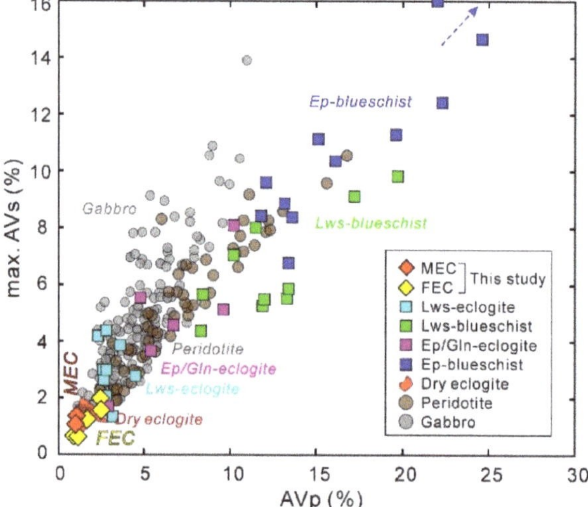

Figure 10. Seismic anisotropies (AVp and max. AVs) of the Flem eclogite (MEC and FEC) and other rock types in ambient conditions. The blue dashed arrow indicates the trend of even higher seismic anisotropies of the epidote bluseschist, which are outside the diagram. Data source: lawsonite eclogite [13,32], lawsonite blueschist [13,69,70], epidote/glaucophane eclogite [14,73,74], epidote blueschist [14,70,73,74], dry eclogite [22], peridotite [75–78], and gabbro [79].

Unlike the seismic anisotropies, the average seismic velocities (approximating the isotropic seismic velocities) and Vp/Vs ratio of Flem eclogite were controlled only by the mineral assemblage. Specifically, both Vp and Vs correlated positively and negatively with the volume proportions of Omp and Amp, respectively (Figure 11a,b). These salient relations were the manifestation of the faster seismic velocities of Omp than the Amp single crystal and the formation of Amp by replacing mainly Omp and Ol. As suggested by Cao et al. [42], the Amp in MEC and FEM has both low- and high-pressure origins, where the high-pressure Amp is the most abundant and is formed by deformation-enhanced hydration reactions under peak-to-early-retrograde UHP metamorphic conditions. Furthermore, despite narrow ranges of seismic velocities and densities, weak positive correlations between Vp, Vs, and bulk rock densities were still displayed (Figure 12), which obeyed Birch's law [9,29,80]. The seismic velocities of Flem eclogite were lower than those of peridotites and dry eclogite; larger than those of Ep-, Gln-, and Lws-eclogite; much greater than those of blueschist and gabbro (Figure 13). The Vp and Vs of Flem eclogite were plotted in a straight line, yielding an average Vp/Vs ratio of 1.765, which was lower and higher than the Ep-/Gln- and Lws-bearing eclogites, respectively (Figure 13). Despite the narrow range of the Vp/Vs ratio ($\approx$1.758–1.770; see Table 2), a strong positive correlation ($R^2 = 0.922$) between the Vp/Vs ratio and Ol volume proportion suggested that Ol could significantly increase the Vp/Vs ratio of eclogite, whereas Amp and Grt decreased the bulk rock Vp/Vs ratio, as indicated by the negative correlations between the Vp/Vs ratio and their volume proportions (Figure 11c). The Vp/Vs ratio of the Flem eclogite was much larger than the non-retrograded eclogite (1.69–1.71) and slightly larger than the retrograded eclogites (1.73–1.77) that contain appreciable amounts of retrograde Amp and Pl [12,28]. This Vp/Vs ratio was also slightly higher than the dry eclogites (1.73–1.75) that are free of Ol and Opx and similar to those of peridotite [81], further indicating that Ol was the source of high Vp/Vs ratio in the Flem eclogite (Figure 13). Owing to the abundance of Gln, which was characterized by the low Vp/Vs ratio, Ep-blueschist mostly had lower Vp/Vs ratios (1.70–1.73) than the Flem eclogite. However, the occurrence of a high-Vp/Vs-ratio Lws can increased the Vp/Vs ratio of Lws-blueschist to a wide range (1.71–1.81). The enrichment of high Vp/Vs ratio Pl caused the Vp/Vs ratios of gabbro (1.75–1.85) to be higher than the Flem eclogite. Therefore, we propose that the much larger Vp and Vs and relatively smaller Vp/Vs ratios of the olivine-rich eclogite (e.g., Flem eclogite) can be used to differentiate them from the country rocks (e.g., metastable and non-eclogized gabbro) in the deep continental crust or subduction channel when high-resolution seismic waves data are available since the detectable body size depends on the resolution (i.e., a wavelength that is determined by velocity and frequency) of seismic waves at different depths.

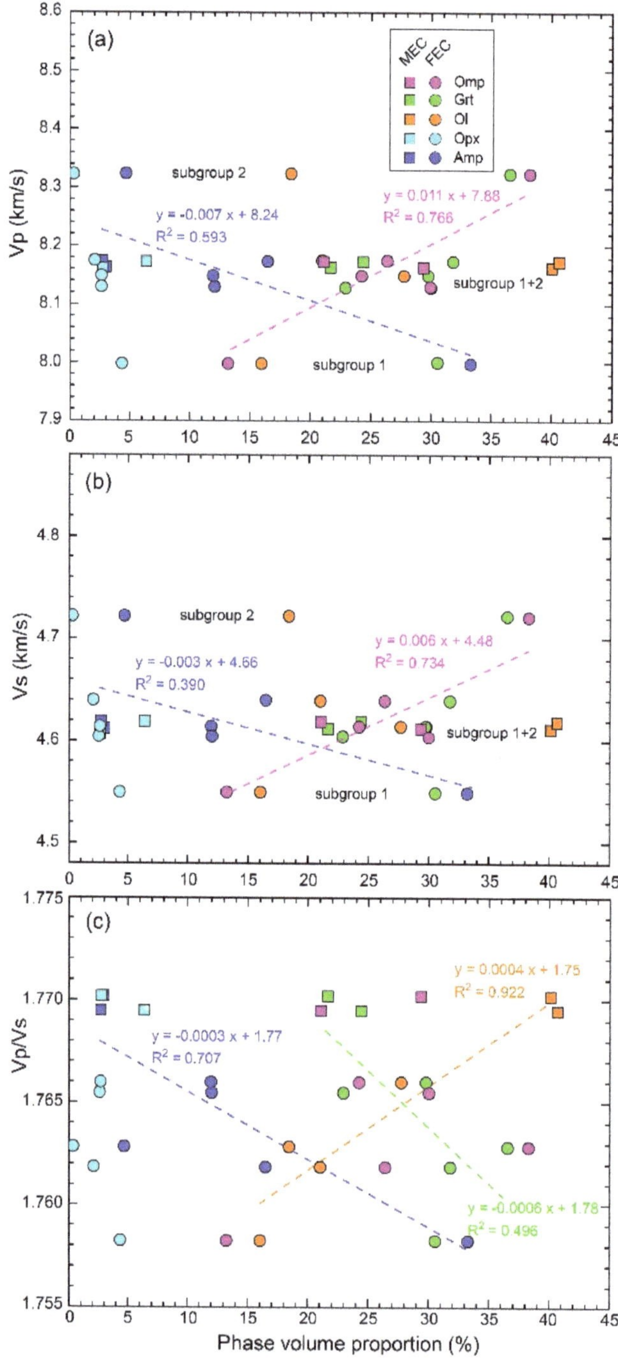

Figure 11. Relations between the phase volume proportions and the bulk rock (**a**) Vp, (**b**) Vs, and (**c**) Vp/Vs ratio in the MEC and FEC. The blue, magenta, orange, and green dashed lines are the best-fitting curves, along with their fitting functions, for Amp, Omp, Ol, and Grt, respectively.

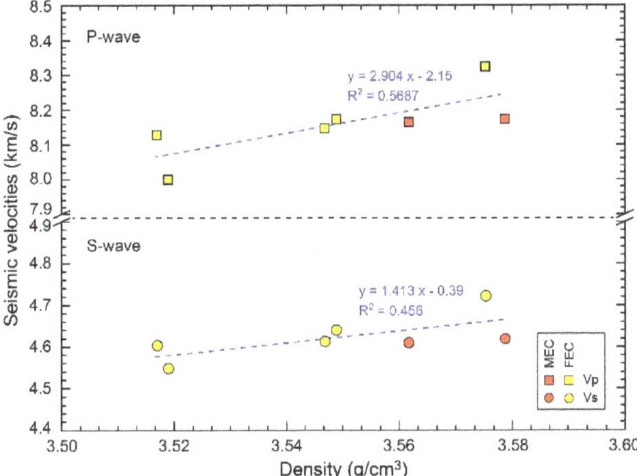

Figure 12. Relations between the bulk rock averaged seismic velocities (Vp and Vs) and density.

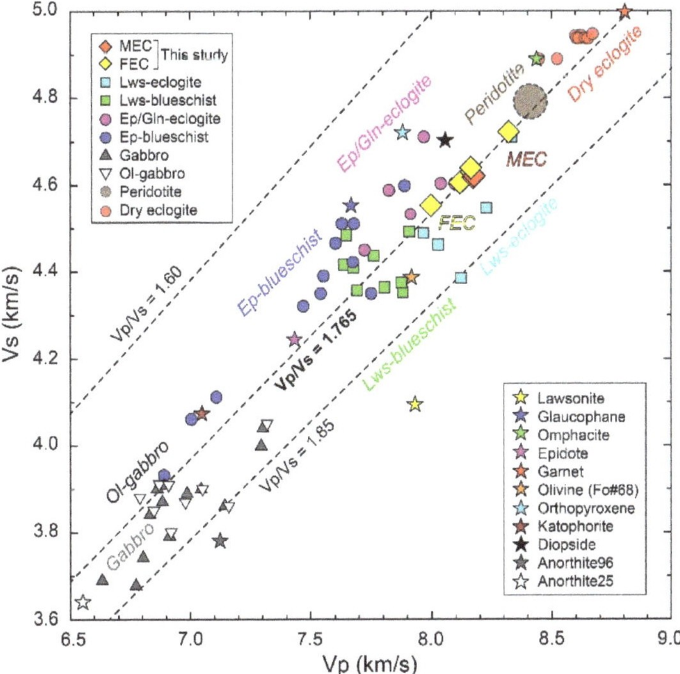

Figure 13. Seismic velocities (Vp and Vs) of the Flem eclogite (MEC and FEC) and other rock types in ambient conditions. The star symbols denote the seismic velocities of isotropic mono-mineralic polycrystals. Data source:s lawsonite eclogite [13], lawsonite blueschsit [13,69,70], epidote/glaucophane eclogite [14,73,74], epidote blueschsit [14,70,73,74], dry eclogite [22], peridotite [81], (olivine) gabbro [82].

6. Conclusions

The Flem eclogite from NW Flemsøya in the Nordøyane UHP domain of the Western Gneiss Region in Norway is a unique type of eclogite that was enriched with olivine and orthopyroxene. Based on the detailed examinations on their seismic properties, several main findings were as follows.

- The seismic anisotropies of the Flem eclogite were largely controlled by the CPO strength and significant destructive effects of CPO interactions on weakening the seismic anisotropies were manifested in both weakly and strongly foliated eclogite. Amphibole CPOs were the main contributors to the higher seismic anisotropies in some amphibole-rich and strongly foliated eclogite.

- The average seismic velocities correlated positively with the rock density and omphacite content, but negatively with the amphibole volume proportion, which is indicative of the dominant role of a mineral assemblage in governing the bulk rock seismic velocities.

- The V_p/V_s ratios were almost constant ($\approx$1.765) between the Flem eclogite samples. These values were similar to the V_p/V_s ratios of peridotite and slightly higher than the dry eclogite that was free of olivine, suggesting that the abundance of olivine was the source of their high V_p/V_s ratios.

- Compared to other common rock types, including hydrous-phase-bearing eclogite, blueschist, peridotite, and gabbro, the AV_p and max. AV_s of Flem eclogite were very low and similar to those of dry eclogite. In contrast, the average V_p and V_s of the Flem eclogite were remarkably larger than the hydrous-phase-bearing eclogite, blueschist, and gabbro, but lower than the dry eclogite and peridotite. The V_p/V_s ratios of the Flem eclogite were relatively higher than those of dry, non-retrograded, and retrograded eclogite, but significantly lower than gabbro. These results thus highlight the feasibilies of utilizing seismic anisotropies, velocities, and V_p/V_s ratios for differentiating olivine-rich eclogite (e.g., Flem eclogite) bodies from their country rocks (e.g., gabbro) in the deep continental crust or subduction channel when high-resolution seismic wave data are available.

Supplementary Materials: The following are available online at http://www.mdpi.com/2075-163X/10/9/774/s1, Figure S1: Seismic properties of the olivine aggregates in the MEC and FEC and olivine single crystal at ambient condition; Figure S2: Seismic properties of the orthopyroxene aggregates in the MEC and FEC and orthopyroxene single crystal at ambient condition; Figure S3: Seismic properties of the garnet aggregates in the MEC and FEC and garnet single crystals in MEC and FEC at ambient condition; Figure S4: Seismic properties of the omphacite aggregates in the MEC and FEC and omphacite single crystal at ambient condition; Figure S5: Seismic properties of the amphibole aggregates in the MEC and FEC and amphibole single crystal at ambient condition.

Author Contributions: Conceptualization, Y.C.; methodology, Y.C.; formal analysis, Y.C.; investigation, Y.C. and J.M.; resources, H.J.; data curation, Y.C.; writing—original draft preparation, Y.C.; writing—review and editing, H.J. and J.M.; visualization, Y.C.; funding acquisition, Y.C. and H.J. All authors have read and agreed to the published version of the manuscript.

Funding: This research was funded by the National Natural Science Foundation of China (grant no. 41902222 to Y.C.), National Research Foundation of Korea (grant no. 2020R1A2CZ003765 to H.J.), and 111 Project of China (grant no. BP0719022).

Acknowledgments: The authors thank Håkon Austrheim for his assistance in collecting eclogite samples for this study. The authors are also grateful to Ruth Keppler and three anonymous reviewers for their helpful comments.

Conflicts of Interest: The authors declare no conflict of interest.

References

1. Godard, G. Eclogites and their geodynamic interpretation: A history. *J. Geodyn.* **2001**, *32*, 165–203. [CrossRef]
2. Austrheim, H. Fluid and deformation induced metamorphic processes around Moho beneath continent collision zones: Examples from the exposed root zone of the Caledonian mountain belt, W-Norway. *Tectonophysics* **2013**, *609*, 620–635. [CrossRef]
3. Peacock, S.M. The importance of blueschist—Eclogite dehydration reactions in subducting oceanic crust. *Geol. Soc. Am. Bull.* **1993**, *105*, 684–694. [CrossRef]
4. Carswell, D.A. *Eclogite Facies Rocks*; Springer: Heidelberg, The Netherlands, 1990.

5. Liou, J.G.; Zhang, R.Y.; Ernst, W.G.; Rumble, D.; Maruyama, S. High-pressure minerals from deeply subducted metamorphic rocks. *Rev. Mineral. Geochem.* **1998**, *37*, 33–96.

6. Wei, C.J.; Yang, Y.; Su, X.L.; Song, S.G.; Zhang, L.F. Metamorphic evolution of low-*T* eclogite from the North Qilian orogen, NW China: Evidence from petrology and calculated phase equilibria in the system NCKFMASHO. *J. Metamorph. Geol.* **2009**, *27*, 55–70. [CrossRef]

7. Hacker, B.R.; Abers, G.A.; Peacock, S.M. Subduction factory 1. Theoretical mineralogy, densities, seismic wave speeds, and H_2O contents. *J. Geophys. Res. Solid Earth* **2003**, *108*, 1–26. [CrossRef]

8. Mørk, M.B.E. Coronite and eclogite formation in olivine gabbro (Western Norway): Reaction paths and garnet zoning. *Mineral. Mag.* **1986**, *50*, 417–426. [CrossRef]

9. Christensen, N.I.; Mooney, W.D. Seismic velocity structure and composition of the continental crust: A global view. *J. Geophys. Res.* **1995**, *100*, 9761–9788. [CrossRef]

10. Fountain, D.M.; Boundy, T.M.; Austrheim, H.; Rey, P. Eclogite-facies shear zones–deep crustal reflectors? *Tectonophysics* **1994**, *232*, 411–424. [CrossRef]

11. Warner, M.; McGeary, S. Seismic reflection coefficients from mantle fault zones. *Geophys. J. R. Astron. Soc.* **1987**, *89*, 223–230. [CrossRef]

12. Keppler, R.; Behrmann, J.H.; Stipp, M. Textures of eclogites and blueschists from Syros island, Greece: Inferences for elastic anisotropy of subducted oceanic crust. *J. Geophys. Res. Solid Earth* **2017**, *122*, 5306–5324. [CrossRef]

13. Cao, Y.; Jung, H. Seismic properties of subducting oceanic crust: Constraints from natural lawsonite-bearing blueschist and eclogite in Sivrihisar Massif, Turkey. *Phys. Earth Planet. Inter.* **2016**, *250*, 12–30. [CrossRef]

14. Cao, Y.; Jung, H.; Song, S.G. Petro-fabrics and seismic properties of blueschist and eclogite in the North Qilian suture zone, NW China: Implications for the low-velocity upper layer in subducting slab, trench-parallel seismic anisotropy, and eclogite detectability in the subduction zone. *J. Geophys. Res. Solid Earth* **2013**, *118*, 3037–3058. [CrossRef]

15. Garber, J.M.; Maurya, S.; Hernandez, J.-A.; Duncan, M.S.; Zeng, L.; Zhang, H.L.; Faul, U.; McCammon, C.; Montagner, J.-P.; Moresi, L.; et al. Multidisciplinary constraints on the abundance of diamond and eclogite in the cratonic lithosphere. *Geochem. Geophys. Geosyst.* **2018**, *19*, 2062–2086. [CrossRef]

16. Puelles, P.; Beranoaguirre, A.; Ábalos, B.; Gil Ibarguchi, J.I.; García de Madinabeitia, S.; Rodríguez, J.; Fernández-Armas, S. Eclogite inclusions from subducted metaigneous continental crust (Malpica-Tui Allochthonous Complex, NW Spain): Petrofabric, geochronology, and calculated seismic properties. *Tectonics* **2017**, *36*, 1376–1406. [CrossRef]

17. Brownlee, S.J.; Hacker, B.R.; Salisbury, M.; Seward, G.; Little, T.A.; Baldwin, S.L.; Abers, G.A. Predicted velocity and density structure of the exhuming Papua New Guinea ultrahigh-pressure terrane. *J. Geophys. Res. Solid Earth* **2011**, *116*. [CrossRef]

18. Haugland, S.M.; Ritsema, J.; Kaneshima, S.; Thorne, M.S. Estimate of the rigidity of eclogite in the lower mantle from waveform modeling of broadband *S*-to-*P* wave conversions. *Geophys. Res. Lett. Geophys. Res. Lett.* **2017**, *44*, 11778–11784. [CrossRef]

19. Park, M.; Jung, H. Relationships between eclogite-facies mineral assemblages, deformation microstructures, and seismic properties in the Yuka terrane, North Qaidam ultrahigh-pressure metamorphic belt, NW China. *J. Geophys. Res. Solid Earth* **2019**, *124*, 13168–13191. [CrossRef]

20. Abalos, B.; Fountain, D.M.; Ibarguchi, J.I.G.; Puelles, P. Eclogite as a seismic marker in subduction channels: Seismic velocities, anisotropy, and petrofabric of Cabo Ortegal eclogite tectonites (Spain). *Geol. Soc. Am. Bull.* **2011**, *123*, 439–456. [CrossRef]

21. Sun, S.S.; Ji, S.C.; Wang, Q.; Xu, Z.Q.; Salisbury, M.; Long, C.X. Seismic velocities and anisotropy of core samples from the Chinese Continental Scientific Drilling borehole in the Sulu UHP terrane, eastern China. *J. Geophys. Res. Solid Earth* **2012**, *117*, 1–24. [CrossRef]

22. Bascou, J.; Barruol, G.; Vauchez, A.; Mainprice, D.; Egydio-Silva, M. EBSD-measured lattice-preferred orientations and seismic properties of eclogites. *Tectonophysics* **2001**, *342*, 61–80. [CrossRef]

23. Wang, Q.; Ji, S.; Salisbury, M.H.; Xia, B.; Pan, M.; Xu, Z. Shear wave properties and Poisson's ratios of ultrahigh-pressure metamorphic rocks from the Dabie-Sulu orogenic belt, China: Implications for crustal composition. *J. Geophys. Res.* **2005**, *110*, B08208. [CrossRef]

24. Llana-Funez, S.; Brown, D. Contribution of crystallographic preferred orientation to seismic anisotropy across a surface analog of the continental Moho at Cabo Ortegal, Spain. *Geol. Soc. Am. Bull.* **2012**, *124*, 1495–1513. [CrossRef]

25. Wang, Q.; Burlini, L.; Mainprice, D.; Xu, Z.Q. Geochemistry, petrofabrics and seismic properties of eclogites from the Chinese Continental Scientific Drilling boreholes in the Sulu UHP terrane, eastern China. *Tectonophysics* **2009**, *475*, 251–266. [CrossRef]

26. Kim, D.; Kim, T.; Lee, J.; Kim, Y.; Kim, H.; Lee, J.I. Microfabrics of omphacite and garnet in eclogite from the Lanterman Range, northern Victoria Land, Antarctica. *Geosci. J.* **2018**, *22*, 939–953. [CrossRef]

27. Zhang, J.F.; Wang, Y.F.; Jin, Z.M. CPO-induced seismic anisotropy in UHP eclogites. *Sci. China Ser. D* **2008**, *51*, 11–21. [CrossRef]

28. Keppler, R.; Ullemeyer, K.; Behrmann, J.H.; Stipp, M.; Kurzawski, R.M.; Lokajíček, T. Crystallographic preferred orientations of exhumed subduction channel rocks from the Eclogite Zone of the Tauern Window (Eastern Alps, Austria), and implications on rock elastic anisotropies at great depths. *Tectonophysics* **2015**, *647–648*, 89–104. [CrossRef]

29. Kern, H.; Jin, Z.; Gao, S.; Popp, T.; Xu, Z. Physical properties of ultrahigh-pressure metamorphic rocks from the Sulu terrain, eastern central China: Implications for the seismic structure at the Donghai (CCSD) drilling site. *Tectonophysics* **2002**, *354*, 315–330. [CrossRef]

30. Kern, H.; Gao, S.; Jin, Z.M.; Popp, T.; Jin, S.Y. Petrophysical studies on rocks from the Dabie ultrahigh-pressure (UHP) metamorphic belt, Central China: Implications for the composition and delamination of the lower crust. *Tectonophysics* **1999**, *301*, 191–215. [CrossRef]

31. Brown, D.; Llana-Funez, S.; Carbonell, R.; Alvarez-Marron, J.; Marti, D.; Salisbury, M. Laboratory measurements of P-wave and S-wave velocities across a surface analog of the continental crust-mantle boundary: Cabo Ortegal, Spain. *Earth Planet. Sci. Lett.* **2009**, *285*, 27–38. [CrossRef]

32. Kim, D.; Wallis, S.; Endo, S.; Ree, J.-H. Seismic properties of lawsonite eclogites from the southern Motagua fault zone, Guatemala. *Tectonophysics* **2016**, *677–678*, 88–98. [CrossRef]

33. Mørk, M.B.E. A gabbro to eclogite transition on Flemsøy, Sunnmøre, western Norway. *Chem. Geol.* **1985**, *50*, 283–310. [CrossRef]

34. Hacker, B.R.; Gans, P.B. Continental collisions and the creation of ultrahigh-pressure terranes: Petrology and thermochronology of nappes in the central Scandinavian Caledonides. *Geol. Soc. Am. Bull.* **2005**, *117*, 117–134. [CrossRef]

35. Andersen, T.B.; Jamtveit, B.; Dewey, J.F.; Swensson, E. Subduction and eduction of continental crust: Major mechanisms during continent-continent collision and orogenic extensional collapse, a model based on the south Norwegian Caledonides. *Terra Nova* **1991**, *3*, 303–310. [CrossRef]

36. Hacker, B.R.; Andersen, T.B.; Johnston, S.; Kylander-Clark, A.R.C.; Peterman, E.M.; Walsh, E.O.; Young, D. High-temperature deformation during continental-margin subduction & exhumation: The ultrahigh-pressure Western Gneiss Region of Norway. *Tectonophysics* **2010**, *480*, 149–171. [CrossRef]

37. Kylander-Clark, A.R.C.; Hacker, B.R.; Johnson, C.M.; Beard, B.L.; Mahlen, N.J.; Lapen, T.J. Coupled Lu–Hf and Sm–Nd geochronology constrains prograde and exhumation histories of high- and ultrahigh-pressure eclogites from western Norway. *Chem. Geol.* **2007**, *242*, 137–154. [CrossRef]

38. Terry, M.P.; Robinson, P. Geometry of eclogite-facies structural features: Implications for production and exhumation of ultrahigh-pressure and high-pressure rocks, Western Gneiss Region, Norway. *Tectonics* **2004**, *23*, 1–23. [CrossRef]

39. Terry, M.P.; Robinson, P.; Ravna, E.J.K. Kyanite eclogite thermobarometry and evidence for thrusting of UHP over HP metamorphic rocks, Nordøyane, Western Gneiss Region, Norway. *Am. Mineral.* **2000**, *85*, 1637–1650. [CrossRef]

40. Carswell, D.A.; Van Roermund, H.L.M.; De Vries, D.F.W. Scandian ultrahigh-pressure metamorphism of Proterozoic basement rocks on Fjørtoft and Otrøy, Western Gneiss Region, Norway. *Int. Geol. Rev.* **2006**, *48*, 957–977. [CrossRef]

41. Butler, J.P.; Jamieson, R.A.; Steenkamp, H.M.; Robinson, P. Discovery of coesite–eclogite from the Nordøyane UHP domain, Western Gneiss Region, Norway: Field relations, metamorphic history, and tectonic significance. *J. Metamorph. Geol.* **2013**, *31*, 147–163. [CrossRef]

42. Cao, Y.; Du, J.; Park, M.; Jung, S.; Park, Y.; Kim, D.; Choi, S.; Jung, H.; Austrheim, H. Metastability and nondislocation-based deformation mechanisms of the Flem eclogite in the Western Gneiss Region, Norway. *J. Geophys. Res. Solid Earth* **2020**, *125*, e2020JB019375. [CrossRef]

43. Renedo, R.N.; Nachlas, W.O.; Whitney, D.L.; Teyssier, C.; Piazolo, S.; Gordon, S.M.; Fossen, H. Fabric development during exhumation from ultrahigh-pressure in an eclogite-bearing shear zone, Western Gneiss Region, Norway. *J. Struct. Geol.* **2015**, *71*, 58–70. [CrossRef]

44. Bunge, H. *Texture Analysis in Materials Science: Mathematical Models*; Butterworths: London, UK, 1982.

45. Skemer, P.; Katayama, B.; Jiang, Z.T.; Karato, S. The misorientation index: Development of a new method for calculating the strength of lattice-preferred orientation. *Tectonophysics* **2005**, *411*, 157–167. [CrossRef]

46. Hielscher, R.; Schaeben, H. A novel pole figure inversion method: Specification of the MTEX algorithm. *J. Appl. Crystallogr.* **2008**, *41*, 1024–1037. [CrossRef]

47. Bachmann, F.; Hielscher, R.; Schaeben, H. Texture Analysis with MTEX–Free and Open Source Software Toolbox. *Solid State Phenom.* **2010**, *160*, 63–68. [CrossRef]

48. Mainprice, D. A FORTRAN program to calculate seismic anisotropy from the lattice preferred orientation of minerals. *Comput. Geosci.* **1990**, *16*, 385–393. [CrossRef]

49. Brown, J.M.; Abramson, E.H. Elasticity of calcium and calcium-sodium amphiboles. *Phys. Earth Planet. Inter.* **2016**, *261*, 161–171. [CrossRef]

50. Webb, S.L.; Jackson, I. The pressure dependence of the elastic moduli of single-crystal orthopyroxene $(Mg_{0.8}Fe_{0.2})SiO_3$. *Eur. J. Mineral.* **1993**, *5*, 1111–1120. [CrossRef]

51. Alexandrov, K.S.; Ryzhova, T.V. Elastic properties of rock-forming minerals. II. Layered silicates. *Bull. Acad. Sci. USSR Geophys. Ser.* **1961**, *9*, 1165–1168.

52. Brown, J.M.; Angel, R.J.; Ross, N.L. Elasticity of plagioclase feldspars. *J. Geophys. Res. Solid Earth* **2016**, *121*, 663–675. [CrossRef]

53. Duan, Y.; Li, X.; Sun, N.; Ni, H.; Tkachev, S.N.; Mao, Z. Single-crystal elasticity of $MgAl_2O_4$-spinel up to 10.9 GPa and 1000 K: Implication for the velocity structure of the top upper mantle. *Earth Planet. Sci. Lett.* **2018**, *481*, 41–47. [CrossRef]

54. Hao, M.; Zhang, J.S.; Pierotti, C.E.; Ren, Z.; Zhang, D. High-pressure single-crystal elasticity and thermal equation of state of omphacite and their implications for the seismic properties of eclogite in the Earth's interior. *J. Geophys. Res. Solid Earth* **2019**, *124*, 2368–2377. [CrossRef]

55. Isaak, D.G.; Anderson, O.L.; Goto, T.; Suzuki, I. Elasticity of single-crystal forsterite measured to 1700 K. *J. Geophys. Res. Solid Earth* **1989**, *94*, 5895–5906. [CrossRef]

56. Speziale, S.; Duffy, T.S.; Angel, R.J. Single-crystal elasticity of fayalite to 12 GPa. *J. Geophys. Res. Solid Earth* **2004**, *109*. [CrossRef]

57. Jiang, F.; Speziale, S.; Duffy, T.S. Single-crystal elasticity of grossular- and almandine-rich garnets to 11 GPa by Brillouin scattering. *J. Geophys. Res. Solid Earth* **2004**, *109*, 1–10. [CrossRef]

58. Chai, M.; Brown, J.M.; Slutsky, L.J. The elastic constants of a pyrope-grossular-almandine garnet to 20 GPa. *Geophys. Res. Lett.* **1997**, *24*, 523–526. [CrossRef]

59. Wang, H.; Simmons, G. Elasticity of some mantle crystal structures 3. spessartite-almandine garnet. *J. Geophys. Res. Solid Earth* **1974**, *79*, 2607–2613. [CrossRef]

60. Liebermann, R.C. Elasticity of ilmenites. *Phys. Earth Planet. Inter.* **1976**, *12*, 5–10. [CrossRef]

61. Almqvist, B.S.G.; Mainprice, D. Seismic properties and anisotropy of the continental crust: Predictions based on mineral texture and rock microstructure. *Rev. Geophys.* **2017**, *55*, 367–433. [CrossRef]

62. Mainprice, D.; Hielscher, R.; Schaeben, H. Calculating anisotropic physical properties from texture data using the MTEX open-source package. *Geol. Soc. Spec. Publ.* **2011**, *360*, 175–192. [CrossRef]

63. Vasin, R.N.; Kern, H.; Lokajíček, T.; Svitek, T.; Lehmann, E.; Mannes, D.C.; Chaouche, M.; Wenk, H.R. Elastic anisotropy of Tambo gneiss from Promontogno, Switzerland: A comparison of crystal orientation and microstructure-based modelling and experimental measurements. *Geophys. J. Int.* **2017**, *209*, 1–20. [CrossRef]

64. Kern, H.; Ivankina, T.I.; Nikitin, A.N.; Lokajíček, T.; Pros, Z. The effect of oriented microcracks and crystallographic and shape preferred orientation on bulk elastic anisotropy of a foliated biotite gneiss from Outokumpu. *Tectonophysics* **2008**, *457*, 143–149. [CrossRef]

65. Ullemeyer, K.; Siegesmund, S.; Rasolofosaon, P.N.J.; Behrmann, J.H. Experimental and texture-derived P-wave anisotropy of principal rocks from the TRANSALP traverse: An aid for the interpretation of seismic field data. *Tectonophysics* **2006**, *414*, 97–116. [CrossRef]

66. Ivankina, T.I.; Kern, H.M.; Nikitin, A.N. Directional dependence of P- and S-wave propagation and polarization in foliated rocks from the Kola superdeep well: Evidence from laboratory measurements and calculations based on TOF neutron diffraction. *Tectonophysics* **2005**, *407*, 25–42. [CrossRef]

67. Ullemeyer, K.; Lokajíček, T.; Vasin, R.N.; Keppler, R.; Behrmann, J.H. Extrapolation of bulk rock elastic moduli of different rock types to high pressure conditions and comparison with texture-derived elastic moduli. *Phys. Earth Planet. Inter.* **2018**, *275*, 32–43. [CrossRef]

68. Ullemeyer, K.; Nikolayev, D.I.; Christensen, N.I.; Behrmann, J.H. Evaluation of intrinsic velocity—Pressure trends from low-pressure P-wave velocity measurements in rocks containing microcracks. *Geophys. J. Int.* **2011**, *185*, 1312–1320. [CrossRef]

69. Cao, Y.; Jung, H.; Song, S.G. Microstructures and petro-fabrics of lawsonite blueschist in the North Qilian suture zone, NW China: Implications for seismic anisotropy of subducting oceanic crust. *Tectonophysics* **2014**, *628*, 140–157. [CrossRef]

70. Kim, D.; Katayama, I.; Michibayashi, K.; Tsujimori, T. Deformation fabrics of natural blueschists and implications for seismic anisotropy in subducting oceanic crust. *Phys. Earth Planet. Inter.* **2013**, *222*, 8–21. [CrossRef]

71. Worthington, J.R.; Hacker, B.R.; Zandt, G. Distinguishing eclogite from peridotite: EBSD-based calculations of seismic velocities. *Geophys. J. Int.* **2013**. [CrossRef]

72. Ji, S.; Shao, T.; Salisbury, M.H.; Sun, S.; Michibayashi, K.; Zhao, W.; Long, C.; Liang, F.; Satsukawa, T. Plagioclase preferred orientation and induced seismic anisotropy in mafic igneous rocks. *J. Geophys. Res. Solid Earth* **2014**. [CrossRef]

73. Ha, Y.; Jung, H.; Raymond, L.A. Deformation fabrics of glaucophane schists and implications for seismic anisotropy: The importance of lattice preferred orientation of phengite. *Int. Geol. Rev.* **2018**, *61*, 720–737. [CrossRef]

74. Bezacier, L.; Reynard, B.; Bass, J.D.; Wang, J.; Mainprice, D. Elasticity of glaucophane, seismic velocities and anisotropy of the subducted oceanic crust. *Tectonophysics* **2010**, *494*, 201–210. [CrossRef]

75. Kang, H.; Jung, H. Lattice-preferred orientation of amphibole, chlorite, and olivine found in hydrated mantle peridotites from Bjørkedalen, southwestern Norway, and implications for seismic anisotropy. *Tectonophysics* **2019**, *750*, 137–152. [CrossRef]

76. Kourim, F.; Beinlich, A.; Wang, K.-L.; Michibayashi, K.; O'Reilly, S.Y.; Pearson, N.J. Feedback of mantle metasomatism on olivine micro–fabric and seismic properties of the deep lithosphere. *Lithos* **2019**, *328–329*, 43–57. [CrossRef]

77. Demouchy, S.; Tommasi, A.; Ionov, D.; Higgie, K.; Carlson, R.W. Microstructures, water contents, and seismic properties of the mantle lithosphere beneath the northern limit of the Hangay Dome, Mongolia. *Geochem. Geophys. Geosyst.* **2019**, *20*, 183–207. [CrossRef]

78. Tommasi, A.; Baptiste, V.; Vauchez, A.; Holtzman, B. Deformation, annealing, reactive melt percolation, and seismic anisotropy in the lithospheric mantle beneath the southeastern Ethiopian rift: Constraints from mantle xenoliths from Mega. *Tectonophysics* **2016**, *682*, 186–205. [CrossRef]

79. Satsukawa, T.; Ildefonse, B.; Mainprice, D.; Morales, L.F.G.; Michibayashi, K.; Barou, F. A database of plagioclase crystal preferred orientations (CPO) and microstructures-implications for CPO origin, strength, symmetry and seismic anisotropy in gabbroic rocks. *Solid Earth* **2013**, *4*, 511–542. [CrossRef]

80. Birch, F. The velocity of compressional waves in rocks to 10 kilobars: Part 2. *J. Geophys. Res.* **1961**, *66*, 2199–2224. [CrossRef]

81. Pera, E.; Mainprice, D.; Burlini, L. Anisotropic seismic properties of the upper mantle beneath the Torre Alfina area (Northern Appennines, Central Italy). *Tectonophysics* **2003**, *370*, 11–30. [CrossRef]

82. Iturrino, G.; Christensen, N.I.; Kirby, S.; Salisbury, M.H. Seismic velocities and elastic properties of oceanic gabbroic rocks from Hole 735B. *PANGAEA* **1991**, *118*. [CrossRef]

 *minerals*

Article

Lattice Preferred Orientation and Deformation Microstructures of Glaucophane and Epidote in Experimentally Deformed Epidote Blueschist at High Pressure

Yong Park, Sejin Jung and Haemyeong Jung *

Tectonophysics Laboratory, School of Earth and Environmental Sciences, Seoul National University, Seoul 08826, Korea; dark2444@snu.ac.kr (Y.P.); shazabi7@snu.ac.kr (S.J.)
* Correspondence: hjung@snu.ac.kr; Tel.: +82-2-880-6733

Received: 30 July 2020; Accepted: 9 September 2020; Published: 11 September 2020

Abstract: To understand the lattice preferred orientation (LPO) and deformation microstructures at the top of a subducting slab in a warm subduction zone, deformation experiments of epidote blueschist were conducted in simple shear under high pressure (0.9–1.5 GPa) and temperature (400–500 °C). At low shear strain ($\gamma \leq 1$), the [001] axes of glaucophane were in subparallel alignment with the shear direction, and the (010) poles were subnormally aligned with the shear plane. At high shear strain ($\gamma > 2$), the [001] axes of glaucophane were in subparallel alignment with the shear direction, and the [100] axes were subnormally aligned with the shear plane. At a shear strain between $2< \gamma <4$, the (010) poles of epidote were in subparallel alignment with the shear direction, and the [100] axes were subnormally aligned with the shear plane. At a shear strain where $\gamma > 4$, the alignment of the (010) epidote poles had altered from subparallel to subnormal to the shear plane, while the [001] axes were in subparallel alignment with the shear direction. The experimental results indicate that the magnitude of shear strain and rheological contrast between component minerals plays an important role in the formation of LPOs for glaucophane and epidote.

Keywords: lattice preferred orientation; glaucophane; epidote; deformation experiment; simple shear; dislocation glide; cataclastic flow

---

1. Introduction

Several previous studies have recognized blueschist as one of the representative rocks in the subduction zone at increased depths. This is because the subducting oceanic crust is considered to transform to blueschist-facies metamorphic rock under high pressure (0.5–2.5 GPa) and relatively low temperature (150–550 °C) conditions [1–3]. The blueschist has been reported worldwide in the paleo-subduction zone or active subduction zone [4–6]. Previous geophysical studies have reported that the low-velocity layer of the subduction zone almost coincides with the upper plane of the double seismic zone at the top of the subducting oceanic crust [7–9]. Glaucophane and lawsonite or epidote in blueschist is considered to potentially affect seismic velocity at the top of the subducting oceanic crust [10,11], as these minerals contain a high H_2O content (as hydroxyl) in their molecular structure [2,12]. Other studies have reported that the tremors and low-frequency earthquakes (LFEs) occur near the top of the low-velocity layers [13,14]. In addition, researchers have suggested that the lattice preferred orientation (LPO) of glaucophane, one of the elastically anisotropic minerals in blueschist, may affect the trench-parallel seismic anisotropy of the forearc region where the subducting slab has a high dip angle [15–18].

To better understand the characteristics of this subduction zone, it is necessary to understand the deformation behavior of the constituent minerals in blueschist. Among these minerals, glaucophane and epidote, recognized as the principal minerals of epidote blueschist, are important at the top of the slab in warm subduction zones. Previous studies on the deformation of natural rocks have suggested several deformation mechanisms for glaucophane; (1) the rigid behavior of glaucophane in eclogitic micaschist [19], (2) rigid body rotation and dynamic recrystallization by dislocation creep in glaucophanite [20], (3) dislocation glide (or slip) in eclogitic micaschist and glaucophane schist [21], (4) dynamic recrystallization by dislocation creep in natural blueschist [17,18], and (5) dissolution–precipitation creep in rocks deformed at high P/T conditions with aqueous fluids [22]. For epidote, five deformation mechanisms have been proposed; (1) rigid body rotation for epidote (zoisite) in metabasite rock [23], (2) dislocation glide for epidote (clinozoisite) in eclogite [24,25], (3) either dislocation glide or sliding on the cleavage by fracturing in naturally deformed rocks [26], and (4) rigid body rotation at relatively low shear strain ($\gamma = 2$) and (5) granular flow and a diffusion-assisted grain boundary sliding at relatively high shear strain ($\gamma = 4.5$ and 7.5) for epidote (zoisite) product in an experimentally deformed plagioclase matrix [27]. Thus, glaucophane and epidote are likely to be deformed through brittle and ductile behavior based on the results from these previous studies. However, the dominant mechanism in the subducting slab in the subduction zone continues to be debated in the literature.

It is possible to develop the LPOs of glaucophane and epidote in blueschist through these deformation mechanisms. Previous studies have reported several LPOs or slip systems of glaucophane. Several dislocation slip systems of glaucophane were first reported by a transmission electron microscopy (TEM)/high-resolution electron microscopy (HREM) study [21]. Recent studies on natural blueschists have reported three types of LPOs for glaucophane; (1) LPO indicating a (100)[001] slip system (or, point/SL-type LPO) [15–18,28–32], (2) LPO indicating a {110}[001] slip system (or, girdle/L-type LPO) [15,16,18,28,31], and (3) LPO indicating a (010)[001] slip system [33]. Although there has been little description on the slip systems of epidote, certain TEM/HREM studies have suggested the easiest slip plane (e.g., [24,34]). There are two types of LPOs for epidote that have recently been reported: one is the LPO indicating a (001)[010] slip system [16,17,28,29,35], and the other is the LPO indicating a {101}[010] slip system [16,30,33,35]. Although LPOs of glaucophane and epidote have been reported, there is little clarity as to the deformation mechanisms that developed their LPOs.

This study conducts deformation experiments for epidote blueschist in simple shear under high pressure (0.9–1.5 GPa) and temperature (400–500 °C), to understand the development of LPOs and deformation mechanisms of glaucophane and epidote in the conditions prevailing over the subducting slab in a warm subduction zone.

2. Methods

2.1. Starting Material

The starting material was a natural epidote blueschist-facies rock collected from the Voltri massif in the western Alps, Italy. This was a massive fine-grained rock with a small grain size (average ~30 μm, Figure 1a). There was no clear foliation and lineation in the hand specimen and thin section. The blueschist was mainly composed of Na-amphibole (glaucophane, ~55%), epidote (~15%), albite (~15%), titanite (~5%), chlorite (~5%), and garnet (almandine, ~5%). The starting material underwent peak metamorphic conditions at a pressure of 22–28 kbar and a temperature of 460–500 °C and was re-equilibrated at a pressure of 10–15 kbar and a temperature of 450–500 °C [36]. The mineral composition of the starting material was measured with a JEOL JXA-8100 electron probe X-ray microanalyzer at the Center for Research Facilities at Gyeongsang National University (GNU), South Korea. The measurement conditions included an accelerating voltage of 15 kV, a current of 10 nA, and a 5×5 μm^2 beam size. The Na-amphibole was classified as pure glaucophane, represented by the formula $Na_{2.0}[(Mg_{1.7}Fe^{2+}_{1.3})(Ca_{0.04}Mn_{0.01})Al_{1.8}Fe^{3+}_{0.09}]Si_{8.1}O_{22}(OH)_2$. The epidote was classified as $Ep_{50.4}Czo_{49.6}$, represented by the formula $Ca_{2.0}Al_{2.0}(Fe^{3+}_{0.50}Al^{VI}_{0.49})(Si_{1.0}O_4)(Si_{2.0}O_7)O(OH)$ (Table 1). The backscattered electron (BSE) image of the starting material illustrates the subhedral shape of the constituent minerals with an absence of fractures and few cleavages (Figure 1b). Most grain boundaries of glaucophane are irregular and curved, while some glaucophane grains represent the retrograded rim from Na-rich to NaCa/Ca-rich, after having undergone somewhat retrograde metamorphism [37]. Epidote minerals also possess irregular grain boundaries and are relatively larger than glaucophane minerals.

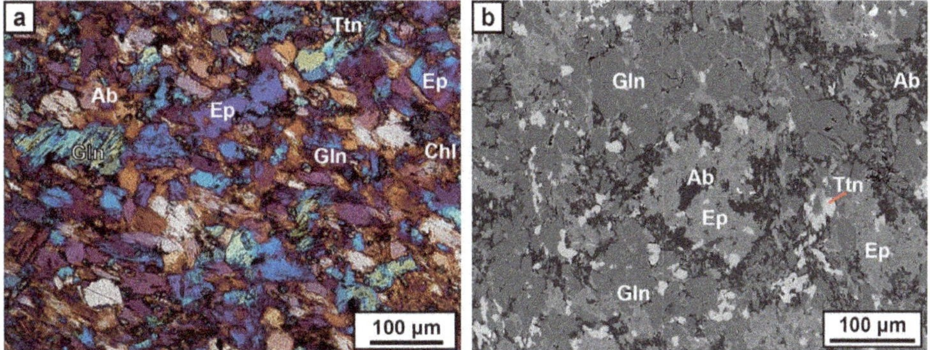

Figure 1. (a) Optical photomicrograph (XPL) with retardation plate (λ = 530 nm) and (b) backscattered electron (BSE) image of the starting material (massive fine-grained epidote blueschist-facies rock). The average grain size of constituent minerals was ~30 µm. Gln: glaucophane, Ep: epidote, Ttn: titanite, Ab: albite, and Chl: chlorite.

2.2. Deformation Experiment in Simple Shear

Deformation experiments of epidote blueschist were conducted using a modified Griggs apparatus at the Tectonophysics Laboratory, School of Earth and Environmental Sciences (SEES) at Seoul National University (SNU), Seoul, South Korea. Figure 2 presents the sample assembly designed for the deformation experiment in simple shear. As there was no clear foliation and lineation in the starting material, the sample was core-drilled in an arbitrary orientation in the same direction as a cylindrical rod with a 3.15 mm diameter and cut at 45° to ~400 µm thick, for the simple shear experiment. The sample was sandwiched between alumina pistons cut at 45° in the maximum principal stress (σ_1) direction, keeping the same orientation for all experiments. Weak CsCl or NaCl was used as a pressure medium. A thin Ni foil, a strain marker, was inserted in the middle of the sample perpendicular to the shear plane. Temperature was monitored using two thermocouples (Pt_{70}-Rh_{30} and Pt_{94}-Rh_6) near the top and bottom of the specimen. The confining pressure and temperature were raised to the target pressure (0.9–1.5 GPa) and temperature (400–500 °C) within ~10 h and ~30 min, respectively. The sample was deformed by moving the tungsten carbide (WC) and alumina pistons at a constant speed (~8.7 × 10^{-5} mm·s^{-1}). Following the deformation experiment, the sample was quenched to room temperature by shutting off the power to preserve microstructures developed during experiments. The confining pressure was decreased to room pressure in the same length of time that it took to increase and reach the target pressure.

Table 1. The representative composition of glaucophane and epidote in the starting material.

Element	SiO$_2$	TiO$_2$	Al$_2$O$_3$	Cr$_2$O$_3$	FeO	MgO	CaO	MnO	Na$_2$O	K$_2$O	NiO	Total
Gln	57.31	0.03	10.68	0.03	11.92	8.22	0.26	0.08	7.30	n.d.	n.d.	95.84
Ep	38.02	0.00	26.76	n.d.	7.61	0.03	23.37	0.07	0.03	0.01	n.d.	95.90
Cation	**Si**	**Ti**	**Al**	**Cr**	**Fe^{2+}**	**Fe^{3+}**	**Mg**	**Ca**	**Mn**	**Na**	**K**	**Sum**
Gln	8.06	0.00	1.77	0.00	1.31	0.09	1.73	0.04	0.01	1.99	-	15.01
Ep	3.01	0.00	2.49	-	0.00	0.50	0.00	1.98	0.00	0.00	0.00	7.99
Gln	Si [T]	Ti [T]	Al [T]	Ti [C]	Al [C]	Cr [C]	Fe^{2+} [C]	Fe^{3+} [C]	Mg [C]	Ca [C]	Mn [C]	-
Cal. #	8.06	0.00	0.00	0.00	1.77	0.00	1.31	0.09	1.72	0.04	0.01	-
Gln	Fe^{2+} [B]	Mg [B]	Ca [B]	Mn [B]	Na [B]	Na [A]	Ca [A]	K [A]	-	-	-	-
Cal. #	0.00	0.00	0.00	0.00	1.99	0.00	0.00	0.00	-	-	-	-
Ep	Si [T]	Al^{IV} [T]	Ti [M]	Cr [M]	Al^{VI} [M]	Fe^{3+} [M]	Mg [M]	Mn [M]	Na [M]	K [M]	Ca [A]	-
Cal. #	3.01	0.00	0.00	0.00	2.49	0.50	0.00	0.00	0.00	0.00	1.98	-

Gln: glaucophane, Ep: epidote, n.d.: not detected, Cal. #: the calculated cation values occupying each site in the crystal structure, [T] and [A]: T-site and A-site in amphibole (glaucophane) and epidote crystal structures, [C] and [B]: C-site and B-site in amphibole (glaucophane) structure, and [M]: M-site in epidote structure.

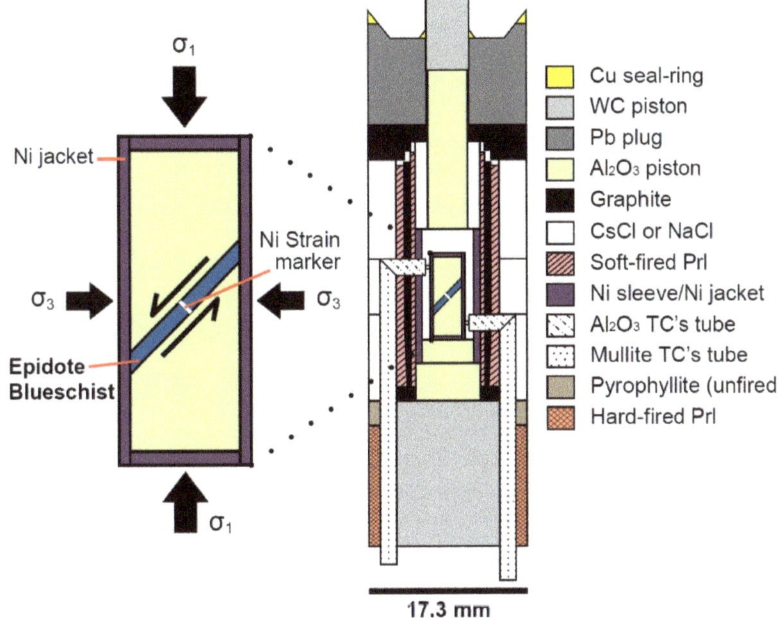

Figure 2. Schematic of sample assembly for the deformation experiment in simple shear. The sample was positioned at 45° to the maximum principal stress (σ_1). A strain marker was placed perpendicular to the shear plane in the middle of the sample.

2.3. Determination of LPOs of Minerals

The LPOs of minerals in epidote blueschist were determined using the electron backscattered diffraction (EBSD) technique [38], prior to and after the deformation experiment. To determine the LPOs, we used HKL Technology's EBSD system (Channel 5 software) with a Nordlys II detector attached to the scanning electron microscope (SEM; JEOL JSM 6380) at the SEES in SNU. The LPOs of glaucophane and epidote were measured in the XZ plane of the deformed samples; the X- and Z-directions represent the shear direction and direction normal to the shear plane, respectively. Samples were polished using alumina powders, diamond paste of 1 μm, and Syton (0.06 μm colloidal silica slurry) because the measurement of the LPO of minerals using SEM involved removing the mechanical surface damage using a chemical–mechanical polishing technique [39]. The polished plane was coated with a ~3 nm thick carbon to prevent charging in the SEM, and the surface was tilted 70° to the incident electron beam in the chamber. EBSD analysis was conducted under a 20 kV accelerating voltage, a 15 mm working distance, and a spot size of 60. For each grain, all EBSD patterns were manually indexed point by point to determine the LPO of the mineral accurately.

2.4. Observation of the Deformation Microstructures in Minerals

The observation of deformed samples following the experiments was conducted using field-emission SEM (FE-SEM; JEOL JSM 7100F) at the SEES in SNU. To observe the intracrystalline deformation microstructures in samples, EBSD mapping of deformed grains was conducted using the Oxford Instruments' EBSD system (AZtec v. 3.4 software), with a Symmetry detector attached to the FE-SEM at the SEES in SNU. EBSD analysis was conducted under a 15 kV accelerating voltage, a 25 mm working distance, a ~3 nA probe current, and a 0.1–0.2 μm step size. Data were subsequently processed using MTEX, a MATLAB toolbox [40], to calculate the local misorientations for textural analysis. This study used the grain reference orientation deviation (GROD) angle map and the kernel angle misorientation (KAM) map. The former is a map generated based on the deviation between the mean orientation of a reference point and those of other points. The KAM map is the misorientation less than the predefined threshold value calculated by the mean orientation between a point and its neighbors. The TEM technique was also used to observe dislocation structures in deformed grains. TEM observations under a 300 kV accelerating voltage were carried out using a JEOL JEM-3010 at the National Center for Inter-university Research Facilities (NCIRF) in SNU. Focused ion beam (FIB) foils for TEM investigations were prepared using the FEI Helios 650 at the NCIRF in SNU. The FIB mineral foils were extracted subparallel to the shear direction and shear plane.

3. Results

3.1. Deformation Microstructures after Experiments

Table 2 summarizes the experimental conditions and results for the deformation experiments on the epidote blueschist. The deformation microstructures of representative samples are shown in Figure 3. Even though we designed and used the sample assembly (Figure 2) for simple shear, there was a small proportion of the pure shear component (~5–9%) for all experiments depending on the shear strain magnitude. We quantified the magnitude of the shear strain (γ) by the rotational angle of the thin Ni strain marker and/or elongated shapes of grains. In the sample deformed with low shear strain ($\gamma \leq 1$), most minerals exhibited a similar shape to that of the starting material (Figure 3a,b). Glaucophane grains had slightly elongated obliquely to the shear direction, with some fractures present subparallel to the shear plane inside a grain. In contrast, epidote grains exhibited kink bands and bookshelf gliding textures [26] (Figure 3b). In the sample deformed with an intermediate shear strain ($1 < \gamma \leq 2$), glaucophane grains had elongated, demonstrating cataclastic mosaic fragmented texture. Their rotation was also more oblique to the shear direction than those deformed with low shear strain. However, epidote grains had reduced elongation than glaucophane, exhibiting shear-band-type fragmented texture in this region (Figure 3c,d). In the sample deformed with high shear strain ($\gamma > 2$), all minerals appeared to be elongated and/or flowed subparallel to the shear direction (Figure 3e,f). Relatively large glaucophanes (~30–50 μm) were highly elongated, consisting of many small-size grains from 0.5 to 5 μm. They were in subparallel alignment with the shear direction, exhibiting a cataclastic/granular flow texture. The microstructure of epidote and titanite also presented with similar cataclastic/granular flow textures.

Table 2. Experimental conditions and results for the deformation experiment on epidote blueschist.

Run #	P (GPa)	T (°C)	Peak σ_d (MPa)	Shear Strain (γ)	Shear Strain Rate (s^{-1})	LPO of Gln	LPO of Ep
JH148	1.2	430 ± 10	700 ± 20	0.4 ± 0.02	1.5×10^{-5}	Type-1	Weak fabric
JH98b	1.5	500 ± 10	710 ± 20	0.6 ± 0.03	3.4×10^{-5}	Type-1	Weak fabric
JH88b	1.5	400 ± 10	800 ± 20	0.8 ± 0.03	4.4×10^{-5}	Type-1	Weak fabric
JH112b	0.9	430 ± 10	720 ± 20	0.9 ± 0.03	4.2×10^{-5}	Type-1	Weak fabric
JH94b	1.5	400 ± 10	900 ± 20	1.5 ± 0.06	6.0×10^{-5}	Transitional	Weak fabric
JH88a	1.5	400 ± 10	800 ± 20	2.1 ± 0.12	1.2×10^{-4}	Type-2	Type-1
JH98a	1.5	500 ± 10	710 ± 20	2.4 ± 0.12	1.4×10^{-4}	Type-2	Type-1
JH150	1.2	480 ± 10	680 ± 20	2.7 ± 0.15	1.2×10^{-4}	Type-2	Type-1
JH112a	0.9	430 ± 10	720 ± 20	2.9 ± 0.20	1.4×10^{-4}	Type-2	Type-1
JH94a	1.5	400 ± 10	900 ± 20	4.5 ± 0.38	1.8×10^{-4}	Type-2	Type-2

σ_d: differential stress, Gln: glaucophane, and Ep: epidote.

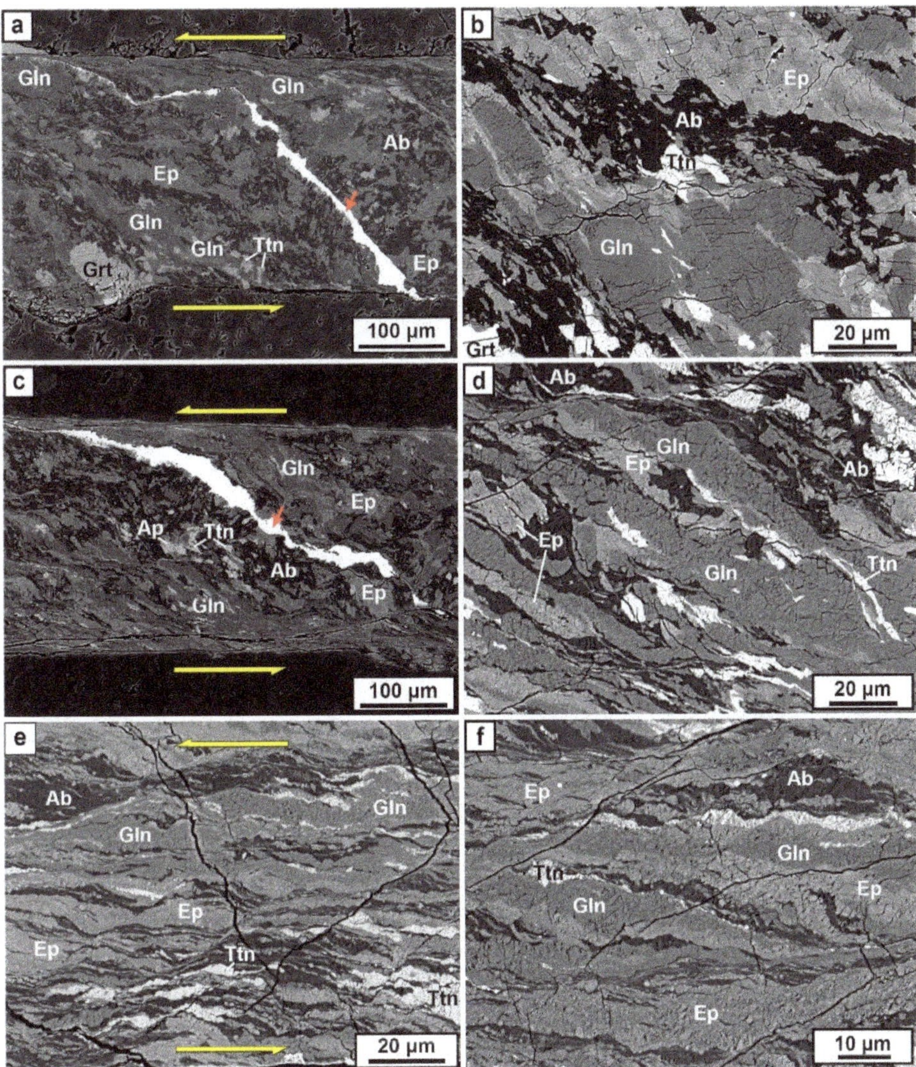

Figure 3. BSE images of deformed samples after simple shear experiments. (**a**,**b**) sample deformed with low shear strain (JH88b, $\gamma = 0.8$, P = 1.5 GPa, and T = 400 °C); (**c**,**d**) sample deformed with intermediate shear strain (JH94b, $\gamma = 1.5$, P = 1.5 GPa, and T = 400 °C); (**e**,**f**) sample deformed with high shear strain (JH94a, $\gamma = 4.5$, P = 1.5 GPa, and T = 400 °C). Long yellow arrows indicate the left-lateral (sinistral) shear sense, and red arrows indicate the Ni strain marker. All samples in the images show the left-lateral (sinistral) shear sense. Gln: glaucophane, Ep: epidote, Ttn: titanite, Ab: albite, Chl: chlorite, Grt: garnet, and Ap: apatite; P: pressure; T = temperature.

In contrast, small grains, originally less than 5 μm in size prior to the experiment, appeared to flow exhibiting a thin ductile banded shape (Figure 4). In this region, the ductile shear band representing the C-C′ structure and the cataclastic flow textures coexist (Figure 4a). The brittle microfault, indicated by a small offset of the thin banded titanite in the glaucophane, and the ductile elongated shear band of small glaucophane minerals was also observed to coexist (Figure 4b).

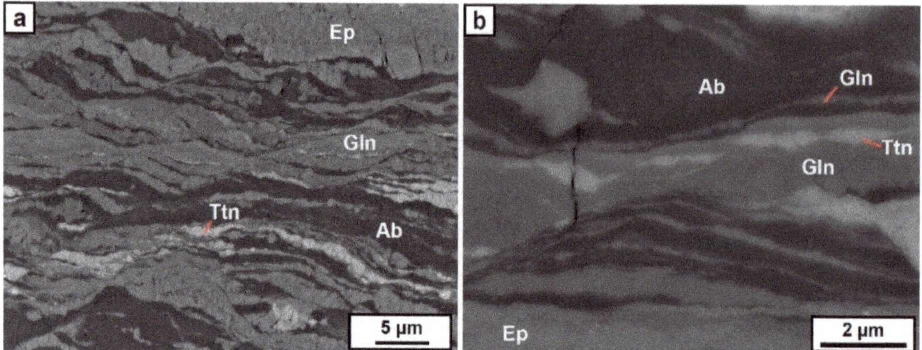

Figure 4. BSE images of sample deformed with high shear strain (JH94a, γ = 4.5). (**a**) Ductile shear band showing C-C′ structure with cataclastic flow texture; (**b**) brittle microfault indicated by the titanite offset and ductile elongated shear band of small glaucophane. All images show the left-lateral (sinistral) shear sense. Gln: glaucophane, Ep: epidote, Ttn: titanite, and Ab: albite.

3.2. Lattice Preferred Orientations of Glaucophane and Epidote

Figure 5 illustrates the LPOs of glaucophane and epidote. The starting material was set up in the same orientation for all experiments. The initial fabric of glaucophane in the starting material is shown in Figure 5a. It exhibited a weak girdle distribution of the crystallographic [100] axes and the (110) and (010) poles, subnormally aligned with the direction becoming a shear direction, and the [001] axes aligned obliquely to that direction. Based on the pole figures of glaucophane deformed after the experiments (Figure 5b–k), the LPO of glaucophane had altered with increasing shear strain (γ). In samples deformed under low shear strain ($\gamma \leq 1$), the (010) poles of glaucophane had subnormally aligned with the shear plane and the [001] axes were in subparallel alignment with the shear direction; this is defined here as type-1 LPO for glaucophane (Figure 5b–e). With increasing shear strain between $1 < \gamma \leq 2$, the [100] axes, and (010) poles were in subparallel alignment with the shear plane and subnormal to the shear direction, appearing to be a transitional fabric (Figure 5f). In samples deformed under high shear strain ($\gamma > 2$), the [100] axes were in subnormally aligned with the shear plane and the [001] axes were in subparallel alignment with the shear direction; this is defined here as type-2 LPO for glaucophane (Figure 5g–k).

The initial fabric of epidote in the starting material showed that the crystallographic [100] axes were subnormally aligned in the direction becoming a shear direction forming a girdle shape, and the (010) poles were in subparallel alignment with that direction (Figure 5a). After deformation experiments, the LPO of epidote had also altered with increasing shear strain (γ). The LPOs of epidote in deformed samples that had undergone low shear strain ($\gamma < 2$) exhibited a weak nonsystematic fabric (Figure 5b–f). However, in samples deformed under a shear strain between $2 < \gamma < 4$, the [100] axes of epidote were subnormally aligned with the shear plane, and the (010) poles were in subparallel alignment with the shear direction; this is defined here as type-1 LPO for epidote (Figure 5g–j). In the sample deformed under a high shear strain of $\gamma = 4.5$, the (010) poles of epidote were subnormally aligned with the shear plane, and the [001] axes were in subparallel alignment with the shear direction; this is defined here as type-2 LPO for epidote (Figure 5k).

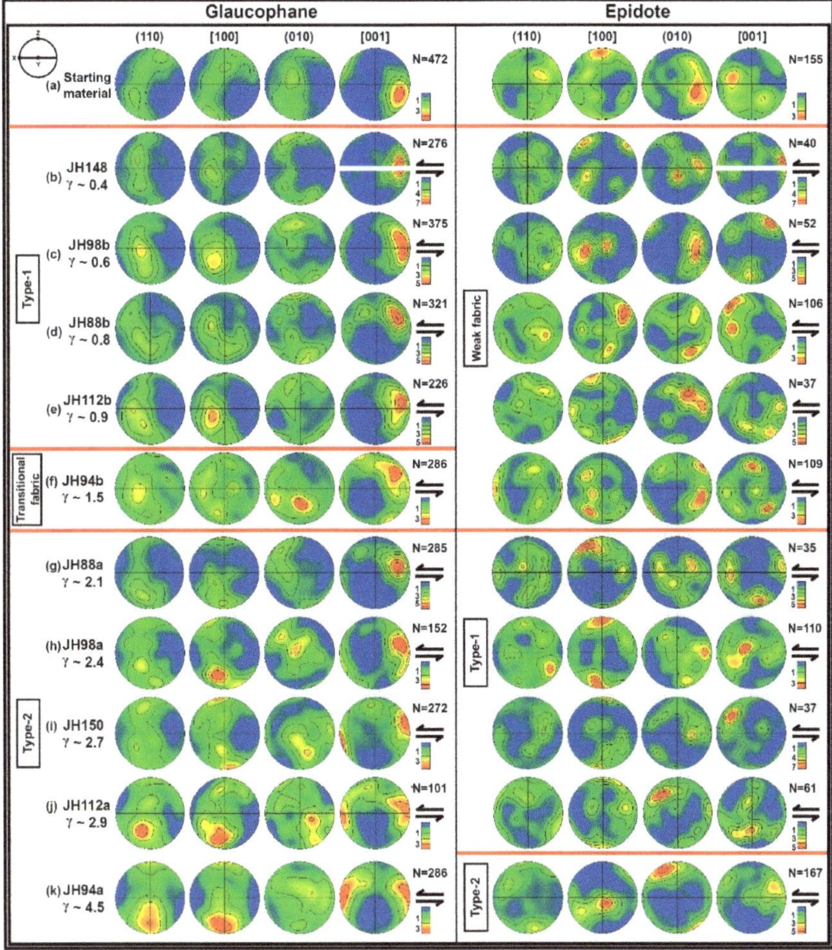

Figure 5. Pole figures of glaucophane and epidote in (**a**) starting material and (**b–k**) deformed samples showing LPO presented in the lower hemisphere using an equal-area projection. A half-scatter width of 30° was used. The X- and Z-directions correspond to the shear direction and the direction normal to the shear plane in the experiment, respectively. White line in pole figure: shear plane; black arrows: shear direction; and N: the number of grains.

3.3. Observations of Intracrystalline Deformation Microstructures in Deformed Glaucophane

EBSD mapping in single glaucophane was undertaken for samples deformed under different shear strains (Figure 6), to observe the intracrystalline deformation microstructures of glaucophane. In glaucophane deformed under a low shear strain of $\gamma = 0.8$ at a pressure of 1.5 GPa and a temperature of 400 °C (JH88b), the GROD angle map showed that the local misorientation had gradually increased toward the edge of the grain (Figure 6b). In contrast, the KAM map indicated that intracrystalline misorientations were present on the inside and edge of the grain (black arrows in Figure 6c).

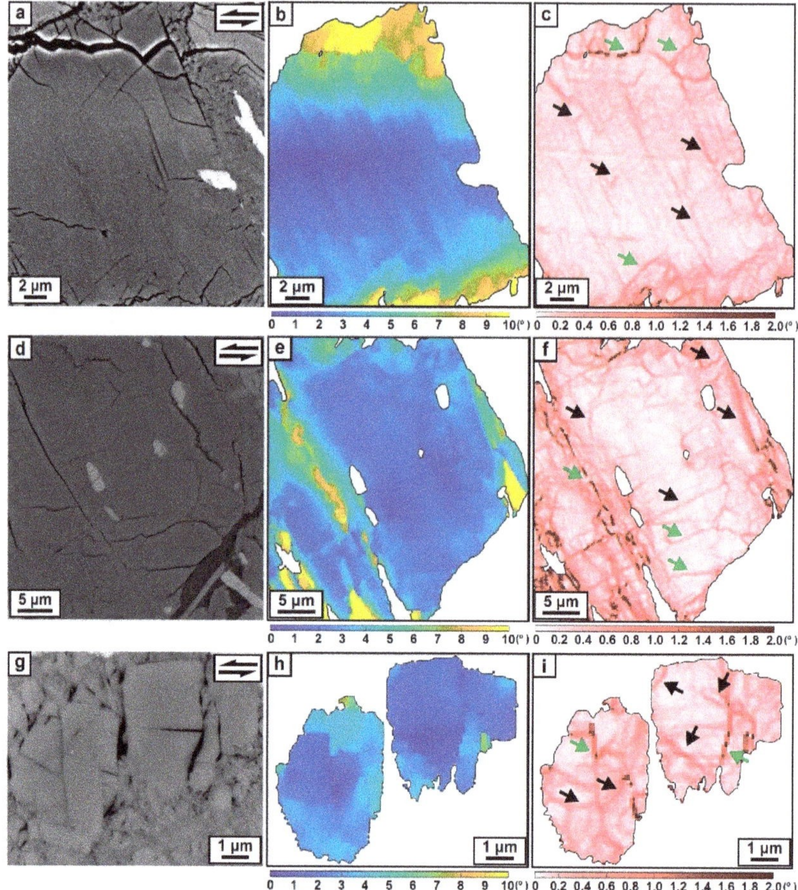

Figure 6. BSE image, grain reference orientation deviation (GROD) angle map (mis2mean), and kernel angle misorientation (KAM) map showing the intracrystalline deformation microstructure of glaucophane deformed under a shear strain of (**a–c**) $\gamma = 0.8$ (in JH88b, P = 1.5 GPa, and T = 400 °C), (**d–f**) $\gamma = 0.6$ (in JH98b, P = 1.5 GPa, and T = 500 °C), and (**g–i**) $\gamma = 4.5$ (in JH94a, P = 1.5 GPa, and T = 400 °C). Black arrows in the BSE images indicate the left-lateral (sinistral) shear sense. Black and green arrows in the KAM map indicate the local plastic strain (i.e., dislocation density or rotation of the lattice) and the tilted crystal orientation by brittle fracture, respectively. Step size of mapping: (**b,c**) 0.15 μm, (**e,f**) 0.2 μm, and (**h,i**) 0.1 μm.

In glaucophane deformed under a low shear strain of $\gamma = 0.6$ at a pressure of 1.5 GPa and high temperature of 500 °C (JH98b), the GROD angle map illustrated that the high misorientation focused on the fracture in the grain (Figure 6e). However, the KAM map presented well-preserved misorientation lines, irrelevant to the fractures (Figure 6f). For glaucophane deformed under a high shear strain of $\gamma = 4.5$ at a pressure of 1.5 GPa and temperature of 400 °C (JH94a), regardless of the cataclastically comminuted grain, the KAM map showed that several misorientation lines were present in the remaining relatively large grains (Figure 6i).

3.4. Observations of Dislocation Microstructures in Deformed Glaucophane and Epidote Using TEM

Weak-beam dark field (WBDF) and bright-field (BF) images were captured using TEM to observe dislocation microstructures in glaucophane and epidote following the experiments; these are shown in Figures 7 and 8. In glaucophane deformed under a low shear strain ($\gamma \leq 1$; $\gamma = 0.6$, in JH98b), WBDF images showed the presence of several dislocation structures. Stacking faults aligned perpendicular to the shear direction (yellow arrow) were observed, in addition to a dislocation loop with a curve toward the shear direction for diffraction vector of $g = 101$ (Figure 7a). Straight dislocations aligned perpendicular or oblique to the shear direction were also observed for $g = 101$ (Figure 7b). In the glaucophane deformed under a high shear strain ($\gamma > 2$; $\gamma = 4.5$, in JH94a), WBDF images of relatively large grains showed the presence of several distorted lattice structures (moiré fringe textures) and straight dislocations aligned obliquely to the shear direction (yellow arrow; Figure 7c,d).

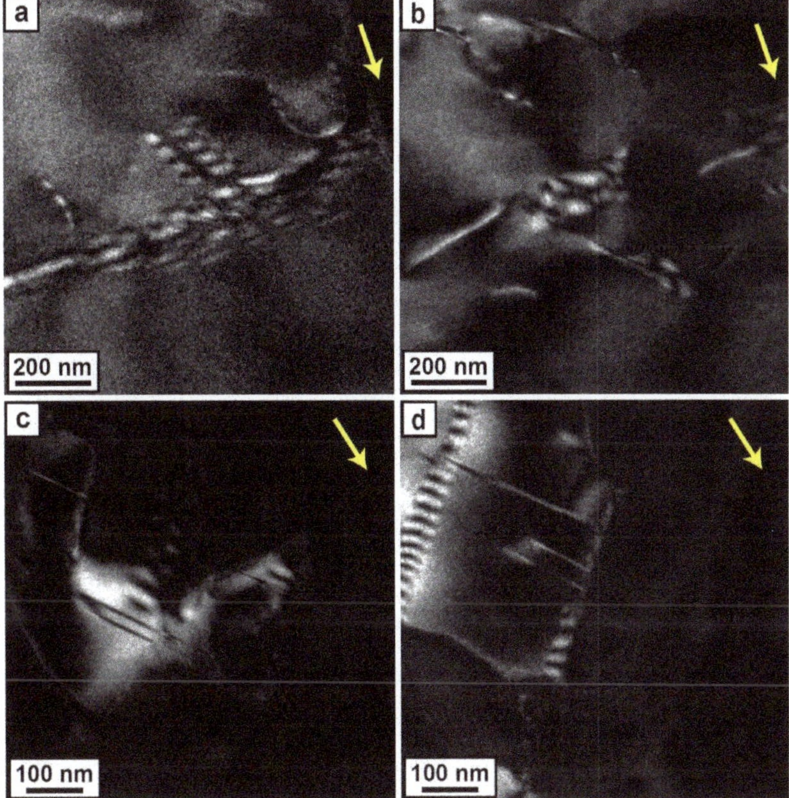

Figure 7. Weak-beam dark field (WBDF) images of glaucophane experimentally deformed with (**a**,**b**) low shear strain ($\gamma = 0.6$) at P = 1.5 GPa, and T = 500 °C (JH98b) and with (**c**,**d**) high shear strain ($\gamma = 4.5$) at P = 1.5 GPa, and T = 400 °C (JH94a). Yellow arrows indicate shear direction in the images. P: pressure; and T: temperature.

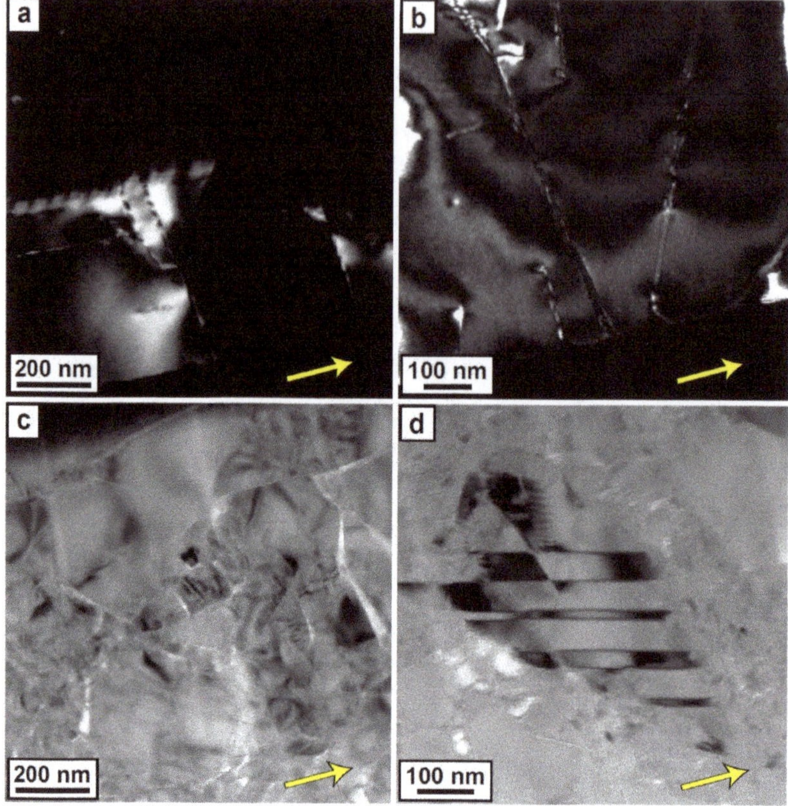

Figure 8. (**a,b**) Weak-beam dark field (WBDF) images of epidote experimentally deformed under a shear strain of $\gamma = 2.4$ at P = 1.5 GPa and T = 500 °C (JH98a); (**c,d**) bright-field (BF) images of epidote experimentally deformed under a high shear strain of $\gamma = 4.5$ at P = 1.5 GPa and T = 400 °C (JH94a). Yellow arrows indicate the shear direction in the images. P: pressure; and T: temperature.

In epidote deformed under a shear strain between $2 < \gamma < 4$ ($\gamma = 2.4$, JH98a), WBDF images illustrated the presence of several twin boundaries as well as subgrain boundaries aligned parallel or perpendicular to the shear direction (yellow arrow; Figure 8a,b). The BF image of epidote deformed under a high shear strain of $\gamma = 4.5$ (JH94a) showed the presence of many distorted lattice structures with brittle failures, similar in texture to glaucophane deformed with a high shear strain (Figure 8c). In addition, a relatively large grain of epidote conserving deformation twins aligned oblique to the shear direction (yellow arrow) was surrounded by grains that were of tens of nanometers size (Figure 8d).

4. Discussion

4.1. LPO Formation and Deformation Mechanisms of Glaucophane

Although the starting material did not show a clear foliation and lineation in the hand specimen, there was an initial fabric, as shown in Figure 5a. However, the initial LPO of minerals in the starting material was obviously altered with increasing shear strain (Figure 5b–k). Type-1 LPOs for glaucophane in samples deformed under a low shear strain ($\gamma \leq 1$) exhibited (010) poles subnormally aligned with the shear plane and the [001] axes in subparallel alignment with the shear direction (Figure 5b–e); this is rarely observed in natural blueschist. A recent study reported this type-1 LPO for glaucophane in weakly deformed (folded) blueschist-facies rock [33]. Previous TEM/HREM study also reported a dislocation slip system of (010)[001] in glaucophane from a blueschist-facies micaschist at a pressure < 1.0 GPa and temperature between 350 and 450 °C [21]. They suggested that the dislocation glide (or slip) may operate at relatively low temperatures as the Na occupying the $M4$ site is able to jump into the neighboring empty A-site in glaucophane [21]. As the easiest slip systems in chain–silicate structures tend to avoid breaking the Si-O bond [41,42], they concluded that the dislocation glide needs to operate between the Si-O tetrahedral chains; this means the (010) in glaucophane (Figure 9a,b) [21]. This suggestion coincides with the observed dislocation in a single crystal of glaucophane deformed under a low shear strain (γ = 0.6) in this study (Figure 7a,b). Various dislocations observed with different diffraction vectors indicate that differing dislocation slip systems were activated by shear deformation. Although this glaucophane was deformed at a relatively high pressure and temperature (i.e., 1.5 GPa and 500 °C, JH98b), the KAM map for glaucophane deformed at a pressure of 1.5 GPa and temperature of 400 °C showed that intracrystalline deformation (i.e., dislocation glide), also occurs at relatively lower temperature (Figure 6c). The glaucophane shape slightly elongated oblique to the shear direction is considered a result of the rigid body rotation in a weak plagioclase matrix (Figure 3b). Thus, these results suggest that glaucophane was deformed by the simultaneous rotation of a relatively rigid body with intracrystalline dislocation glide (or slip) forming a strong type-1 LPO under a pressure between 0.9 and 1.5 GPa and a temperature between 400 and 500 °C. This result is consistent with the suggestion from a previous study that rigid body rotation and dynamic recrystallization by dislocation creep are the deformation mechanisms for glaucophane [20]. The glaucophane that had deformed under a shear strain between $1 < \gamma \leq 2$, was elongated showing cataclastic and mosaic fragmented texture. This microstructure may indicate that the predominant deformation of glaucophane had converted to brittle behavior as cataclastic flow and rigid body rotation, exhibiting a transitional LPO with a decrease in the fabric strength of glaucophane (Figure 5f).

For deformed samples under a high shear strain ($\gamma > 2$), type-2 LPOs of glaucophane exhibited the [100] axes subnormally aligned with the shear plane and the [001] axes in subparallel alignment the shear direction (Figure 5g–k), which have been reported in many natural blueschists [15–18,28–32]. This type-2 LPO of glaucophane is also similar to the type-1 LPO of hornblende [43,44]. A previous TEM/HREM study reported major dislocation glides for (100)[001] and {110}[001] in glaucophane from an eclogitic micaschist (at 1.5–1.8 GPa and 550–600 °C). These are sufficiently active independent slip systems that ensure its ductile behavior during progressive deformation [21]. Other previous studies on natural blueschist have also suggested that dislocation creep was a dominant mechanism in the formation of type-2 glaucophane LPO [17,18]. Based on the TEM investigation in this study, many dislocations and distorted lattice structures were present in a single crystal of glaucophane deformed under a high shear strain (γ = 4.5) under a pressure of 1.5 GPa and temperature of 400 °C (JH94a; Figure 7c,d). The KAM map of the deformed grain illustrated many intracrystalline misorientations, irrelevant to the fracture, indicating the activation of dislocations (Figure 6i). High shear strain energy is likely to form a (100) slip plane by breaking the octahedral Mg-O bond, weak $M4$-site (occupied by Na), and A-site (empty space; Figure 9a,c). As such, these results suggest that the magnitude of shear strain may be a major cause for changes to dominant slip systems and the formation of different LPOs of glaucophane in the blueschist-facies metamorphic condition. In addition, elongated aggregates of small-size grains with cataclastic/granular flow texture for glaucophane deformed under a strain of $\gamma > 2$ (Figure 3e,f), suggest that the type-2 LPO formation may be influenced by simultaneous cataclastic flow deformation of glaucophane. However, originally small grains of glaucophane (~5 μm) prior to deformation, appear to flow exhibiting a ductile thin banded shape (Figure 4). These microstructures imply that high shear strain has a different influence on relatively large grains ($\geq$30 μm) and very fine-grains of glaucophane, causing cataclastic flow deformation and a crystal plastic deformation via dynamic recrystallization [45], respectively. As a result, the rheological contrast between component minerals, shear strain, and grain size may be important factors influencing the deformation mechanisms of glaucophane.

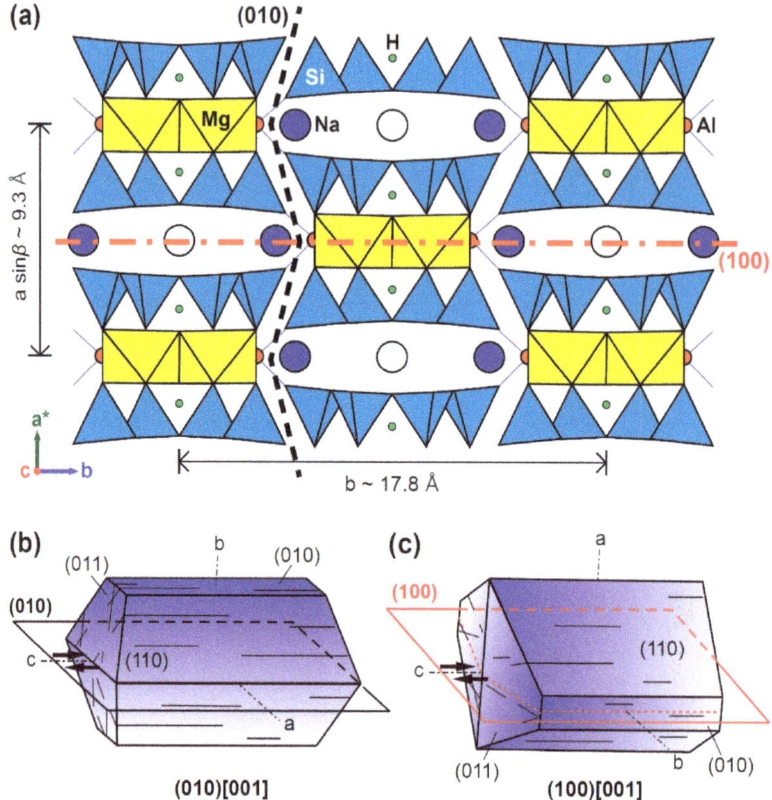

Figure 9. (**a**) Simplified glaucophane crystal structure projected onto the (001) plane. Black dashed and red dot-dashed lines indicate the tracks of the easiest slip system planes at low and high shear strain, respectively. Blue tetrahedron: Si-O structure, yellow octahedron: Mg-O structure (*M1* site), red circle: Al (*M2* site), green circle: H, blue circle: Na (*M4* site), and white circle: empty space (*A*-site). The Mg-O octahedron in the *M3* site has been omitted; (**b**,**c**) schematic of the glaucophane crystal form and different dominant slip systems in glaucophane. The transparent truncate sheet and black arrows indicate a dominant slip plane and a dominant slip direction, respectively.

4.2. LPO Formation and Deformation Mechanisms of Epidote

Our experimental results showed that epidote in samples deformed under a low shear strain of $\gamma \leq 2$ exhibited mostly weak and nonsystematic LPOs (Figure 5b–f). This may be due to insufficient strain developing an intracrystalline deformation in epidote as a rigid material. A previous study also reported similar results, revealing that low shear strain ($\gamma = 2$) was insufficient to produce a strong preferred orientation of experimentally deformed epidote mineral as a relatively rigid particle in a deformed plagioclase matrix at a pressure of 1.5 GPa and a temperature of 750 °C [27]. In contrast, the type-1 LPO of epidote in samples deformed under a shear strain between $2 < \gamma < 4$ (Figure 5g–j) and type-2 LPO of epidote deformed under a shear strain of $\gamma = 4.5$ (Figure 5k) were different from previously reported LPOs [16,17,28–30,33,35]. However, a previous TEM and HREM investigation reported stacking faults on (100) and lamellar twins on (100) as the easiest shearing planes of epidote mineral (clinozoisite and zoisite) [34]. Other studies have also observed deformation twin lamellae on (100) of clinozoisite in eclogite [24,25]. This plane was consistent with a relatively weak bond for the *A*-site (occupied by Ca) and *M3*-site (occupied by Al-Fe^{3+}; Figure 10a,b) [46]. On TEM observations of the epidote crystal deformed under a shear strain between $2 < \gamma < 4$ ($\gamma = 2.4$, JH98a), several aligned dislocations, subgrain boundaries, and deformation twins were present (Figure 8a,b). This suggests that dislocation glide or slip may play a major role in the formation of type-1 LPO for epidote. This result is consistent with deformation processes of epidote group minerals as suggested by Franz and Liebscher [26]; these are either dislocation glide or sliding on the cleavage by fracturing in naturally deformed rocks. On the contrary, another previous study suggested a rigid body rotation as an LPO formation mechanism for epidote group minerals in metabasite rocks through two potential processes. Firstly, where zoisite minerals are deformed by shearing parallel to the (100) cleavage planes and rotated to align with the (100) plane subparallel to foliation. Secondly, where they are deformed by stretching in the [010] axes, resulting in boudinaged crystals in subparallel alignment with lineation, and the (100) plane is simultaneously aligned, subparallel to foliation [23]. As the epidote is a relatively rigid mineral in this blueschist sample, a rigid body rotation may also be a major process for the development of type-1 LPO.

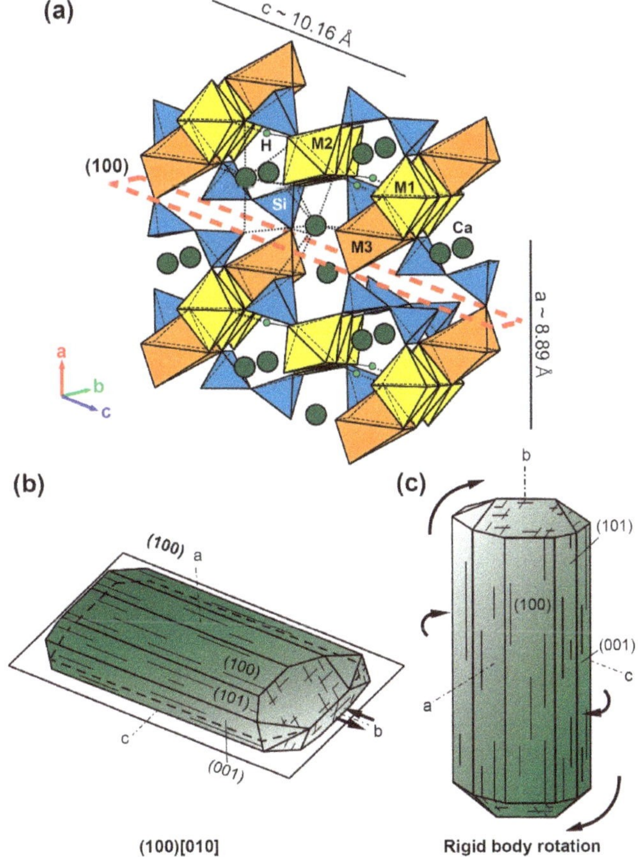

Figure 10. (**a**) Simplified epidote crystal structure. Red dashed rectangular indicate the tracks of the easiest (100) slip plane. Blue tetrahedron: Si-O structure, yellow octahedron: Al-O structure (*M1* and *M2* site), orange octahedron: Al-Fe^{3+} substitution-O (*M3* site), blue circle: H, and green circle: Ca (*A1* and *A2* site); (**b,c**) schematic of the epidote crystal form with different behavior by slip. The transparent truncate sheet and black arrows in (**b**) indicate a dominant slip plane and a dominant slip direction, respectively. Black arrows in (**c**) indicate a direction of rigid body rotation by shear strain.

The type-2 LPO of epidote has not yet been reported. A previous experimental study has suggested that epidote minerals are deformed by granular flow and diffusion-assisted grain boundary sliding at relatively high shear strain (γ = 4.5 and 7.5), thereby developing shear band textures at a pressure of 1.5 GPa and temperature of 750 °C [27]. Based on the TEM investigation in this study, relatively large grains of epidote conserving deformation twins surrounded by nanocrystalline grains were observed in the sample deformed under a shear strain of γ > 4 (γ = 4.5, JH94a; Figure 8c,d). These microstructures indicate that cataclastic/granular flow with rigid body rotation may play an important role in the formation of type-2 LPO of epidote (Figure 10c). In contrast, a previous study on naturally bent and kinked epidote crystal suggested that all {100}, {010}, {001} planes play a major role in the deformation of epidote group minerals, either as dislocation slip planes or as cleavage planes [26]. Thus, type-1 and type-2 LPOs of epidote were likely to have been developed by dislocation creep and cataclastic flow with rigid body rotation due to rheological contrasts with other minerals. Further studies on the slip system of epidote are required at various pressure and temperature conditions to better understand the epidote LPOs/slip systems transition.

4.3. Implications for Deformation Mechanisms of Epidote Blueschist in a Warm Subduction Zone

The EBSD analysis and TEM investigation of experimental specimens revealed that glaucophane and epidote, major constituent minerals of epidote blueschist, had deformed simultaneously through brittle and ductile behavior at a pressure between 0.9 and 1.5 GPa and a temperature between 400 and 500 °C. These results indicate that epidote blueschist is likely to undergo deformation in the brittle–ductile regime. The nonlocalized cataclastic flow in the deformed sample may be attributed to the deformation microstructures in the semi-brittle regime following the brittle–ductile transition [47,48]. Thus, a brittle–ductile transition of glaucophane is likely to occur under a pressure of 1.0–1.5 GPa at 400–500 °C in epidote blueschist. This result is relatively consistent with previous suggestions that glaucophane may be deformed under a ductile regime (i.e., dislocation creep), in natural epidote blueschist at the pressures of ~0.5–2.0 GPa [16,17,35]. A previous experimental study on lawsonite blueschist also suggested that the brittle–ductile transition of glaucophane is likely to occur at a pressure of ~2 GPa and temperatures between 400 and 500 °C [49]. One may argue that the cataclastic deformation microstructure in this study was produced by the deformation of the sample at a fast strain rate. However, there are also examples showing cataclastic deformation microstructures in natural blueschists and blueschist-facies metamorphic rocks that were deformed at low strain rates [50,51]. In addition, experimental results have suggested that while the microstructure of epidote is similar to glaucophane, the shear strain required to change the LPO of epidote appears to be larger than glaucophane (Figure 11). This is likely a result of the strength contrast between glaucophane and epidote, as deformation is concentrated and strain is localized on relatively weaker minerals [52]. As a result, epidote blueschist that exists in a warm subduction zone may be deformed under the coexistence of brittle and ductile regimes due to the rheological contrast between constituent minerals under stable pressure and temperature conditions.

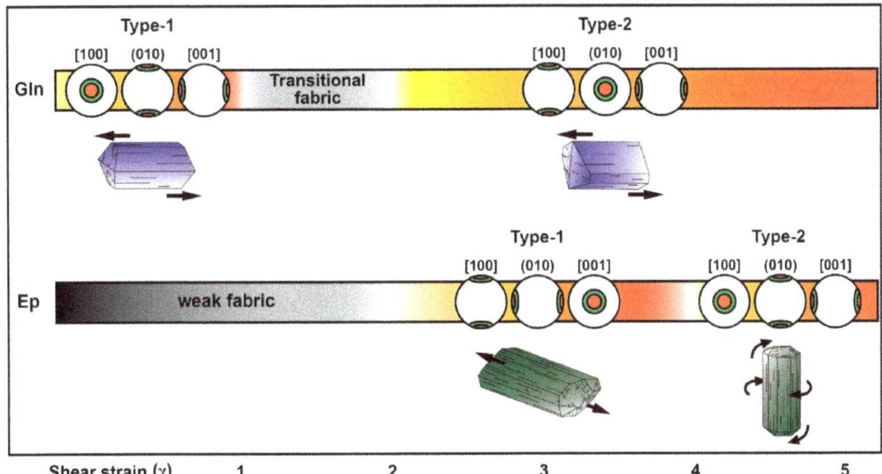

Figure 11. Schematic of the development of LPOs of glaucophane and epidote with increasing shear strain. Gln: glaucophane; and Ep: epidote.

4.4. Implications for Seismic Anisotropy of Subducting Slab in a Subduction Zone

The seismic anisotropy of shear waves in the forearc region of the subduction zone has been mainly observed by SKS waves propagating through the anisotropic mediums and throughout the area subperpendicular to the Earth's surface [53]. As SKS anisotropic data are measured by path integration, the source for the splitting of S-waves has been regarded as multiple anisotropic layers. Among these layers, the mantle wedge and subslab mantle have been considered as the main anisotropic mediums because of the large volume proportions in the lithosphere [54]. Specifically, the B-type LPO of olivine in the mantle wedge has been suggested as a cause for the change from the trench-normal to trench-parallel polarization direction of the fast shear wave in the forearc region [55,56]. However, subducting oceanic crust is also a potential anisotropic layer where many elastically anisotropic minerals are stable [3,12]. Therefore, recent studies have suggested that type-2 LPO of glaucophane may affect the trench-parallel seismic anisotropy of the forearc region, where the subducting slab has a high dip angle [15–18]. This is because the propagation velocity of the fast S-wave (Vs1) through the b- and c-axes of the glaucophane is much faster than that through an a-axis in glaucophane. In contrast, the type-1 LPO of glaucophane may contribute to the trench-normal polarization direction regardless of the subducting angle because the b- and c-axes are continuously aligned parallel to the shear direction in the subducting slab with different dip angles. The type-1 and type-2 LPOs of epidote in this study have not yet been reported in natural samples. However, if epidote with type-1 LPO exists in the subducting slab, this LPO may also influence the trench-normal seismic anisotropy regardless of whether the subducting angle due to the Vs1 through the a- and b-axes of the epidote is much faster than that through the c-axis. In contrast, if epidote with type-2 LPO exists in the subducting slab, this LPO may contribute to the trench-parallel seismic anisotropy where the subducting slab has a low dip angle. However, further research is needed to determine whether these LPO types of epidote could exist in natural blueschist. Consequently, different types of LPOs for glaucophane and epidote may contribute to the different seismic anisotropies of the forearc region in the subduction zone: type-1 LPOs of glaucophane and epidote can contribute to the trench-normal seismic anisotropy in the subducting slab, and type-2 LPOs of glaucophane and epidote may affect the trench-parallel seismic anisotropy of the forearc region, where the subducting slab has a high dip angle and a low dip angle, respectively.

5. Conclusions

We conducted deformation experiments of epidote blueschist in simple shear using the modified Griggs apparatus to understand deformation microstructures, the development of LPOs, and deformation mechanisms of glaucophane and epidote over a warm subducting slab. After the experiments, two different types of glaucophane and epidote LPOs were developed by different mechanisms depending on the magnitude of shear strain and grain size. TEM observations and EBSD mapping of glaucophane suggest that the LPO of glaucophane was developed by dislocation creep under a shear strain of $0.4 \leq \gamma \leq 4.5$. At a shear strain of $\gamma > 2$, cataclastic/granular flow microstructures were observed in large grains. At a high shear strain of $\gamma = 4.5$, thin ductile shear bands were observed in the small grains. Our experimental results suggest that the brittle–ductile transition of glaucophane likely occurs at epidote blueschist-facies metamorphic conditions. On the other hand, intracrystalline microstructures in epidote such as subgrain boundaries and twins revealed by TEM observations suggest that the LPO of epidote was developed by dislocation creep under a shear strain of $2 < \gamma < 4$. However, the development of the LPO of epidote under a high shear strain ($\gamma > 4$) is considered to be influenced by cataclastic flow with rigid body rotation because epidote is more rigid than other minerals in the matrix. Therefore, our data suggest that the magnitude of shear strain, grain size, and rheological contrast between component minerals are important factors influencing the deformation mechanisms and development of LPOs of glaucophane and epidote. However, there is a need for further studies on the fabric transition of epidote under various pressure and temperature conditions.

Author Contributions: Conceptualization, Y.P. and H.J.; methodology, Y.P. and S.J.; software, Y.P.; validation, Y.P. and H.J.; formal analysis, Y.P.; investigation, Y.P. and S.J.; resources, H.J.; data curation, Y.P.; writing—original draft preparation, Y.P.; writing—review and editing, Y.P., H.J., and S.J.; visualization, Y.P.; supervision, H.J.; project administration, H.J.; funding acquisition, H.J. All authors have read and agreed to the published version of the manuscript.

Funding: This research was funded by the Mid-career Research Program through the National Research Foundation of Korea (NRF: 2020R1A2C2003765) to H.J.

Acknowledgments: The authors are grateful to the anonymous reviewers whose suggestions and comments have notably improved the manuscript.

Conflicts of Interest: The authors declare no conflict of interest.

References

1. Evans, B.W. Phase relations of epidote-blueschists. *Lithos* **1990**, *25*, 3–23. [CrossRef]
2. Peacock, S.M. The importance of blueschist eclogite dehydration reactions in subducting oceanic crust. *Geol. Soc. Am. Bull.* **1993**, *105*, 684–694. [CrossRef]
3. Schmidt, M.W.; Poli, S. Experimentally based water budgets for dehydrating slabs and consequences for arc magma generation. *Earth Planet. Sci. Lett.* **1998**, *163*, 361–379. [CrossRef]
4. Agard, P.; Yamato, P.; Jolivet, L.; Burov, E. Exhumation of oceanic blueschists and eclogites in subduction zones: Timing and mechanisms. *Earth-Sci. Rev.* **2009**, *92*, 53–79. [CrossRef]
5. Ernst, W. Tectonic history of subduction zones inferred from retrograde blueschist PT paths. *Geology* **1988**, *16*, 1081–1084. [CrossRef]
6. Tsujimori, T.; Ernst, W.G. Lawsonite blueschists and lawsonite eclogites as proxies for palaeo-subduction zone processes: A review. *J. Metamorph. Geol.* **2014**, *32*, 437–454. [CrossRef]
7. Hasegawa, A.; Nakajima, J.; Kita, S.; Okada, T.; Matsuzawa, T.; Kirby, S.H. Anomalous deepening of a belt of intraslab earthquakes in the Pacific slab crust under Kanto, central Japan: Possible anomalous thermal shielding, dehydration reactions, and seismicity caused by shallower cold slab material. *Geophys. Res. Lett.* **2007**, *34*. [CrossRef]
8. Kawakatsu, H.; Watada, S. Seismic evidence for deep-water transportation in the mantle. *Science* **2007**, *316*, 1468–1471. [CrossRef]
9. Tsuji, Y.; Nakajima, J.; Hasegawa, A. Tomographic evidence for hydrated oceanic crust of the Pacific slab beneath northeastern Japan: Implications for water transportation in subduction zones. *Geophys. Res. Lett.* **2008**, *35*. [CrossRef]

10. Abers, G.A.; Nakajima, J.; van Keken, P.E.; Kita, S.; Hacker, B.R. Thermal–petrological controls on the location of earthquakes within subducting plates. *Earth Planet. Sci. Lett.* **2013**, *369–370*, 178–187. [CrossRef]

11. Hirose, F.; Nakajima, J.; Hasegawa, A. Three-dimensional seismic velocity structure and configuration of the Philippine Sea slab in southwestern Japan estimated by double-difference tomography. *J. Geophys. Res.* **2008**, *113*, 09311–09326. [CrossRef]

12. Hacker, B.R.; Abers, G.A.; Peacock, S.M. Subduction factory 1. Theoretical mineralogy, densities, seismic wave speeds, and H2O contents. *J. Geophys. Res. Solid Earth* **2003**, *108*, 2021–2026, 2029. [CrossRef]

13. Audet, P.; Kim, Y. Teleseismic constraints on the geological environment of deep episodic slow earthquakes in subduction zone forearcs: A review. *Tectonophysics* **2016**, *670*, 1–15. [CrossRef]

14. Audet, P.; Bostock, M.G.; Boyarko, D.C.; Brudzinski, M.R.; Allen, R.M. Slab morphology in the Cascadia fore arc and its relation to episodic tremor and slip. *J. Geophys. Res.* **2010**, *115*. [CrossRef]

15. Cao, Y.; Jung, H. Seismic properties of subducting oceanic crust: Constraints from natural lawsonite-bearing blueschist and eclogite in Sivrihisar Massif, Turkey. *Phys. Earth Planet. Inter.* **2016**, *250*, 12–30. [CrossRef]

16. Cao, Y.; Jung, H.; Song, S. Petro-fabrics and seismic properties of blueschist and eclogite in the North Qilian suture zone, NW China: Implications for the low-velocity upper layer in subducting slab, trench-parallel seismic anisotropy, and eclogite detectability in the subduction z. *J. Geophys. Res. Solid Earth* **2013**, *118*, 3037–3058. [CrossRef]

17. Kim, D.; Katayama, I.; Michibayashi, K.; Tsujimori, T. Deformation fabrics of natural blueschists and implications for seismic anisotropy in subducting oceanic crust. *Phys. Earth Planet. Inter.* **2013**, *222*, 8–21. [CrossRef]

18. Cao, Y.; Jung, H.; Song, S. Microstructures and petro-fabrics of lawsonite blueschist in the North Qilian suture zone, NW China: Implications for seismic anisotropy of subducting oceanic crust. *Tectonophysics* **2014**, *628*, 140–157. [CrossRef]

19. Ildefonse, B.; Lardeaux, J.-M.; Caron, J.-M. The behavior of shape preferred orientations in metamorphic rocks: Amphiboles and jadeites from the Monte Mucrone area (Sesia-Lanzo zone, Italian Western Alps). *J. Struct. Geol.* **1990**, *12*, 1005–1011. [CrossRef]

20. Zucali, M.; Chateigner, D.; Dugnani, M.; Lutterotti, L.; Ouladdiaf, B. Quantitative texture analysis of glaucophanite deformed under eclogite facies conditions (Sesia-Lanzo Zone, Western Alps): Comparison between X-ray and neutron diffraction analysis. *Geol. Soc. Lond. Spec. Publ.* **2002**, *200*, 239–253. [CrossRef]

21. Reynard, B.; Gillet, P.; Willaime, C. Deformation mechanisms in naturally deformed glaucophanes: A TEM and HREM study. *Eur. J. Mineral.* **1989**, *1*, 611–624. [CrossRef]

22. Wassmann, S.; Stöckhert, B. Rheology of the plate interface—Dissolution precipitation creep in high pressure metamorphic rocks. *Tectonophysics* **2013**, *608*, 1–29. [CrossRef]

23. Brunsmann, A.; Franz, G.; Erzinger, J.; Landwehr, D. Zoisite-and clinozoisite-segregations in metabasites (Tauern Window, Austria) as evidence for high-pressure fluid-rock interaction. *J. Metamorph. Geol.* **2000**, *18*, 1–22. [CrossRef]

24. Müller, W.F.; Franz, G. Unusual deformation microstructures in garnet, titanite and clinozoisite from an eclogite of the Lower Schist Cover, Tauern Window, Austria. *Eur. J. Mineral.* **2004**, *16*, 939–944. [CrossRef]

25. Müller, W.F.; Franz, G. TEM-microstructures in omphacite and other minerals from eclogite near to a thrust zone; the Eclogite Zone–Venediger nappe area, Tauern Window, Austria. *Neues Jahrb. Mineral.-Abh. J. Mineral. Geochem.* **2008**, *184*, 285–298.

26. Franz, G.; Liebscher, A. Physical and Chemical Properties of the Epidote Minerals—An Introduction–. *Rev. Mineral. Geochem.* **2004**, *56*, 1–81. [CrossRef]

27. Stünitz, H.; Tullis, J. Weakening and strain localization produced by syn-deformational reaction of plagioclase. *Int. J. Earth Sci.* **2001**, *90*, 136–148. [CrossRef]

28. Bezacier, L.; Reynard, B.; Bass, J.D.; Wang, J.; Mainprice, D. Elasticity of glaucophane, seismic velocities and anisotropy of the subducted oceanic crust. *Tectonophysics* **2010**, *494*, 201–210. [CrossRef]

29. Ha, Y.; Jung, H.; Raymond, L.A. Deformation fabrics of glaucophane schists and implications for seismic anisotropy: The importance of lattice preferred orientation of phengite. *Int. Geol. Rev.* **2018**, *61*, 720–737. [CrossRef]

30. Fujimoto, Y.; Kono, Y.; Hirajima, T.; Kanagawa, K.; Ishikawa, M.; Arima, M. P-wave velocity and anisotropy of lawsonite and epidote blueschists: Constraints on water transportation along subducting oceanic crust. *Phys. Earth Planet. Inter.* **2010**, *183*, 219–228. [CrossRef]

31. Kim, D.; Katayama, I.; Michibayashi, K.; Tsujimori, T. Rheological contrast between glaucophane and lawsonite in naturally deformed blueschist from Diablo Range, California. *Isl. Arc* **2013**, *22*, 63–73. [CrossRef]

32. Teyssier, C.; Whitney, D.L.; Toraman, E.; Seaton, N.C.A. Lawsonite vorticity and subduction kinematics. *Geology* **2010**, *38*, 1123–1126. [CrossRef]

33. Cossette, É.; Schneider, D.; Audet, P.; Grasemann, B.; Habler, G. Seismic properties and mineral crystallographic preferred orientations from EBSD data: Results from a crustal-scale detachment system, Aegean region. *Tectonophysics* **2015**, *651–652*, 66–78. [CrossRef]

34. Ray, N.J.; Putnis, A.; Gillet, P. Polytypic relationship between clinozoisite and zoisite. *Bull. Mineral.* **1986**, *109*, 667–685. [CrossRef]

35. Cao, Y.; Song, S.; Niu, Y.L.; Jung, H.; Jin, Z.M. Variation of mineral composition, fabric and oxygen fugacity from massive to foliated eclogites during exhumation of subducted ocean crust in the North Qilian suture zone, NW China. *J. Metamorph. Geol.* **2011**, *29*, 699–720. [CrossRef]

36. Malatesta, C.; Crispini, L.; Federico, L.; Capponi, G.; Scambelluri, M. The exhumation of high pressure ophiolites (Voltri Massif, Western Alps): Insights from structural and petrologic data on metagabbro bodies. *Tectonophysics* **2012**, *568–569*, 102–123. [CrossRef]

37. Vignaroli, G.; Rossetti, F.; Bouybaouene, M.; Massonne, H.J.; Theye, T.; Faccenna, C.; Funiciello, R. A counter-clockwise P–T path for the Voltri Massif eclogites (Ligurian Alps, Italy). *J. Metamorph. Geol.* **2005**, *23*, 533–555. [CrossRef]

38. Prior, D.J.; Boyle, A.P.; Brenker, F.; Cheadle, M.C.; Day, A.; Lopez, G.; Peruzzo, L.; Potts, G.J.; Reddy, S.; Spiess, R. The application of electron backscatter diffraction and orientation contrast imaging in the SEM to textural problems in rocks. *Am. Mineral.* **1999**, *84*, 1741–1759. [CrossRef]

39. Lloyd, G.E. Atomic number and crystallographic contrast images with the SEM: A review of backscattered electron techniques. *Mineral. Mag.* **1987**, *51*, 3–19. [CrossRef]

40. Bachmann, F.; Hielscher, R.; Schaeben, H. Grain detection from 2d and 3d EBSD data—Specification of the MTEX algorithm. *Ultramicroscopy* **2011**, *111*, 1720–1733. [CrossRef]

41. Nicolas, A.; Poirier, J.P. *Crystalline Plasticity and Solid State Flow in Metamorphic Rocks*; John Wiley & Sons: Hoboken, NJ, USA, 1976; p. 444.

42. van Duysen, J.C.; Doukhan, J.C. Room temperature microplasticity of a spodumene LiAlSi2O6. *Phys. Chem. Miner.* **1984**, *10*, 125–132. [CrossRef]

43. Ko, B.; Jung, H. Crystal preferred orientation of an amphibole experimentally deformed by simple shear. *Nat. Commun.* **2015**, *6*, 1–10. [CrossRef] [PubMed]

44. Jung, H. Crystal preferred orientations of olivine, orthopyroxene, serpentine, chlorite, and amphibole, and implications for seismic anisotropy in subduction zones: A review. *Geosci. J.* **2017**, *21*, 985–1011. [CrossRef]

45. Drury, M.R.; Urai, J.L. Deformation-related recrystallization processes. *Tectonophysics* **1990**, *172*, 235–253. [CrossRef]

46. Deer, W.A.; Howie, R.A.; Zussman, J. *Rock-Forming Minerals: Disilicates and Ring Silicates*, 2nd ed.; Geological Society of London: London, UK, 1986; p. 629.

47. Kohlstedt, D.; Evans, B.; Mackwell, S. Strength of the lithosphere: Constraints imposed by laboratory experiments. *J. Geophys. Res. Solid Earth* **1995**, *100*, 17587–17602. [CrossRef]

48. Karato, S.-I. *Deformation of Earth Materials: An Introduction to the Rheology of Solid Earth*; Cambridge University Press: Cambridge, UK, 2008; p. 463.

49. Kim, D.; Katayama, I.; Wallis, S.; Michibayashi, K.; Miyake, A.; Seto, Y.; Azuma, S. Deformation microstructures of glaucophane and lawsonite in experimentally deformed blueschists: Implications for intermediate-depth intraplate earthquakes. *J. Geophys. Res. Solid Earth* **2015**, *120*, 1229–1242. [CrossRef]

50. Balestro, G.; Festa, A.; Tartarotti, P. Tectonic significance of different block-in-matrix structures in exhumed convergent plate margins: Examples from oceanic and continental HP rocks in Inner Western Alps (Northwest Italy). *Int. Geol. Rev.* **2014**, *57*, 581–605. [CrossRef]

51. Ukar, E.; Cloos, M. Cataclastic deformation and metasomatism in the subduction zone of mafic blocks-in-mélange, San Simeon, California. *Lithos* **2019**, *346–347*, 105116. [CrossRef]

52. Ebert, A.; Herwegh, M.; Pfiffner, A. Cooling induced strain localization in carbonate mylonites within a large-scale shear zone (Glarus thrust, Switzerland). *J. Struct. Geol.* **2007**, *29*, 1164–1184. [CrossRef]

53. Long, M.D. Constraints on Subduction Geodynamics from Seismic Anisotropy. *Rev. Geophys.* **2013**, *51*, 76–112. [CrossRef]
54. Long, M.D.; Becker, T.W. Mantle dynamics and seismic anisotropy. *Earth Planet. Sci. Lett.* **2010**, *297*, 341–354. [CrossRef]
55. Jung, H.; Karato, S.-I. Water-induced fabric transitions in olivine. *Science* **2001**, *293*, 1460–1463. [CrossRef] [PubMed]
56. Karato, S.-I.; Jung, H.; Katayama, I.; Skemer, P. Geodynamic significance of seismic anisotropy of the upper mantle: New insights from laboratory studies. *Annu. Rev. Earth Planet. Sci.* **2008**, *36*, 59–95. [CrossRef]

Article

Evolution of Deformation Fabrics Related to Petrogenesis of Upper Mantle Xenoliths Beneath the Baekdusan Volcano

Munjae Park [1,2], Youngwoo Kil [3,*] and Haemyeong Jung [1]

[1] Tectonophysics Laboratory, School of Earth and Environmental Sciences, Seoul National University, Seoul 08826, Korea; munjaepark@korea.ac.kr (M.P.); hjung@snu.ac.kr (H.J.)
[2] Department of Earth and Environmental Sciences, Korea University, Seoul 02841, Korea
[3] Department of Energy and Resources Engineering, Chonnam National University, Gwangju 61186, Korea
* Correspondence: ykil@jnu.ac.kr; Tel.: +82-62-530-1731

Received: 31 July 2020; Accepted: 20 September 2020; Published: 21 September 2020

Abstract: Knowledge of the formation and evolution of cratonic subcontinental lithospheric mantle is critical to our understanding of the processes responsible for continental development. Here, we report the deformation microstructures and lattice preferred orientations (LPOs) of olivine and pyroxenes alongside petrological data from spinel peridotite xenoliths beneath the Baekdusan volcano. We have used these datasets to constrain the evolution of deformation fabrics related to petrogenesis from the Baekdusan peridotites. Based on petrographic features and deformation microstructures, we have identified two textural categories for these peridotites: coarse- and fine-granular harzburgites (CG and FG Hzb). We found that mineral composition, equilibrium temperature, olivine LPO, stress, and extraction depth vary considerably with the texture. We suggest that the A-type olivine LPO in the CG Hzb may be related to the preexisting Archean cratonic mantle fabric (i.e., old frozen LPO) formed under high-temperature, low-stress, and dry conditions. Conversely, we suggest that the D-type olivine LPOs in the FG Hzb samples likely originated from later localized deformation events under low-temperature, high-stress, and dry conditions after a high degree of partial melting. Moreover, we consider the Baekdusan peridotite xenoliths to have been derived from a compositionally and texturally heterogeneous vertical mantle section beneath the Baekdusan volcano.

Keywords: spinel peridotite xenoliths; deformation microstructures; lattice preferred orientation; petrogenesis; mantle heterogeneity; Baekdusan volcano

---

1. Introduction

Cratons are the oldest and thickest domains of the Earth's lithosphere [1,2] and are commonly associated with subcontinental lithospheric mantle (SCLM), which is characterized by a cold lithospheric mantle root extending to depths >250 km [3]. Knowledge of the formation and evolution of SCLM is critical to our understanding of the processes responsible for continental development [4]. Cratonic SCLM typically comprises of highly refractory residues produced by high degrees of partial melting [5–8]. This refractory mantle is unlikely to melt again subsequently unless heated by a hot mantle plume [2] because the cratonic lithosphere is relatively cold and rigid. However, some cratons, such as the North China Craton (NCC), did not remain stable following their formation; this kind of craton is characterized by extensive ductile deformation of deep crustal rocks and an abundance of crust-derived felsic magmatism [2].

Fragments of deep cratonic lithosphere, such as peridotite xenoliths, can be brought to the surface by basaltic magmas. These xenoliths can provide valuable insights into deformation processes in the upper mantle [9,10]; thus, enhancing our understanding of physical and chemical characteristics and deformation conditions in the upper mantle [11–13]. Previous experimental studies considering deformation fabrics have focused on the lattice-preferred orientations (LPOs) of olivine at high pressures and temperatures and have demonstrated that the LPOs of olivine can be controlled significantly by water and stress [14–16], temperature [17], and pressure [18–20]. The Baekdusan (also called as Changbaishan in China) volcano is a prominent active volcano located on the border between China and North Korea. To date, few petrological and geochemical studies have been undertaken to investigate peridotite xenoliths hosted in basaltic rocks associated with this volcano [21,22]. Accordingly, the physicochemical heterogeneity of cratonic mantle in this region, and its relationship to local deformation fabrics and petrological characteristics (i.e., modal and chemical compositions), remains poorly constrained. Moreover, the detailed microstructures of the deformed peridotites beneath Baekdusan have not yet been reported in the literature. To address these gaps in knowledge, we have studied spinel peridotite xenoliths beneath the Baekdusan volcano. Here, we report our results and discuss their significance for constraining the evolution of deformation fabrics produced in olivine and pyroxenes in relation to petrogenesis in the cratonic mantle.

2. Geological Background and Sampling

The Chinese portion of the Sino-Korean craton is known as the NCC and is one of the oldest cratonic blocks globally [2]; it contains Eoarchean rocks with ages up to 3.8 Ga [25,26]. Previously, the NCC has been divided into eastern and western blocks and the intervening Trans-North China Orogen based on its age, lithological assemblages, and tectonic evolution [27]. The Archean basement of the eastern NCC is composed primarily of the following: tonalite–trondhjemite–granodiorite (TTG) gneisses (2.6–2.5 Ga); mafic to ultramafic igneous rocks and syntectonic granites (~2.5 Ga); and a variety of supracrustal rocks [27,28]. The NCC remained stable from its amalgamation (~1.85 Ga) until the initiation of massive circum-craton Phanerozoic subduction and collisional orogenies [3]. However, the eastern part of the NCC experienced widespread tectonothermal reactivation during the Phanerozoic subduction, as revealed by the presence of voluminous Late Mesozoic mafic and felsic igneous rocks and extensive Tertiary alkali basaltic volcanism [29–31].

The Baekdusan volcanic field is located along the northeastern margin of the NCC (Figure 1). Moreover, it is located far away from the Japanese subduction zone (ca. 900 km) (Figure 1), indicating that volcanism here is of intraplate origin [32]. However, recent seismotomographic studies have suggested the presence of a stagnant slab in the mantle transition zone and the possible presence of a hydrous mantle upwelling beneath Baekdusan [33–35]; this may have produced subduction-induced hydrous mantle upwelling. The Baekdusan volcano consists of an early-stage basaltic plateau, a middle-stage trachytic cone, and a late-stage explosive comenditic ignimbrite [36,37]. Previous studies have shown that the main early shield-forming eruptions and cone construction occurred ~22.6 to 1.5 Ma and between ~1.0 Ma and 20 ka, respectively, followed by Quaternary volcanic events including the Millennium eruption [36,38]. We collected peridotite xenoliths hosted in massive basaltic rocks from the northern part (42°12′30.98″ N, 128°11′41.70″ E) of the Baekdusan volcano.

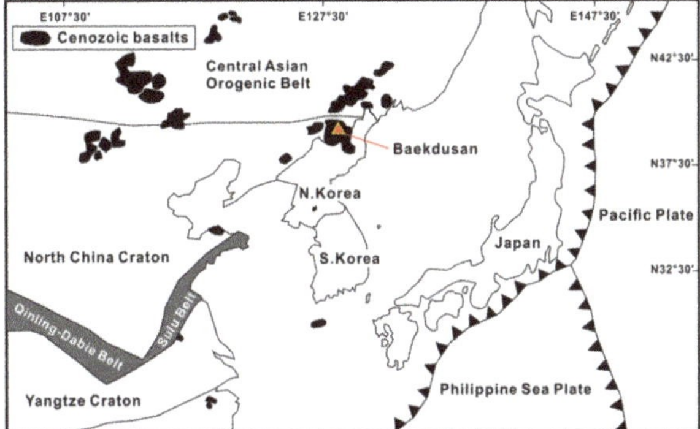

Figure 1. Simplified geologic map of the Baekdusan area showing the distribution of Cenozoic intraplate basalts in northeastern China and Korea (modified from [21,23,24]) and the sampled location at Baekdusan.

3. Methods

3.1. Mineral Chemistry

The major element compositions of target minerals (olivine, orthopyroxene, clinopyroxene, and spinel) were determined using a field emission electron probe microanalyzer (FE-EPMA; JEOL JXA-8530F) at the Korea Polar Research Institute. The accelerating voltage, beam current, beam diameter, and counting time were 15 kV, 10 nA, 3 μm, and 20 s, respectively. The detection limits (3σ) in wt% are as follows: 0.01 for CaO and K_2O, 0.02 for MgO and Na_2O, 0.03 for SiO_2, TiO_2, Al_2O_3, and Cr_2O_3, and 0.04 for FeO and MnO. Three to four grains were measured for each mineral and we represent here averages. There was no clear distinction in element concentrations between the cores and rims of the constituent minerals. Based on the obtained mineral compositions, we calculated equilibrium temperatures using the two-pyroxene thermometer by Brey and Köhler [39].

3.2. Measurement of LPO

For petrographic and microstructural observations, the foliation and lineation of rock samples were determined by observing the compositional layering and alignment of olivine, pyroxenes, and spinel grains. Rock sections were cut parallel to the XZ plane of the XYZ structural reference frame (where XY represents the macroscopic foliation and the X direction represents the direction of the stretching lineation), and thin sections were produced for electron backscatter diffraction (EBSD) analysis according to previously published methods [40,41]. The LPOs of the minerals were measured using an automated EBSD system and a symmetry detector (Oxford Instruments, Abingdon, UK) attached to a field emission scanning electron microscope (FE-SEM; JEOL JSM-7100F) at the School of Earth and Environmental Sciences at Seoul National University. The working parameters for the EBSD analysis included accelerating voltage, working distance, and stage tilt of 15.0 kV, 25.0 mm, and 70° from the horizontal, respectively. Kikuchi diffraction patterns were obtained automatically using the Aztec software (Oxford Instruments HKL) with a sampling step size of 25 μm; this is significantly smaller than the average grain size of the constituent minerals. The HKL CHANNEL 5 software was used to process the EBSD data as described in previous studies [40,41]. To avoid oversampling of large grains and to permit comparison of fabrics between samples, pole figures were constructed from one point per grain [40,41]. Contoured pole figures for each mineral were plotted using lower-hemisphere equal-area stereographic projection. Fabric strength of the LPOs was quantified using the M-index [42]. The LPOs of clinopyroxenes obtained here were compared with the conventional LPO of clinopyroxene, where the [001] axes are commonly aligned subparallel to the lineation and the (010) planes are aligned subparallel to the foliation in naturally deformed rocks [43,44].

4. Results

4.1. Petrographic Features and Deformation Microstructures

We found the spinel harzburgite xenoliths from Baekdusan to consist of olivine (63–75 vol.%), orthopyroxene (21–34 vol.%), clinopyroxene (up to 5 vol.%), and minor spinel (up to 0.6 vol.%), based on the area fraction of each mineral in the EBSD maps (Table 1). Generally, we found all peridotite xenoliths to exhibit weak foliation. We identified two main textural categories of peridotites based on petrographic features (including grain size): coarse-granular and fine-granular (Table 1).

Table 1. Summarized results of the studied peridotite xenoliths from Baekdusan.

Sample	Rock	Texture	Modal Composition [1] (%)				T [2]	GS [3]	Stress [4]	Olivine LPO	
			Ol	Opx	Cpx	Sp	(°C)	(μm)	(MPa)	LPO	M [5]
BD-19	Hzb	CG	71.0	23.4	5.0	0.6	992	864 ± 86	9	A-type	0.24
BD-17	Hzb	FG	62.0	33.7	4.0	0.3	855	384 ± 52	16	D-type	0.23
BD-21	Hzb	FG	74.8	21.0	4.0	0.3	845	452 ± 45	14	D-type	0.22

Hzb (harzburgite), CG and FG (coarse- and fine-granular), Ol (olivine), Opx (orthopyroxene), Cpx (clinopyroxene), and Sp (spinel). [1] Based on the area fraction of each mineral in the EBSD maps. [2] Equilibrium temperature was calculated by the two-pyroxene thermometer [39] at 15 kbar, st. dev. ±16 °C. [3] Grain-size (GS) of olivine was estimated by EBSD mapping analysis. [4] Stress was estimated using recrystallized grain size piezometer [45], st. dev. ±1 MPa. [5] M-index (M): the misorientation index which shows the fabric strength of olivine [42].

The coarse-granular peridotite (BD-19) is characterized by large grains (2–8 mm) with abundant 120° triple junctions at grain boundaries (Figure 2a,b). It represents roughly a protogranular texture. The mean grain size of olivine grains in this peridotite is approximately 860 μm (Table 1) and olivine grains in the sample typically exhibit subgrains and undulose extinction (Figure 3b).

Figure 2. Optical photomicrographs of large-scale thin sections in plane-polarized light (**a,c**) and cross-polarized light (**b,d**) illustrating the typical microstructures from the Baekdusan peridotite xenoliths. (**a,b**) Coarse-granular peridotite (BD-19). (**c,d**) Fine-granular peridotite (BD-17). Ol (olivine), Opx (orthopyroxene), Cpx (clinopyroxene), and Sp (spinel).

The fine-granular peridotites (BD-17 and BD-21) are characterized by relatively small grains (1–3 mm) with equigranular textures (Figure 2c,d). We commonly observed 120° triple junctions at grain boundaries in these samples, indicating a well-equilibrated microstructure. The mean grain sizes of olivine crystals in BD-17 and BD-21 were approximately 380 μm and 450 μm (Table 1), respectively. We found subgrains and undulose extinctions to be more common in these olivine grains than in the coarse-granular sample (Figure 3d,f).

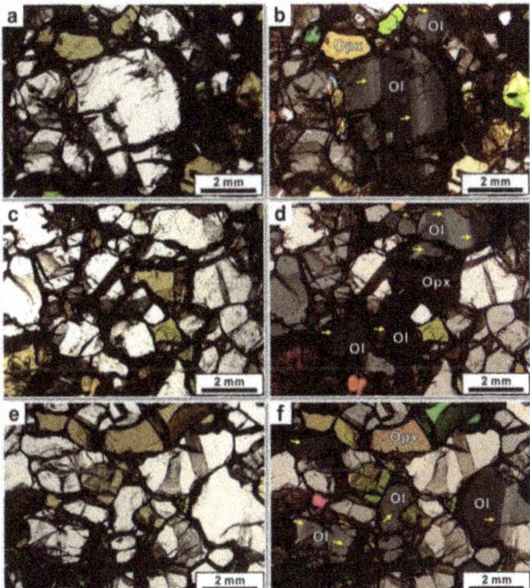

Figure 3. Optical photomicrographs of small-scale thin sections in plane-polarized light (**a,c,e**) and cross-polarized light (**b,d,f**) illustrating the typical microstructures from the Baekdusan peridotite xenoliths. (**a,b**) Coarse-granular peridotite (BD-19). (**c,d**) Fine-granular peridotite (BD-17). (**e,f**) Fine-granular peridotite (BD-21). Ol (olivine), Opx (orthopyroxene). Yellow arrows denote subgrain boundaries and/or undulose extinction in olivine.

Both the coarse- and fine-granular peridotites typically display straight and gently curved grain boundary morphologies (Figures 2 and 3). Orthopyroxenes typically exhibit clinopyroxene exsolution lamellae, whereas clinopyroxenes have orthopyroxene lamellae or exsolved spinel blobs. Clinopyroxene is present only as small lobate forms in the matrix (Figure 2a,c), often with spongy rims. Spinels are present as small and anhedral grains forming intergrowths with orthopyroxene or as interstitial grains with amoeboid shapes (Figure 2a,c).

4.2. Mineral Compositions and Equilibrium Temperatures

The mineral compositions of spinel peridotite xenoliths from Baekdusan are presented in Table 2. We found the olivine Fo contents to be similar in the coarse- and fine-granular peridotites (90.7–91.1) and found no clear distinction in element concentrations between the cores and rims of the olivine crystals. The orthopyroxenes are enstatite-rich, with Mg# [=100Mg/(Mg + Fe)] of 90.6 and 91.1–91.4 in the coarse- and fine-granular peridotites, respectively. The Al_2O_3 contents of orthopyroxenes in the coarse-granular peridotite (3.92 wt%) are slightly higher than those in the fine-granular peridotites (2.63–2.65 wt%). Similarly, we found the Cr_2O_3 contents of orthopyroxenes to be 0.60 wt% and 0.47 wt% in the coarse- and fine-granular peridotites, respectively. The Cr# [=100Cr/(Cr + Al)] of orthopyroxenes is 9.2 and 10.6–10.8 in the coarse-and fine-granular peridotites, respectively. For clinopyroxenes, we found Mg# to be 91.3 and 93.5–93.7 in the coarse- and fine-granular peridotites, respectively. The Al_2O_3 contents of clinopyroxenes in the coarse-granular peridotite (4.09 wt%) are also slightly higher than those in the fine-granular peridotites (2.71–2.72 wt%). The Mg# and Cr# of pyroxenes in the fine-granular peridotites are higher than those in the coarse-granular peridotite (Table 2 and Figure 4). For spinels, we found a clear distinction in Al_2O_3 and Cr_2O_3 contents between coarse- and fine-granular peridotites. In particular, we found spinels in fine-granular peridotites to be characterized by much higher Cr# (37.6–38.3) and lower Mg# (66.4–66.5) than those in the coarse-granular peridotite, for which we found Cr# and Mg# to be 24.1 and 73.3, respectively. Spinel Mg# exhibits a negative correlation with spinel Cr# (Figure 4a), whereas spinel Cr# exhibits a positive correlation with olivine Fo content (Figure 4b). Based on the spinel Cr# versus Mg# relationship and the relationship between olivine Fo# and spinel Cr#, the Baekdusan peridotites plot between abyssal and suprasubduction zone (SSZ) peridotites (Figure 4). This is broadly consistent with a previous geochemical study [21], in which the olivine and spinel compositions of Baekdusan harzburgites extended toward more fertile (SSZ) compositions.

Table 2. Compositions of minerals of studied peridotite xenoliths from Baekdusan.

Sample	BD-19				BD-17				BD-21			
Rock	Harzburgite				Harzburgite				Harzburgite			
Texture	Coarse-Granular				Fine-Granular				Fine-Granular			
Mineral	Ol	Opx	Cpx	Sp	Ol	Opx	Cpx	Sp	Ol	Opx	Cpx	Sp
SiO_2 (wt%)	41.26	55.09	52.60	-	41.20	56.65	53.69	-	40.75	56.11	53.53	-
TiO_2	-	0.04	0.09	0.10	-	0.04	0.11	0.06	-	-	0.08	0.09
Al_2O_3	-	3.92	4.09	46.56	-	2.63	2.71	35.53	-	2.65	2.72	35.90
Cr_2O_3	-	0.60	0.81	21.98	-	0.47	0.82	32.91	-	0.47	0.87	32.22
FeO*	8.73	6.08	2.88	12.13	9.02	5.87	2.13	14.48	8.99	5.69	2.03	14.69
MnO	0.10	0.21	0.12	0.21	0.09	0.16	0.06	0.22	0.15	0.11	0.03	0.18
MgO	50.09	32.86	16.99	18.67	49.66	33.92	17.21	16.12	49.11	33.77	16.98	16.26
CaO	0.04	0.91	21.67	-	0.04	0.56	23.07	-	0.05	0.55	23.18	-
Na_2O	-	0.05	0.55	-	-	-	0.47	-	-	-	0.39	-
K_2O	-	-	-	-	-	-	-	-	-	-	-	-
Total	100.26	99.75	99.82	99.69	100.08	100.34	100.26	99.33	99.10	99.36	99.80	99.34
Si	1.002	1.909	1.911	0.000	1.004	1.944	1.941	0.000	1.003	1.943	1.944	0.000
Ti	0.000	0.001	0.003	0.002	0.000	0.001	0.003	0.001	0.001	0.000	0.002	0.002
Al	0.000	0.160	0.175	1.496	0.000	0.106	0.115	1.211	0.000	0.108	0.116	1.222
Cr	0.000	0.016	0.023	0.474	0.000	0.013	0.023	0.753	0.000	0.013	0.025	0.736
Fe	0.177	0.176	0.088	0.277	0.184	0.169	0.064	0.350	0.185	0.165	0.062	0.355
Mn	0.002	0.006	0.004	0.005	0.002	0.005	0.002	0.005	0.003	0.003	0.001	0.004
Mg	1.814	1.698	0.920	0.759	1.804	1.736	0.927	0.695	1.802	1.743	0.919	0.700
Ca	0.001	0.034	0.843	0.000	0.001	0.021	0.894	0.000	0.001	0.020	0.902	0.000
Na	0.000	0.003	0.039	0.001	0.000	0.002	0.033	0.001	0.000	0.000	0.027	0.000
K	0.000	0.000	0.000	0.000	0.000	0.000	0.000	0.000	0.000	0.000	0.000	0.000
Sum	2.997	4.003	4.007	3.014	2.996	3.996	4.003	3.017	2.996	3.996	3.998	3.019
Mg#	91.1	90.6	91.3	73.3	90.7	91.1	93.5	66.5	90.7	91.4	93.7	66.4
Cr#		9.2	11.7	24.1		10.8	16.8	38.3		10.6	17.6	37.6

Total Fe as FeO*; Mg# = 100Mg/(Mg + Fe); Cr# = 100Cr/(Cr + Al); - (below the detection limits); Ol (olivine); Opx (orthopyroxene); Cpx (clinopyroxene); and Sp (spinel).

We estimated the equilibrium temperatures of the Baekdusan peridotites using the two-pyroxene geothermometer by Brey and Köhler [39]; this is a well-known method for calculating the solubility of the enstatite component in diopside, coexisting with orthopyroxene. We assumed a pressure of 15 kbar for our temperature calculations because the Cr-spinel in the peridotite can be stable at pressures of 8–25 kbar [46,47]. A pressure change of ±10 kbar would result in estimated temperature variations of only ±20 °C because the two-pyroxene geothermometer is relatively insensitive to pressure. Use of this method yielded equilibrium temperatures of 992 °C and 845–855 °C for the coarse- and fine-granular peridotites, respectively (Table 1). This range of equilibrium temperatures is consistent with the results of a previous study that also adopted the two-pyroxene geothermometer of Brey and Köhler (752–997 °C) [21].

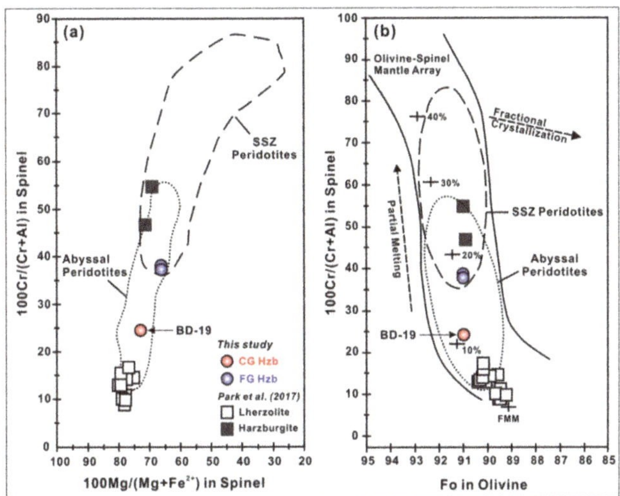

Figure 4. Plots of (**a**) spinel Cr# versus Mg# relationship and (**b**) olivine Fo# versus spinel Cr# relationship. The olivine–spinel mantle array and melting trend (annotated by melting %) given in (**b**) are from Arai [48]. Data sources: abyssal peridotites [49] and suprasubduction zone (SSZ) peridotites [50–52]. FMM = fertile MORB mantle. Previous data from the Baekdusan peridotite xenoliths [21] are also shown as squares for comparison. Coarse-granular (CG; red circles) and fine-granular (FG; blue circles) harzburgite (Hzb) from the Baekdusan peridotites.

4.3. LPOs of Minerals

The LPO of olivine from the coarse-granular peridotite (BD-19) indicates maximum concentration of the [100] axis subparallel to the lineation, with the [010] axis strongly aligned subnormal to the foliation (Figure 5a). This is a characteristic of the A-type olivine LPO [14], indicating a dominant slip system of (010)[100]. In contrast, the LPOs of olivine from the fine-granular peridotites (BD-17 and BD-21) are characterized by a strong alignment of the [100] axis subparallel to the lineation and a weak girdle pattern for both the [010] and [001] axes subnormal to the lineation (Figure 5b,c). This is a characteristic of the D-type olivine LPO and indicates the activation of multiple slip systems of {0kl}[100]. The fabric strength of olivine (i.e., the M-index) in the Baekdusan peridotites ranges between 0.22 and 0.24 (Table 1 and Figure 5).

Generally, the fabrics of orthopyroxenes (Figure 6a–c) are much weaker than those of olivine in our samples (Figure 5). The LPOs of orthopyroxenes from both the coarse- and fine-granular peridotites exhibit alignment of the [100] axis subnormal to the foliation, whereas the [001] axis is aligned subparallel to the lineation (Figure 6a–c). This is a characteristic of the type-AC LPO of orthopyroxene [53]. The LPOs of clinopyroxenes are illustrated in Figure 6d–f, in which the fabrics of clinopyroxenes are relatively weak. The LPOs of clinopyroxene typically indicate alignment of the [010] axis subnormal to the foliation (Figure 6d,e) and alignment of the [001] axis subparallel to the lineation (Figure 6d–f). This is considered a conventional characteristic of the LPO of clinopyroxene in naturally deformed rocks [43,44]. The fabric strengths (i.e., M-index) of orthopyroxenes (0.04–0.07) and clinopyroxenes (0.04–0.05) are weaker than those of olivine in our samples (Figures 5 and 6).

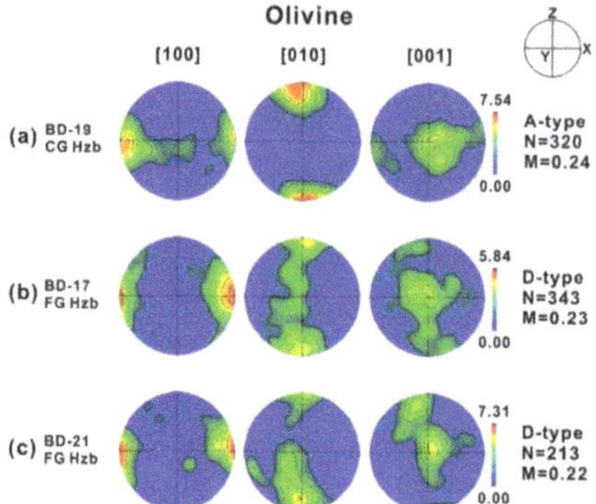

Figure 5. Pole figures showing the lattice preferred orientations (LPOs) of olivine from coarse-granular (CG) and fine-granular (FG) harzburgite (Hzb) from the Baekdusan peridotites. (**a**) The LPO of olivine in coarse-granular peridotite (BD-19). (**b**,**c**) The LPOs of olivine in fine-granular peridotites (BD-17, BD-21). The east–west direction corresponds to the stretching lineation (X), and the north–south direction (Z) is normal to the foliation. Color-coding represents the density of data points. Equal-area lower-hemisphere projection was used with a half scattering width of 20°. N indicates the number of grains measured. Olivine fabric (LPO) strength is denoted by M (M-index) [42].

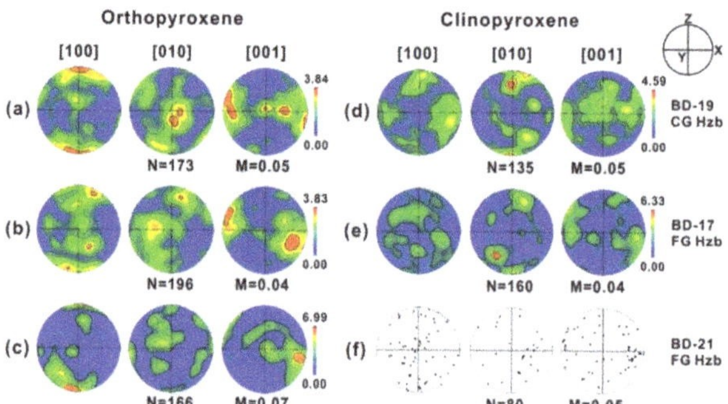

Figure 6. Pole figures showing the lattice preferred orientations (LPOs) of (**a**–**c**) orthopyroxene and (**d**–**f**) clinopyroxene from coarse-granular (CG) and fine-granular (FG) harzburgite (Hzb) from the Baekdusan peridotites. The east–west direction corresponds to the stretching lineation (X) and the north–south direction (Z) is normal to the foliation. Color-coding represents the density of data points. Equal-area lower-hemisphere projection was used with a half scattering width of 20°. N indicates the number of grains measured. Pyroxene fabric (LPO) strength is denoted by M (M-index) [42].

5. Discussion

5.1. The Genesis of Olivine Fabrics in the Upper Mantle

Olivine fabrics (LPOs) developed under lithospheric conditions are well preserved in both coarse- and fine-granular peridotites from the Baekdusan volcano (Figure 5). Our peridotite samples exhibit intracrystalline deformation microstructures such as subgrain boundaries and undulose extinctions in olivine (Figure 3). Together, these deformation microstructures and LPO patterns imply that the Baekdusan peridotites were deformed by a dislocation creep regime.

We found the olivine LPO patterns to vary depending on the sample texture. For the coarse-granular peridotite, the olivine LPO exhibited alignment of the [100] axes subparallel to the lineation and a strong alignment of the [010] axis subnormal to the foliation (Figure 5a); this is a characteristic of the A-type olivine LPO. Previously, the A-type olivine LPOs have been observed under conditions of high temperature, low water content, and low stress, both in experimental studies investigating the plastic deformation of olivine [14,54–57] and in naturally deformed peridotites [12,58–61]. Application of the two-pyroxene geothermometer to the coarse-granular peridotite from the Baekdusan volcano produced a high estimate of equilibrium temperature (approximately 990 °C) (Table 1), and the stress of this sample was estimated to be low (9 MPa, Table 1). Therefore, we consider the A-type LPO of olivine in the coarse-granular peridotite to have been preserved from the original olivine fabric formed under high-temperature (approximately 1000 °C), low-stress, and dry conditions.

The olivine LPOs for our fine-granular peridotites exhibit a strong alignment of the [100] axes subparallel to the lineation and a weak girdle pattern for both the [010] and [001] axes subnormal to the lineation (Figure 5b,c). This is a characteristic of the D-type olivine LPO. Previously, the D-type olivine LPOs have been observed primarily at low temperature and high stress regions in lithospheric shear zones [61–66]. Previous experimental studies, natural observations, and numerical modeling suggest that the D-type olivine LPOs may be formed by the following: (i) activation of multiple {0kl}[100] slip systems under higher stress conditions than those forming the A-type olivine LPOs [16,55,67]; (ii) activation of only the (010)[100] slip system under transtensional deformation regimes [68]; (iii) activation of dominant (010)[100] and (001)[100] slip systems under high-temperature and low-stress conditions, with strain compatibility constraints relaxed by grain-to-grain interactions (numerical simulations) [69]; or (iv) dislocation-accommodated grain boundary sliding (DisGBS) [56,63] under moderate transient strain conditions [70]. The equilibrium temperature we have obtained for the fine-granular peridotites from the Baekdusan volcano is lower (845–855 °C) than that obtained for the coarse-granular peridotite (Table 1). Moreover, the stress estimates obtained using a recrystallized grain size piezometer [45] indicate that stress in the fine-granular peridotites is relatively high (14–16 MPa) compared to that in the coarse-granular peridotite (9 MPa) (Table 1). If the D-type LPO of olivine was formed by the transtensional deformation regime as in case (ii), the LPOs of orthopyroxene and clinopyroxene would exhibit similar characteristics, i.e., the girdle distribution observed for the [010] and [001] axes, because the LPOs of pyroxenes and olivine typically form against the same structural background [71,72]. Furthermore, the mechanisms required for the formation of D-type olivine fabrics in cases (iii) to (iv) are typically related to high temperatures (1200 °C), and thus cases from (ii) to (iv) must be rejected. Consequently, we consider the D-type LPO of olivine in the fine-granular peridotites from the Baekdusan volcano to have been formed under lower temperature (approximately 850 °C), slightly higher stress, and dry conditions at shallower depth, comparing to the A-type LPO of olivine.

5.2. Petrogenesis and Evolution of Deformation Fabrics of the Baekdusan Peridotites

The lack of a reliable geobarometer for the spinel peridotite stability field makes it very difficult to estimate extraction depths (or equilibrium pressure) for the spinel peridotite xenoliths. However, the extraction depths (or equilibrium pressure) of xenoliths can often be estimated by plotting their equilibrium temperatures on a geotherm derived from surface heat flow data, because heat flow is a key parameter used to constrain lithospheric thermal structures [73]. Thus, we estimated the extraction depths (or equilibrium pressures) of the Baekdusan peridotite xenoliths (Figure 7) by plotting their equilibrium temperatures ($T_{\text{Two-Pyx}}$; two-pyroxene geothermometer; Table 1) on the geotherm of the Bohai Bay Basin in the NCC [74]; this is one of the few available datasets in the vicinity of the Baekdusan volcano. For the coarse-granular ($T_{\text{Two-Pyx}}$: 992 °C) and fine-granular ($T_{\text{Two-Pyx}}$: 845–855 °C) peridotites, we obtained extraction depth ranges of approximately 50–55 km and 40–45 km, respectively (Figure 7). These results are consistent with the depth ranges obtained for the Moho (30–35 km) and lithosphere–asthenosphere boundary (50–70 km) in a previous geophysical study [75] and with the spinel–garnet phase transformation boundary [76] as shown in Figure 7. Such consistent variation in equilibrium temperature with the sample texture may indicate that the Baekdusan peridotite xenoliths were derived from a compositionally and texturally heterogeneous vertical mantle section beneath the Baekdusan volcano. However, additional petrochemical and microstructural data relating to the Baekdusan peridotite xenoliths will be required to validate this hypothesis.

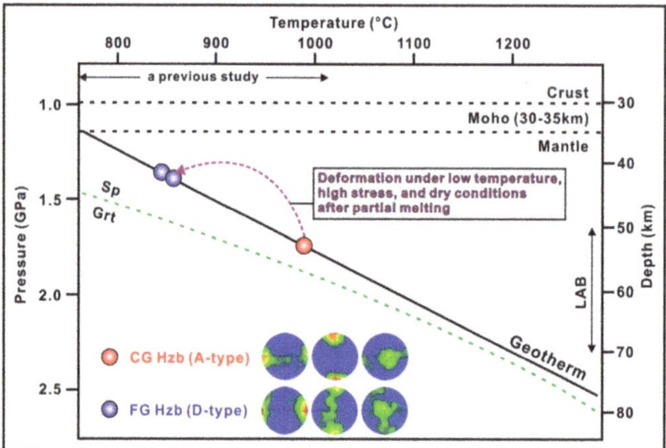

Figure 7. A summary figure showing the relationship between olivine fabrics and deformation conditions in the Baekdusan peridotites. The equilibrium temperatures are projected onto the geotherm (solid line) of the Bohai Bay Basin in the NCC [74]. The spinel (Sp)–garnet (Grt) phase transformation (green dashed line) is from O'Reilly and Griffin [76]. The crust–mantle boundary (Moho) depth (30–35 km; black dashed lines) and lithosphere–asthenosphere boundary (LAB) depth (50–70 km) beneath the Baekdusan volcano are from Kim et al. [75]. The equilibrium temperature range indicated by previous data from the Baekdusan peridotite xenoliths is also shown [21]. Coarse-granular (CG) and fine-granular (FG) harzburgite (Hzb). Representative A-type and D-type LPOs of olivine from BD-19 and BD-17, respectively.

In general, the relationship between the olivine Fo content (Mg#) and spinel Cr# of peridotites is an excellent indicator of the extent of melt extraction from mantle rocks [48,49,77]. Both the olivine Fo content and spinel Cr# increase with increasing degree of partial melting in the peridotite samples studied here (Figure 4b). The major element mineral chemistry (Table 2) and olivine–spinel compositional relationships (Figure 4b) obtained here indicate that the Baekdusan peridotites (i.e., harzburgites) are residues that have experienced partial melting (ca. < 10% and < 20% for the coarse-granular and fine-granular harzburgites, respectively). This result is broadly consistent with the degree of partial melting (ca. ≤ 25%) estimated in a previous geochemical study [21]. Based on petrographic and microstructural observations, we found intracrystalline deformation features such as subgrains, undulose extinctions, and strong olivine LPOs in the fine-granular harzburgites, although these have experienced a high degree of partial melting (ca. < 20%). This explains the relatively weak fabric strength (M-index) observed for the LPOs of clinopyroxene and orthopyroxene compared to olivine (Figures 5 and 6). If a high degree of partial melting had occurred in the fine-granular harzburgites after the main deformation events, the recorded intracrystalline deformation microstructures observed in the present study would have been destroyed by annealing (recovery) processes (e.g., grain growth) under relatively high-temperature conditions during the partial melting process. In contrast, the A-type olivine LPO has been preserved in the coarse-granular peridotite owing to the relatively small degree of partial melting (ca. < 10%). In summary, we infer that a high degree of partial melting occurred in the fine-granular harzburgites before the main deformation event; this resulted in intracrystalline deformation microstructures, forming the D-type olivine LPO under low-temperature, high-stress, and dry conditions (Figure 7). The A-type olivine LPO observed in the coarse-granular peridotite may be related to the preexisting Archean cratonic mantle fabric (i.e., old frozen LPO) that formed under high-temperature and low-stress conditions in a deeper section (50–55 km), whereas the D-type olivine LPO in the fine-granular peridotites likely originated from later localized deformation events under low-temperature and high-stress conditions in a shallower section (40–45 km) after a high degree of partial melting. However, a more precise interpretation of the tectonic history of our study area will require more data regarding the microstructures and geochemistry of these lithologies.

6. Conclusions

We report, for the first time, the deformation microstructures and lattice preferred orientations (LPOs) of minerals alongside petrological data from spinel peridotite xenoliths beneath the Baekdusan volcano. Based on petrographic features and deformation microstructures, we have identified two categories of peridotites: coarse- and fine-granular harzburgites. The olivine LPO in the coarse-granular peridotite exhibits a maximum concentration of [100] axes subparallel to the lineation, whereas the [010] axes are aligned strongly subnormal to the foliation (A-type olivine LPO). In contrast, the olivine LPOs in the fine-granular peridotites exhibit a strong alignment of the [100] axes subparallel to the lineation and a weak girdle pattern for both the [010] and [001] axes subnormal to the lineation (D-type olivine LPO). We found that mineral composition, equilibrium temperature, stress, and extraction depth vary considerably with the sample texture. Based on these differences, we infer that the A-type olivine LPO of the coarse-granular peridotite reflects the preexisting Archean cratonic mantle fabric (i.e., old frozen LPO) formed under high-temperature (approximately 990 °C), low-stress, and dry conditions in a deep section (50–55 km). Conversely, we infer that the D-type olivine LPO of the fine-granular peridotites likely originated from a later localized deformation event under low-temperature (approximately 850 °C), high-stress, and dry conditions in a shallow section (40–45 km) after a high degree of partial melting (ca. < 20%). We propose that the Baekdusan peridotite xenoliths were derived from a compositionally and texturally heterogeneous vertical mantle section beneath the Baekdusan volcano.

Author Contributions: Conceptualization, M.P. and Y.K.; methodology, M.P. and H.J.; software, M.P.; validation, M.P., H.J., and Y.K.; formal analysis, M.P.; investigation, M.P.; resources, Y.K. and H.J.; data curation, M.P.; writing—original draft preparation, M.P.; writing—review and editing, Y.K. and H.J.; visualization, M.P.; supervision, H.J.; project administration, H.J.; and funding acquisition, M.P., Y.K., and H.J. All authors have read and agreed to the published version of the manuscript.

Funding: This study was supported by the National Research Foundation of Korea (NRF: 2018R1D1A3B07048228 to Y.K., NRF: 2017R1A2B2004688, 2020R1A2C2003765 to H.J. and NRF: 2019R1C1C1004402 to M.P.).

Acknowledgments: The authors would like to thank Hwayoung Kim (Korea Polar Research Institute) for the EPMA analyses. The authors are grateful to the Editor and two anonymous reviewers, whose suggestions and comments have notably improved the manuscript.

Conflicts of Interest: The authors declare no conflict of interest.

References

1. Bascou, J.; Doucet, L.S.; Saumet, S.; Ionov, D.A.; Ashchepkov, I.V.; Golovin, A.V. Seismic velocities, anisotropy and deformation in Siberian cratonic mantle: EBSD data on xenoliths from the Udachnaya kimberlite. *Earth Planet. Sci. Lett.* **2011**, *304*, 71–84. [CrossRef]
2. Wu, F.-Y.; Yang, J.-H.; Xu, Y.-G.; Wilde, S.A.; Walker, R.J. Destruction of the North China Craton in the Mesozoic. *Annu. Rev. Earth Planet. Sci.* **2019**, *47*, 173–195. [CrossRef]
3. Liu, J.; Cai, R.; Pearson, D.G.; Scott, J.M. Thinning and destruction of the lithospheric mantle root beneath the North China Craton: A review. *Earth Sci. Rev.* **2019**, *196*, 102873. [CrossRef]
4. Tang, Y.-J.; Zhang, H.-F.; Ying, J.-F.; Su, B.-X. Widespread refertilization of cratonic and circum-cratonic lithospheric mantle. *Earth Sci. Rev.* **2013**, *118*, 45–68. [CrossRef]
5. Frey, F.A.; Green, D.H. The mineralogy, geochemistry and origin of Iherzolite inclusions in Victorian basanites. *Geochim. Cosmochim. Acta* **1974**, *38*, 1023–1059. [CrossRef]
6. Jordan, T.H. The continental tectosphere. *Rev. Geophys.* **1975**, *13*, 1–12. [CrossRef]
7. King, S.D. Archean cratons and mantle dynamics. *Earth Planet. Sci. Lett.* **2005**, *234*, 1–14. [CrossRef]
8. Griffin, W.L.; O'Reilly, S.Y.; Afonso, J.C.; Begg, G.C. The Composition and Evolution of Lithospheric Mantle: A Re-evaluation and its Tectonic Implications. *J. Petrol.* **2009**, *50*, 1185–1204. [CrossRef]
9. Mercier, J.-C.C.; Nicolas, A. Textures and fabrics of upper-mantle peridotites as illustrated by xenoliths from basalts. *J. Petrol.* **1975**, *16*, 454–487. [CrossRef]

10. Jung, H. Crystal preferred orientations of olivine, orthopyroxene, serpentine, chlorite, and amphibole, and implications for seismic anisotropy in subduction zones: A review. *Geosci. J.* **2017**, *21*, 985–1011. [CrossRef]

11. Park, M.; Berkesi, M.; Jung, H.; Kil, Y. Fluid infiltration in the lithospheric mantle beneath the Rio Grande Rift, USA: A fluid-inclusion study. *Eur. J. Mineral.* **2017**. [CrossRef]

12. Park, M.; Jung, H.; Kil, Y. Petrofabrics of olivine in a rift axis and rift shoulder and their implications for seismic anisotropy beneath the Rio Grande rift. *Isl. Arc* **2014**, *23*, 299–311. [CrossRef]

13. Park, Y.; Jung, H. Deformation microstructures of olivine and pyroxene in mantle xenoliths in Shanwang, eastern China, near the convergent plate margin, and implications for seismic anisotropy. *Int. Geol. Rev.* **2015**, *57*, 629–649. [CrossRef]

14. Jung, H.; Karato, S. Water-induced fabric transitions in olivine. *Science* **2001**, *293*, 1460–1463. [CrossRef] [PubMed]

15. Katayama, I.; Jung, H.; Karato, S.I. New type of olivine fabric from deformation experiments at modest water content and low stress. *Geology* **2004**, *32*, 1045–1048. [CrossRef]

16. Jung, H.; Katayama, I.; Jiang, Z.; Hiraga, I.; Karato, S. Effect of water and stress on the lattice-preferred orientation of olivine. *Tectonophysics* **2006**, *421*, 1–22. [CrossRef]

17. Katayama, I.; Karato, S. Effect of temperature on the B- to C-type olivine fabric transition and implication for flow pattern in subduction zones. *Phys. Earth Planet. Inter.* **2006**, *157*, 33–45. [CrossRef]

18. Mainprice, D.; Tommasi, A.; Couvy, H.; Cordier, P.; Frost, D.J. Pressure sensitivity of olivine slip systems and seismic anisotropy of Earth's upper mantle. *Nature* **2005**, *433*, 731–733. [CrossRef]

19. Jung, H.; Mo, W.; Green, H.W. Upper mantle seismic anisotropy resulting from pressure-induced slip transition in olivine. *Nat. Geosci.* **2009**, *2*, 73–77. [CrossRef]

20. Ohuchi, T.; Kawazoe, T.; Nishihara, Y.; Nishiyama, N.; Irifune, T. High pressure and temperature fabric transitions in olivine and variations in upper mantle seismic anisotropy. *Earth Planet. Sci. Lett.* **2011**, *304*, 55–63. [CrossRef]

21. Park, K.; Choi, S.H.; Cho, M.; Lee, D.-C. Evolution of the lithospheric mantle beneath Mt. Baekdu (Changbaishan): Constraints from geochemical and Sr-Nd-Hf isotopic studies on peridotite xenoliths in trachybasalt. *Lithos* **2017**, *286–287*, 330–344. [CrossRef]

22. Choi, H.-O.; Choi, S.H.; Lee, Y.S.; Ryu, J.-S.; Lee, D.-C.; Lee, S.-G.; Sohn, Y.K.; Liu, J.-Q. Petrogenesis and mantle source characteristics of the late Cenozoic Baekdusan (Changbaishan) basalts, North China Craton. *Gondwana Res.* **2020**, *78*, 156–171. [CrossRef]

23. Wu, F.-Y.; Walker, R.J.; Ren, X.-W.; Sun, D.-Y.; Zhou, X.-H. Osmium isotopic constraints on the age of lithospheric mantle beneath northeastern China. *Chem. Geol.* **2003**, *196*, 107–129. [CrossRef]

24. Kuritani, T.; Kimura, J.-I.; Ohtani, E.; Miyamoto, H.; Furuyama, K. Transition zone origin of potassic basalts from Wudalianchi volcano, northeast China. *Lithos* **2013**, *156–159*, 1–12. [CrossRef]

25. Liu, D.Y.; Nutman, A.P.; Compston, W.; Wu, J.S.; Shen, Q.H. Remnants of ≥3800 Ma crust in the Chinese part of the Sino-Korean craton. *Geology* **1992**, *20*, 339–342. [CrossRef]

26. Liu, D.; Wilde, S.A.; Wan, Y.; Wu, J.; Zhou, H.; Dong, C.; Yin, X. New U-Pb and Hf isotopic data confirm Anshan as the oldest preserved segment of the North China Craton. *Am. J. Sci.* **2008**, *308*, 200–231. [CrossRef]

27. Zhao, G.; Wilde, S.A.; Cawood, P.A.; Sun, M. Archean blocks and their boundaries in the North China Craton: Lithological, geochemical, structural and P–T path constraints and tectonic evolution. *Precambrian Res.* **2001**, *107*, 45–73. [CrossRef]

28. Yang, J.-H.; Wu, F.-Y.; Wilde, S.A.; Belousova, E.; Griffin, W.L. Mesozoic decratonization of the North China block. *Geology* **2008**, *36*, 467–470. [CrossRef]

29. Wu, F.-Y.; Lin, J.-Q.; Wilde, S.A.; Zhang, X.O.; Yang, J.-H. Nature and significance of the Early Cretaceous giant igneous event in eastern China. *Earth Planet. Sci. Lett.* **2005**, *233*, 103–119. [CrossRef]

30. Wu, F.-Y.; Yang, J.-H.; Wilde, S.A.; Zhang, X.-O. Geochronology, petrogenesis and tectonic implications of Jurassic granites in the Liaodong Peninsula, NE China. *Chem. Geol.* **2005**, *221*, 127–156. [CrossRef]

31. Zou, H.; Zindler, A.; Xu, X.; Qi, Q. Major, trace element, and Nd, Sr and Pb isotope studies of Cenozoic basalts in SE China: Mantle sources, regional variations, and tectonic significance. *Chem. Geol.* **2000**, *171*, 33–47. [CrossRef]

32. Cheong, A.C.-S.; Jeong, Y.-J.; Jo, H.J.; Sohn, Y.K. Recurrent Quaternary magma generation at Baekdusan (Changbaishan) volcano: New zircon U-Th ages and Hf isotopic constraints from the Millennium Eruption. *Gondwana Res.* **2019**, *68*, 13–21. [CrossRef]

33. Zhao, D.; Tian, Y.; Lei, J.; Liu, L.; Zheng, S. Seismic image and origin of the Changbai intraplate volcano in East Asia: Role of big mantle wedge above the stagnant Pacific slab. *Phys. Earth Planet. Inter.* **2009**, *173*, 197–206. [CrossRef]

34. Tang, Y.; Obayashi, M.; Niu, F.; Grand, S.P.; Chen, Y.J.; Kawakatsu, H.; Tanaka, S.; Ning, J.; Ni, J.F. Changbaishan volcanism in northeast China linked to subduction-induced mantle upwelling. *Nat. Geosci.* **2014**, *7*, 470–475. [CrossRef]

35. Kuritani, T.; Xia, Q.-K.; Kimura, J.-I.; Liu, J.; Shimizu, K.; Ushikubo, T.; Zhao, D.; Nakagawa, M.; Yoshimura, S. Buoyant hydrous mantle plume from the mantle transition zone. *Sci. Rep.* **2019**, *9*, 6549. [CrossRef]

36. Wei, H.; Liu, G.; Gill, J. Review of eruptive activity at Tianchi volcano, Changbaishan, northeast China: Implications for possible future eruptions. *Bull. Volcanol.* **2013**, *75*, 706. [CrossRef]

37. Wei, H.; Sparks, R.S.J.; Liu, R.; Fan, Q.; Wang, Y.; Hong, H.; Zhang, H.; Chen, H.; Jiang, C.; Dong, J.; et al. Three active volcanoes in China and their hazards. *J. Asian Earth Sci.* **2003**, *21*, 515–526. [CrossRef]

38. Wei, H.; Wang, Y.; Jin, J.; Gao, L.; Yun, S.-H.; Jin, B. Timescale and evolution of the intracontinental Tianchi volcanic shield and ignimbrite-forming eruption, Changbaishan, Northeast China. *Lithos* **2007**, *96*, 315–324. [CrossRef]

39. Brey, G.P.; Köhler, T. Geothermobarometry in four-phase lherzolites II. New thermobarometers, and practical assessment of existing thermobarometers. *J. Petrol.* **1990**, *31*, 1353–1378. [CrossRef]

40. Park, M.; Jung, H. Relationships between eclogite-facies mineral assemblages, deformation microstructures, and seismic properties in the Yuka Terrane, North Qaidam Ultrahigh-Pressure Metamorphic Belt, NW China. *J. Geophys. Res. Solid Earth* **2019**, *124*, 13168–13191. [CrossRef]

41. Park, M.; Jung, H. Analysis of electron backscattered diffraction (EBSD) mapping of geological materials: Precautions for reliably collecting and interpreting data on petro-fabric and seismic anisotropy. *Geosci. J.* **2020**. [CrossRef]

42. Skemer, P.; Katayama, I.; Jiang, Z.; Karato, S.-I. The misorientation index: Development of a new method for calculating the strength of lattice-preferred orientation. *Tectonophysics* **2005**, *411*, 157–167. [CrossRef]

43. Amiguet, E.; Raterron, P.; Cordier, P.; Couvy, H.; Chen, J.C. Deformation of diopside single crystal at mantle pressure 1: Mechanical data. *Phys. Earth Planet. Inter.* **2009**, *177*, 122–129. [CrossRef]

44. Helmstaedt, H.; Anderson, O.L.; Gavasci, A.T. Petrofabric studies of eclogite, spinel-Websterite, and spinel-lherzolite Xenoliths from kimberlite-bearing breccia pipes in southeastern Utah and northeastern Arizona. *J. Geophys. Res.* **1972**, *77*, 4350–4365. [CrossRef]

45. Van der Wal, D.; Chopra, P.; Drury, M.; Gerald, J.F. Relationships between dynamically recrystallized grain size and deformation conditions in experimentally deformed olivine rocks. *Geophys. Res. Lett.* **1993**, *20*, 1479–1482. [CrossRef]

46. Nickel, K.G. Phase equilibria in the system SiO_2-MgO-Al_2O_3-CaO-Cr_2O_3 (SMACCR) and their bearing on spinel/garnet lherzolite relationships. *Neues Jahrb. Mineral. Abh.* **1986**, *155*, 259–287.

47. Webb, S.A.C.; Wood, B.J. Spinel-pyroxene-garnet relationships and their dependence on Cr/Al ratio. *Contrib. Mineral. Petrol.* **1986**, *92*, 471–480. [CrossRef]

48. Arai, S. Characterization of spinel peridotites by olivine-spinel compositional relationships: Review and interpretation. *Chem. Geol.* **1994**, *113*, 191–204. [CrossRef]

49. Dick, H.J.B.; Bullen, T. Chromian spinel as a petrogenetic indicator in abyssal and alpine-type peridotites and spatially associated lavas. *Contrib. Mineral. Petrol.* **1984**, *86*, 54–76. [CrossRef]

50. Ishii, T.; Robinson, P.T.; Maekawa, H.; Fiske, R. Petrological studies of peridotites from diapiric serpentinite seamounts in the Izu-Ogasawara-Mariana forearc, Leg 125. *Proc. Ocean. Drill. Program Sci. Results* **1992**, *125*, 445–485.

51. Parkinson, I.J.; Pearce, J.A. Peridotites from the Izu-Bonin-Mariana Forearc (ODP Leg 125): Evidence for Mantle Melting and Melt—Mantle Interaction in a Supra-Subduction Zone Setting. *J. Petrol.* **1998**, *39*, 1577–1618. [CrossRef]

52. Pearce, J.A.; Barker, P.F.; Edwards, S.J.; Parkinson, I.J.; Leat, P.T. Geochemistry and tectonic significance of peridotites from the South Sandwich arc-basin system, South Atlantic. *Contrib. Mineral. Petrol.* **2000**, *139*, 36–53. [CrossRef]

53. Jung, H.; Park, M.; Jung, S.; Lee, J. Lattice preferred orientation, water content, and seismic anisotropy of orthopyroxene. *J. Earth Sci.* **2010**, *21*, 555–568. [CrossRef]

54. Zhang, S.; Karato, S.-I. Lattice preferred orientation of olivine aggregates deformed in simple shear. *Nature* **1995**, *375*, 774–777. [CrossRef]

55. Zhang, S.; Karato, S.-I.; Fitz Gerald, J.; Faul, U.H.; Zhou, Y. Simple shear deformation of olivine aggregates. *Tectonophysics* **2000**, *316*, 133–152. [CrossRef]

56. Hansen, L.N.; Zimmerman, M.E.; Kohlstedt, D.L. The influence of microstructure on deformation of olivine in the grain-boundary sliding regime. *J. Geophys. Res. Solid Earth* **2012**, *117*, B09201. [CrossRef]

57. Karato, S.; Jung, H.; Katayama, I.; Skemer, P. Geodynamic significance of seismic anisotropy of the upper mantle: New insights from laboratory studies. *Annu. Rev. Earth Planet. Sci.* **2008**, *36*, 59–95. [CrossRef]

58. Michibayashi, K.; Mainprice, D. The Role of Pre-existing Mechanical Anisotropy on Shear Zone Development within Oceanic Mantle Lithosphere: An Example from the Oman Ophiolite. *J. Petrol.* **2004**, *45*, 405–414. [CrossRef]

59. Jung, H.; Mo, W.; Choi, S.H. Deformation microstructures of olivine in peridotite from Spitsbergen, Svalbard and implications for seismic anisotropy. *J. Metamorph. Geol.* **2009**, *27*, 707–720. [CrossRef]

60. Falus, G.; Tommasi, A.; Soustelle, V. The effect of dynamic recrystallization on olivine crystal preferred orientations in mantle xenoliths deformed under varied stress conditions. *J. Struct. Geol.* **2011**, *33*, 1528–1540. [CrossRef]

61. Park, M.; Jung, H. Microstructural evolution of the Yugu peridotites in the Gyeonggi Massif, Korea: Implications for olivine fabric transition in mantle shear zones. *Tectonophysics* **2017**, *709*, 55–68. [CrossRef]

62. Skemer, P.; Warren, J.M.; Kelemen, P.B.; Hirth, G. Microstructural and Rheological Evolution of a Mantle Shear Zone. *J. Petrol.* **2010**, *51*, 43–53. [CrossRef]

63. Warren, J.M.; Hirth, G.; Kelemen, P.B. Evolution of olivine lattice preferred orientation during simple shear in the mantle. *Earth Planet. Sci. Lett.* **2008**, *272*, 501–512. [CrossRef]

64. Linckens, J.; Herwegh, M.; Muntener, O. Linking temperature estimates and microstructures in deformed polymineralic mantle rocks. *Geochem. Geophys. Geosyst.* **2011**, *12*. [CrossRef]

65. Linckens, J.; Herwegh, M.; Muntener, O.; Mercolli, I. Evolution of a polymineralic mantle shear zone and the role of second phases in the localization of deformation. *J. Geophys. Res. Solid Earth* **2011**, *116*. [CrossRef]

66. Kaczmarek, M.-A.; Jonda, L.; Davies, H.L. Evidence of melting, melt percolation and deformation in a supra-subduction zone (Marum ophiolite complex, Papua New Guinea). *Contrib. Mineral. Petrol.* **2015**, *170*, 1–23. [CrossRef]

67. Bystricky, M.; Kunze, K.; Burlini, L.; Burg, J.-P. High Shear Strain of Olivine Aggregates: Rheological and Seismic Consequences. *Science* **2000**, *290*, 1564–1567. [CrossRef]

68. Tommasi, A.; Tikoff, B.; Vauchez, A. Upper mantle tectonics: Three-dimensional deformation, olivine crystallographic fabrics and seismic properties. *Earth Planet. Sci. Lett.* **1999**, *168*, 173–186. [CrossRef]

69. Tommasi, A.; Mainprice, D.; Canova, G.; Chastel, Y. Viscoplastic self-consistent and equilibrium-based modeling of olivine lattice preferred orientations: Implications for the upper mantle seismic anisotropy. *J. Geophys. Res. Solid Earth* **2000**, *105*, 7893–7908. [CrossRef]

70. Hansen, L.N.; Zhao, Y.-H.; Zimmerman, M.E.; Kohlstedt, D.L. Protracted fabric evolution in olivine: Implications for the relationship among strain, crystallographic fabric, and seismic anisotropy. *Earth Planet. Sci. Lett.* **2014**, *387*, 157–168. [CrossRef]

71. Higgie, K.; Tommasi, A. Deformation in a partially molten mantle: Constraints from plagioclase lherzolites from Lanzo, western Alps. *Tectonophysics* **2014**, *615–616*, 167–181. [CrossRef]

72. Yang, Y.; Abart, R.; Yang, X.; Shang, Y.; Ntaflos, T.; Xu, B. Seismic anisotropy in the Tibetan lithosphere inferred from mantle xenoliths. *Earth Planet. Sci. Lett.* **2019**, *515*, 260–270. [CrossRef]

73. Balling, N. Heat flow and thermal structure of the lithosphere across the Baltic Shield and northern Tornquist Zone. *Tectonophysics* **1995**, *244*, 13–50. [CrossRef]

74. Jiang, G.; Hu, S.; Shi, Y.; Zhang, C.; Wang, Z.; Hu, D. Terrestrial heat flow of continental China: Updated dataset and tectonic implications. *Tectonophysics* **2019**, *753*, 36–48. [CrossRef]

75. Kim, S.; Tkalčić, H.; Rhie, J. Seismic constraints on magma evolution beneath Mount Baekdu (Changbai) volcano from transdimensional Bayesian inversion of ambient noise data. *J. Geophys. Res. Solid Earth* **2017**, *122*, 5452–5473. [CrossRef]

76. O'Reilly, S.Y.; Griffin, W.L. 4-D Lithosphere Mapping: Methodology and examples. *Tectonophysics* **1996**, *262*, 3–18. [CrossRef]

77.   Hellebrand, E.; Snow, J.E.; Dick, H.J.B.; Hofmann, A.W. Coupled major and trace elements as indicators of the extent of melting in mid-ocean-ridge peridotites. *Nature* **2001**, *410*, 677–681. [CrossRef]

Article

Microstructures and Fabric Transitions of Natural Ice from the Styx Glacier, Northern Victoria Land, Antarctica

Daeyeong Kim [1,*], David J. Prior [2], Yeongcheol Han [3], Chao Qi [4], Hyangsun Han [5] and Hyeon Tae Ju [6]

[1] Division of Polar Earth-System Sciences, Korea Polar Research Institute, Incheon 21990, Korea
[2] Department of Geology, University of Otago, Dunedin 9054, New Zealand; david.prior@otago.ac.nz
[3] Division of Polar Paleoenvironment, Korea Polar Research Institute, Incheon 21990, Korea; yhan@kopri.re.kr
[4] Key Laboratory of Earth and Planetary Physics, Institute of Geology and Geophysics, Chinese Academy of Sciences, Beijing 100029, China; qichao@mail.iggcas.ac.cn
[5] Department of Geophysics, Kangwon National University, Chuncheon 24341, Korea; hyangsun@kangwon.ac.kr
[6] Antarctic K-Route Expedition Team, Korea Polar Research Institute, Incheon 21990, Korea; hyeontae@kopri.re.kr
* Correspondence: dkim@kopri.re.kr; Tel.: +82-32-760-5443

Received: 10 August 2020; Accepted: 7 October 2020; Published: 8 October 2020

Abstract: We investigated the microstructures of five ice core samples from the Styx Glacier, northern Victoria Land, Antarctica. Evidence of dynamic recrystallization was found in all samples: those at 50 m mainly by polygonization, and those at 170 m, largely by grain boundary migration. Crystallographic preferred orientations of all analyzed samples (view from the surface) typically showed a single cluster of c-axes normal to the surface. A girdle intersecting the single cluster occurs at 140–170 m with a tight cluster of a-axes normal to the girdle. We interpret the change of crystallographic preferred orientations (CPOs) at <140 m as relating to a combination of vertical compression, and shear on a horizontal plane, and the girdle CPOs at depths >140 m, as the result of horizontal extension. Based on the data obtained from the ground penetrating radar, the underlying bedrock topography of a nunatak could have generated the extensional stress regime in the study area. The results imply changeable stress regimes that may occur during burial as a result of external kinematic controls, such as an appearance of a small peak in the bedrock.

Keywords: Ice; microstructure; crystallographic preferred orientation (CPO); Styx Glacier; electron backscatter diffraction (EBSD)

---

1. Introduction

Understanding the rheological behavior of Antarctic ice is essential for predictions of sea-level change in the future. Ice in the polar regions flow continuously from the continents to the oceans driven by gravity. The flow of ice is usually controlled by two processes: frictional sliding on the bedrock, which often occurs where the base is at pressure melting or pre-melting conditions [1–8], and the internal plastic deformation of the polycrystalline ice [9–15].

The plastic deformation of Antarctic ice results in the development of crystallographic preferred orientations (CPOs), which may induce a strong mechanical anisotropy that can enhance or impede the flow of ice and can be used to study the kinematics of flow [16]. The CPOs of ice vary with the kinematics (e.g., compression or shear), stress, temperature, and strain [10,17–19]. In nature, usually three types of ice c-axes fabrics form: single cluster, double clusters, and a girdle. Single and double clusters of ice c-axes can form due to simple shear [20]. Single cluster can also form due to uniaxial compression (e.g., [21]). Girdles have been interpreted as due to uniaxial extension [22]. The ice CPOs are often reorganized by kinematic changes, such as the progressive development of folds, shear zones, and an increased depth of burial (confining pressure) [21–25]; however, studies related to the CPO evolution are limited, especially for natural ice.

Single-maxima clusters are observed in uniaxial compression laboratory experiments only at low temperatures and/or fast strain rates [17,26]. The vast majority of laboratory experiments give an open cone (small circle) CPO in uniaxial compression. Laboratory experiments have shown that a transition from an open cone to a single cluster of *c*-axes [26–28] with increasing stress, decreasing temperature, and increasing strain. This has been interpreted as an increase in pyramidal slip systems by Piazolo et al. [27] and to a change in balance between lattice rotation and grain boundary migration recrystallization by Qi et al. [28]. A series of simple shear experiments have reported a transition of double cluster to single clustered *c*-axes with increasing strain, and have documented different patterns of evolution with strain dependent upon deformation temperature [29–31]. The transition in the CPOs of natural ice is often ascribed to environmental variables, such as the concentration of air bubbles, layers of volcanic tephra, snow accumulation rates, and the structures of bedrock [32–34]. Although recent experimental studies have reported stress and temperature as prevailing factors that lead to the formation CPOs of ice [28,29], microstructural evolution of natural ice, which occurs at lower strain rates, and proceeds to higher strain than experiments, probably involves further complexities that are not adequately understood.

In this study, we report a microstructural evolution of ice in cores drilled from the Styx Glacier in northern Victoria Land, Antarctica, which is a renowned site for snow accumulation. The slow ice flow allows for a high probability of the development of diverse stress regimes. We describe the deformation behavior of the ice with increasing depth based on calculations of physical properties and careful investigations of CPOs using electron backscatter diffraction (EBSD) in a scanning electron microscope. We also compare the deformation features of the sample to conventional ice flow models, which is followed by a discussion of glaciological implications.

2. Materials and Methods

2.1. Geological Outline and Sample Descriptions

The Styx Glacier in northern Victoria Land lies along the western margin of the Ross Sea (Figure 1) and is a renowned site for snow accumulation that is minimally influenced by katabatic winds. Hence, several ice cores have been retrieved from the Styx Glacier in recent decades [35–38]. The Styx Glacier consists of a flat surface with a relief of <100 m over a total area of >100 km^2. The velocity of its ice surface is ca. 2.3 m yr^{-1} [39], which was measured from multiple satellite (Sentinel-1) synthetic aperture radar interferometry data (InSAR) (Figure 2). As a part of the Korean Antarctic Research Program based at the Jang Bogo Station, a drilling program led by the Korea Polar Research Institute, was conducted at the Styx Glacier to obtain an ice core that developed during the past 2 kyr [40]. An ice core with a total depth of 210.5 m was collected from the Styx Glacier at 73°51.10′ S, 163°41.22′ E, at an elevation of 1623 m, during the 2014–2015 austral summer season. The ice base at the sample site was determined to be at 550 m depth by a ground-penetrating radar (GPR) investigation [41]. Bedrock signals in the GPR data confirmed the existence of a nunatak, located ~4 km southeast (SE) of the sample site. The presence of the nunatak is consistent with the reduction in ice flow velocity and divergent change of the ice flow direction, observed from the satellite data (Figures 2 and 3), as ice flows past the site in the south-southeast (SSE) direction.

Due to low densities at <40 m and rich in cracks at >180 m depth, samples of the ice core were selected from depths of about 50, 80, 110, 140, and 170 m, labeled as SG_50, _80, _110, _140, and _170, respectively [40]. The samples were stored at ca. −25 °C for preserving the fabric (e.g., [42]). Characteristics of the cores are summarized in Table 1. The ages were estimated to be 0.26 to 1.83 kyr, based on a combined approach of an empirical firn densification model [43] and the CH$_4$ gas age with the correction for the difference between the gas and ice ages of ~0.3 kyr [44]; densities ranged from

780 to 920 kg/m^3 (Figure 4a,b). The borehole temperature, measured 2 y after the drilling, was −31.7 °C at a depth of 15 m and −28.3 °C at the base of the hole (210 m). A model calculation reported a mean annual temperature of −32.5 °C, and a basal temperature of −20.5 °C at 550 m [44]. Snow accumulation rate is reported as 203 kg m^{-2} yr^{-1} [38].

Vertical longitudinal strain rates can be estimated by assuming that finite longitudinal strain is accommodated by density increase from the surface. The strain rate separating the sample and the surface is calculated by dividing the natural log of the change in ice density ($\Delta \rho_{ice}$) by the age of the sample, with shortening being positive. The age is derived from the depth-age scale [44] (Table 1). Initial density of 350 kg/m^3 was assumed at the surface. Finite longitudinal strain increases to 0.97 at 170 m depth. Longitudinal strain rate decreases from 10^{-10} s^{-1} at shallow depths to 2×10^{-11} s^{-1} at 170 m depth (Figure 4c). An upper bound horizontal shear strain rate is estimated by assuming a linear decrease in velocity from the surface to the base of the ice: a horizontal shear strain rate of ~10^{-10} has been calculated by dividing the ice flow velocity, 2.3 m year^{-1}, by thickness, 550 m (Table 1). A lower bound shear strain rate estimate was made by assuming the driving force for horizontal shear strain is at a maximum at the base of the glacier (550 m) and reduces linearly to zero at the surface. Shear strain rate at any depth is proportional to depth to the power of n, where n is the stress exponent in the flow law. Using this relationship and the fact that the average of shear strain rates for all depths must correspond to the average shear strain rate averaged across the entire ice thickness (2.3 m yr^{-1}/550 m) allows the shear strain rate and shear strain at any given depth to be calculated. A similar approach can be used to account for the warmer temperatures at the base of the glacier using Q = 60 kJmol^{-1}. Using an n value of 3 (appropriate for low strains), lower bound horizontal shear strain rate increases from 10^{-13} at 50 m to 7×10^{-12} at 170 m depth (Table 1).

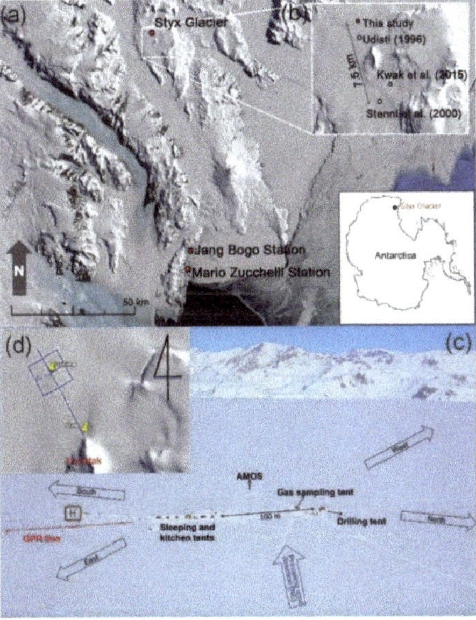

Figure 1. Maps of the drilling site, modified from Han et al. [40]. (**a**) The Styx Glacier in northern Victoria Land lies along the western margin of the Ross Sea. The sample locations of Udisti [45], Stenni et al. [38] and Kwak et al. [36] in (**b**) are marked for comparison. (**c**) The site where samples were collected, showing the various tents during the 2014–2015 summer expedition. GPR: ground-penetrating radar; AMOS: automatic meteorological observation system. (**d**) Investigation lines for ground penetrating radar (GPR). Details of the measurements and results will be presented in Figure 3.

Table 1. Sample descriptions.

Sample Numbers	Location (Latitude, Longitude)	Depth (m)	Density (kg/m³) [a]	Age (kyr) [b]	Temperature (°C) [c]	Finite Longitudinal Strain [d]	Longitudinal Strain Rate (s⁻¹)	Upper Bound Horizontal Shear Strain Rate (s⁻¹) [e]	Lower Bound Horizontal Shear Strain Rate (s⁻¹) [f]	Lower Bound Horizontal Shear Strain
SG_50		50.03–50.08	783	0.26	−31.2	0.81	9.82×10^{-11}		1.02×10^{-13}	0.00
SG_80	Styx Glacier	79.97–80.02	884	0.55	−30.8	0.93	5.34×10^{-11}		5.10×10^{-13}	0.01
SG_110	(73° 51′ 6.0″S,	109.91–109.96	900	0.90	−30.3	0.94	3.33×10^{-11}	1.3×10^{-10}	1.52×10^{-12}	0.04
SG_140	163° 41′ 13.2″E)	140.47–140.52	917	1.31	−29.8	0.96	2.33×10^{-11}		3.49×10^{-12}	0.14
SG_170		170.19–170.24	922	1.83	−28.9	0.97	1.68×10^{-11}		6.90×10^{-12}	0.40

[a] Density is measured using cored ice masses in a range of 0.4~0.8 m length. [b] Ice ages were derived from Yang et al. (2018). [c] Borehole temperature was measured twice with 1 m interval, in November 2016, 2 years after drilling. [d] Finite longitudinal strain was calculated as $\ln(\Delta \rho_{ice})$ where $\Delta \rho_{ice}$ is the difference between the density of the surface and the density of the sample. [e] Upper bound horizontal shear strain rate was calculated by ice flow velocity (2.3 m/year) divided by thickness (550 m). [f] Lower bound horizontal shear strain rate was estimated with an assumption that the driving force for horizontal shear strain is at a maximum at the base of the glacier and reduces linearly to zero at the surface.

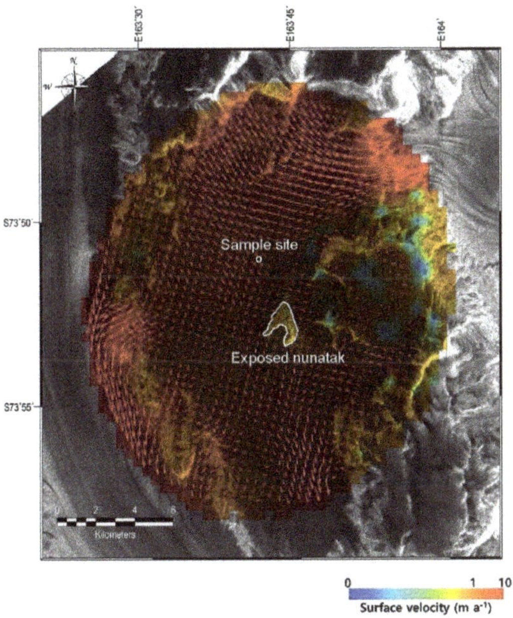

Figure 2. Ice surface velocity and flow direction (red arrows) around the drilling site, derived from satellite radar interferometry data [39]. The background is Sentinel-1 synthetic aperture radar image acquired on 17 June 2017.

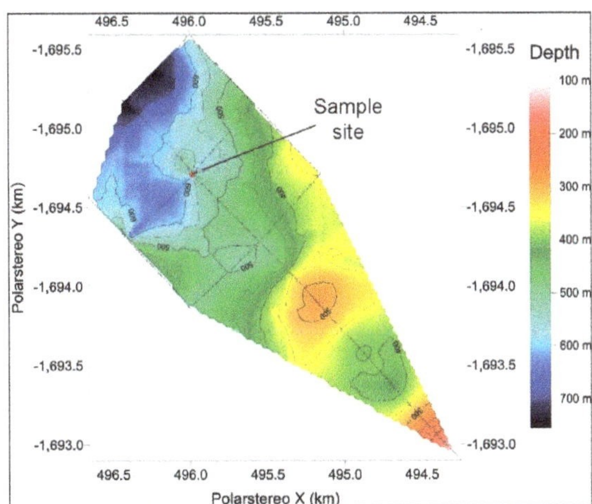

Figure 3. Ground-penetrating radar (GPR) data plotted using the program Surfer. The GPR data were acquired using a MALÅ Rough Terrain Antenna (RTA) system. The control unit linked to global positioning system (GPS) receiver measured signals with 25 MHz center frequency, 4.4 m transmitter-receiver separation, 1 trace s^{-1} record, 13,000 ns time window, and 8 stacks. The raw data were post-processed by background removal, gain, and Kirchhoff migration using the program Reflexw. The radio wave velocity was employed as 0.20 m ns^{-1} for snow (0–20 m depth), 0.19 m ns^{-1} for firn (20–50 m depth), and 0.17 m ns^{-1} for ice (>50 m depth) [46].

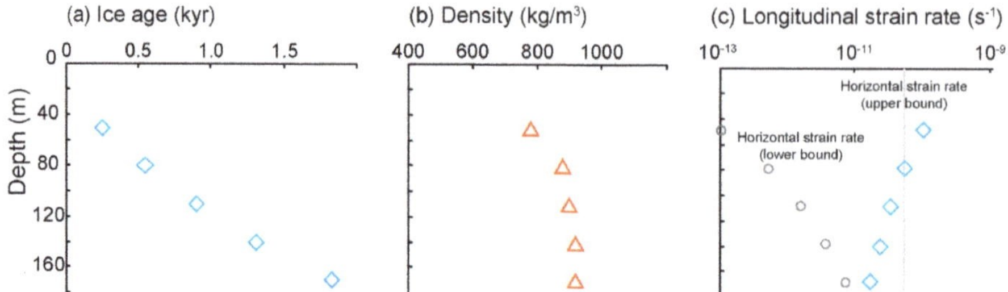

Figure 4. Plots of (**a**) ice age, (**b**) density, and (**c**) longitudinal strain rate as a factor of increasing depth. (**a**) The ice age of the core was derived from Yang et al. [44]. (**b**) Density is computed from the measured weight of the core divided by volume. (**c**) Longitudinal strain rate is calculated from density divided by time. Upper and lower bound horizontal strain rates are presented as a grey line.

2.2. Analytical Methods

The microstructural features of rocks are usually observed in sections cut perpendicular to the foliation and parallel to the lineation (e.g., [47]). The foliation (bedding) of ice, in this setting, can be considered as the plane parallel to the surface, which forms due to the annual snow accumulation and densification during burial. Lineation cannot usually be observed owing to the transparency of ice. Azimuthal orientations are not normally preserved in rotary coring and for this study we cut an arbitrary vertical (normal to foliation) observation plane. The samples were cut by a ring blade (a circular saw) in a cold room at −15 to −20 °C, and mounted on a copper ingot at ~5 °C, thus, producing a bond between the sample and the ingot by melting and re-freezing a thin layer. As soon as a bond was formed, the ingot and the sample were cooled rapidly. To strengthen the bond, water droplets at 0 °C were added into the holes at the back of the ingot which froze onto the bottom of the sample. We used a microtome to cut the samples to thickness of ~1 cm in a freezer at −15 to −20 °C. The final polishing was achieved by using sandpaper (1500 mesh) attached to a polishing plate sit 1–2 cm above liquid nitrogen at the base of a polystyrene box, so that the plate was about −20 to −30 °C (for further details, see [48]).

To study the microstructures of ice, high-quality crystallographic orientation maps were acquired by indexing the Kikuchi bands (diffraction patterns) using an Oxford Instruments EBSD camera on a Zeiss Sigma variable-pressure, field-emission, scanning electron microscope (FE–SEM), housed at the University of Otago, Dunedin, New Zealand. Each polished sample (ca. 30 × 45 mm) was transferred through a nitrogen-filled glove box into the SEM to minimize condensation. After the sample was inserted to the cold stage in the SEM chamber at ≤−80 °C, the chamber was evacuated to a high vacuum and the sample was cooled to ≤−100 °C. A sublimation process was then carried out by bringing the chamber pressure to the atmospheric pressure and warming the sample to approximately −80 °C to remove frost and flatten the surface of the sample, and then the chamber was evacuated again to a high vacuum and the sample was cooled again to less than −100 °C. The orientation maps were generated with an accelerating voltage of 30 kV, a working distance of ~20 mm, a beam current of ~10 nA, and a step size of 30–50 μm with ~10 Pa nitrogen gas pressure to prevent charging (for further details of the process, see [48]). The raw data thus obtained, were processed using the software Channel 5 and the MATLAB-based toolbox MTEX [49]. Contoured pole figures of [0001], <11–20> and <10–10> were generated from all indexed pixels using MTEX. The *J*- and *M*-indices were determined [50–52] as measure of fabric strength using MTEX. Eigenvalues for calculating shape factor, *K*, were computed using the MTEX [53].

3. Results

All results from our microstructural analyses are summarized in Table 2. As illustrated in Figure 4, the age and density of the ice increases, and the longitudinal strain rate decreases with increasing depth. In this section, we describe the microstructures, CPOs, and related data.

Table 2. Results of EBSD analysis.

Sample Numbers	Step Size (μm)	Success Rate (%) [a]	Mean Grain Size with std. dev. (μm)	Fabric Strength		Seismic Anisotropy (%)		Number of Crystals	Number of Indexed Pixels
				J-index	M-index	AV_P	AV_S		
SG_50	40	84.46	1240 ± 550	1.4618	0.0372	0.7	1.92	385	257,290
SG_80	40	93.46	1580 ± 970	1.9663	0.0725	1	2.68	242	349,756
SG_110	40	96.21	1790 ± 1200	2.1067	0.0853	0.7	2.07	185	373,586
SG_140	40	93.12	2000 ± 1280	5.0772	0.3009	2.8	6.87	101	159,714
SG_170	40	97.40	2060 ± 1400	4.2485	0.2327	2.2	5.77	129	343,550

[a] Success rate represents the initial number of indexed grains divided by all the analyzed samples before noise reduction, and zero solution includes the space represented by the bubbles and cracks; therefore, the number is underestimated.

The high-quality EBSD maps, figures for misorientation between each pixel and the mean orientation of the parent grain (mis2mean), and the grain-size distributions are displayed in Figure 5a–c, respectively, for intracrystalline and intercrystalline deformation features. Analyzed sections have different arbitrary azimuthal orientations as azimuthal information was lost in coring. Sample SG_50 contains a large number of air bubbles. The ice crystals in this sample are characterized by relatively small and uniform grain size, equigranular shapes, and straight grain boundaries (Figure 5a,b). Samples from greater depths have; a lower fraction of bubbles, larger grain size, larger variations in grain size, and more angular and irregular grain boundaries. The figures for mis2mean indicate that a few areas with high misorientations are distributed along the grain boundaries, particularly near the triple junctions (Figure 5b). Mean grain size increases with depth and the peak distribution ranged between 1000–2000 µm (Figure 5c). Grain size at the peak frequency of the distribution is about 1.5 mm at 50 m and decreases to ~1 mm at 80 and 110 m.

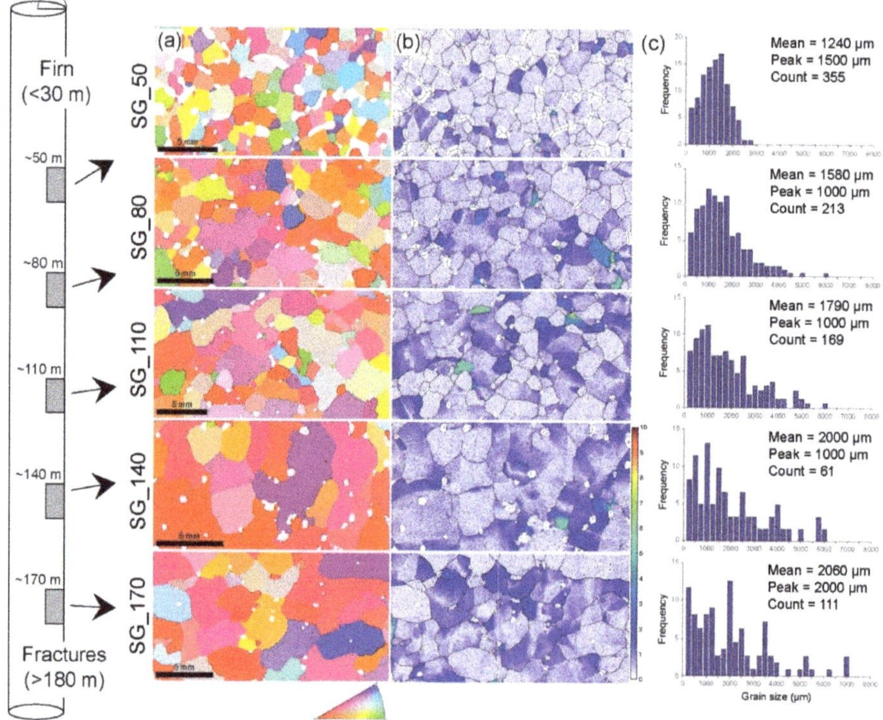

Figure 5. High-quality electron backscatter diffraction (EBSD) data showing (**a**) orientation (Euler angle), (**b**) mis2mean maps, and (**c**) grain sizes of the analyzed samples from the Styx Glacier. The locations of samples in the drilled core are shown to the left of (**a**). White areas in (**a**) and (**b**) present the existence of air bubbles. Color indices for (**a**) and (**b**) are located in the lower right of each. The maps in (**b**) show grain boundaries (>10°) as black lines. The MATLAB toolbox MTEX is employed for all figures.

The crystallographic orientations are shown in a reference frame where the vertical is in the center of the stereonets and the data are rotated around the vertical axis so that the a-axis maxima lie in the east–west (E–W) direction. This approach facilitated a better comparison of data from different depths, but the reader is reminded that the azimuthal orientation is arbitrary (the direction of the *a*-axis maxima is unknown). The *c*-axes become more concentrated with increasing depths (Figure 6a). The CPOs are characterized by a vertical cluster of *c*-axes with weak secondary maxima at ≤110 m. The strong cluster of *c*-axes lies within a broad vertical girdle at depths ≥140 m (Figure 6b). The girdle is partial at 140 m and almost complete at 170 m. The vertical girdle of *c*-axes lies perpendicular to the *a*-axis maxima. The *a*-axes and poles to m-planes lie sub-horizontally. Weaker *a*-axis sub-maxima lie at ±60 degrees to the main maximum (normal to c-axis girdle). Maxima in the poles to m-planes lie at ±30 degrees and at 90 degrees to the a-axis maxima: those at ±30 degrees are strongest.

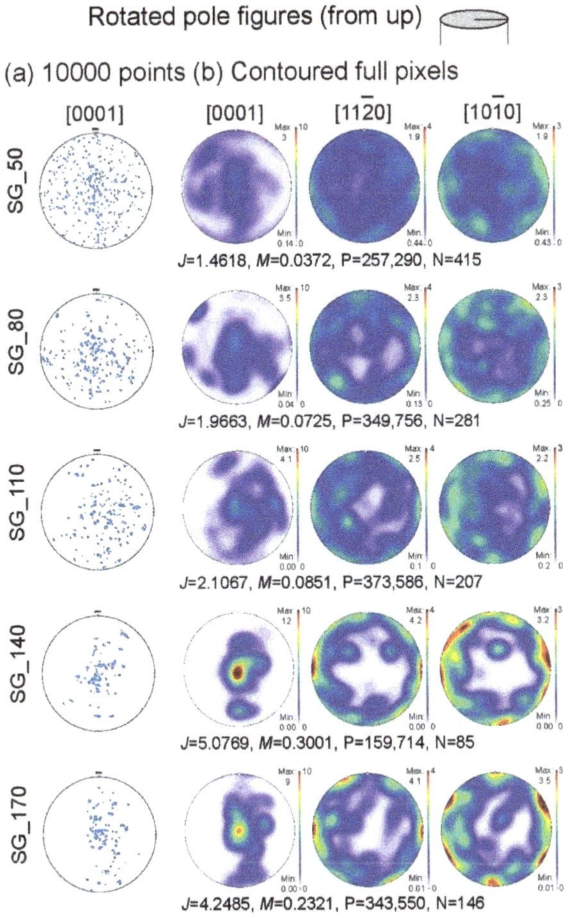

Figure 6. Rotated crystallographic preferred orientations of the ice samples using (**a**) 2000 random points and (**b**) full pixels data. The small cylinder at the top right shows the orientation of the plotted pole figures. The crystallographic data are plotted using the MATLAB-based toolbox MTEX as poles on upper-hemisphere equal-area projections with contours in multiples of uniform distribution. P = Number of pixels; *J* = *J*-index; *M* = *M*-index; N = Number of grains.

The *J*- and *M*-indices, mean grain size and the shape factor, *K*, of each sample are displayed in Figure 7. Fabric indices (*J* = 1.46–2.11 and *M* = 0.04–0.09) are weak at depths ≤110 m, remarkably high at the depth of 140 m (*J* = 5.08 and *M* = 0.30), and decrease again at 170 m (*J* = 4.25 and *M* = 0.23) (Figure 7a and Table 2). Mean grain size and standard deviation increase slightly, corresponding to increasing density, with increasing depth (Figure 7b and Table 2). The shape factor, *K*, computed from eigenvalues of the *c*-axis distribution [53] provides quantitative identification of *c*-axes being more clustered at 50–110 m depth and more girdle-like at 140–170 m.

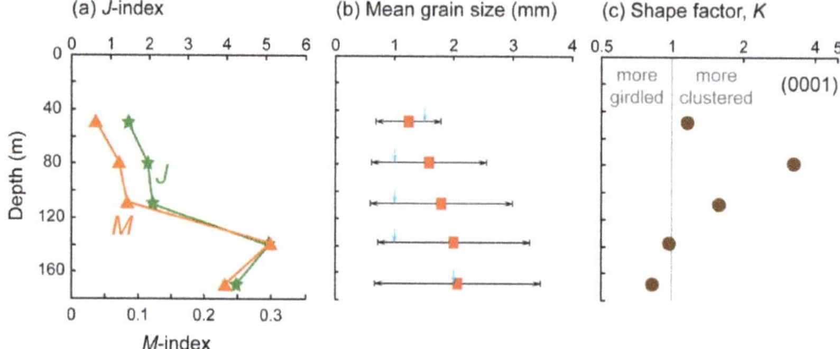

Figure 7. Plots of (**a**) *J*-index, (**b**) mean grain size, and (**c**) shape factor, *K*, as a factor of increasing depth. The range presented as black arrows and a blue arrow and in (**b**) denote the standard deviation and peak grain size, respectively. The shape factor is calculated from eigenvalues [53].

Rotation axes (misorientation axes) of low-angle boundaries (2°–10°) in our samples were analyzed in inverse pole figures (Figure 8). The rotation axes of ice were found to be generally distributed along the basal plane, and/or parallel to *c*-axes. Strong intensity parallel to the *c*-axes is activated strongly at 140 m, and more weakly at 50 and 170 m, while intensity along the basal plane is clear in most samples, except the sample from 140 m.

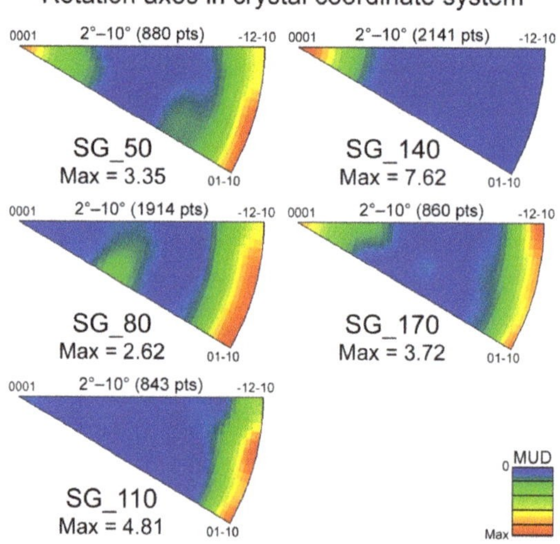

Figure 8. Rotation axes of the analyzed samples computed using the program Channel 5. The term "Max" denotes the maximum of mean uniform density. These are upper hemisphere plots, contoured with a half width of 10°.

4. Discussion

4.1. Stress Regimes and CPO Development

Deformation features can be identified from intracrystalline microstructures, such as the CPOs (e.g., [54–57]). (Table 1). In terms of density, sample SG_50 is classified as firn–ice transition, and the other samples are defined as ice [58,59]. Deformation temperatures between −29.1 to −31.2 °C estimated from the measurements of the bore hole, are colder than the vast majority of published deformation experiments for which CPO and microstructural data have been presented, but are at a similar temperature to −30 °C experiments presented by [17,26,29].

The CPOs of natural ice could be attributed to both vertical shortening and horizontal shear. The upper bounds of the estimated rate of horizontal shear is equivalent or greater than the calculated rate of finite longitudinal shortening at 80–170 m (Table 1). Lower bound estimations of horizontal shear strain at 140–170 m (>0.05) could induce a strong CPO [60]. The weak central cluster with or without multiple maxima of *c*-axes and weak fabric strengths (Figures 6 and 7a) at depths of ≤110 m, can be ascribed to low strains [20,60–63]. Higher longitudinal strain rate than horizontal shear strain rate at 50 m suggests that the single cluster of *c*-axes is related with the vertical compression under firn-ice transition condition (Figure 4c) [18]. The single cluster with multiple maxima of ice *c*-axes from 80–110 m can be attributed to a dominant activation of the horizontal shear, as estimated upper bound horizontal strain rates are higher than estimated vertical strain rates in all samples below the 50 m sample (Figure 4c). All samples have estimated shear strains that are significant, given that strong CPOs are well developed by shear strains of ~0.2 [29]. Similar CPOs have been reported as products of deformation due to recrystallization under a simple shear regime by numerical models [29,64]. Therefore, we infer that the CPOs formed at ≤110 m are influenced by both the finite longitudinal and horizontal shear regimes.

In contrast, the *c*-axes of deeper samples (≥140 m) show a strong cluster with weak maxima aligned vertically (a girdle) with an abrupt increase in fabric indices, followed by more girdle-like ($K < 1$) shape factors, and longitudinal strain rates that are about one to two orders of magnitude lower than the horizontal strain rates (Figure 4c, Figure 7a,c and Table 1). The girdle for *c*-axes is often ascribed to the development of a horizontal extension, based on the orientation of the vertical girdle that intersects the single cluster [19,22,23,65–67]. We suggest that horizontal extension may become important, probably in addition to vertical compression and horizontal shear. The strong alignment *a*-axis at 140–170 m depths may lend support for the horizontal extension origin of the girdle CPOs. Slip on basal planes to facilitate extension would lead to the slip direction aligning with extension direction. The lower *J*-index at 170 m rather than at 140 m, may be attributed to the 'countervailing effect' of the stronger girdle against the presence of a strong cluster or a small number of analyzed crystals at 140 m. The *J*- and *M*-indices are scalars and they become less useful for comparing CPO strengths when CPO symmetry changes. The occurrence of tephra at depths of 97.01, 99.18, and 165.37 m indicates that the change in CPO may be irrelevant with impurities in the ice [40].

4.2. Recrystallization Processes and Slip Systems

The flow of ice can be influenced significantly by microstructural changes, particularly those attributed to recrystallization by grain boundary migration (GBM) and/or by lattice rotation including associated recovery, polygonization, and subgrain rotation (e.g., [28,29,68]). Recrystallization by GBM is driven by reduction in the mean intragranular distortion of the polycrystals. GBM alone results in an increase in grain size as no new orientation (nuclei) are created. Recrystallization by lattice rotations is accompanied by a substantial reduction as defects and associated distortion become localized in subgrain boundaries greating cells that may evolve to become new grains (nuclei) [22,69].

Sample SG_50, representing the firn–ice transition [58,59], shows straight grain boundaries, a relatively uniform grain size, triple junctions of ~120°, weak fabrics, and a very weak single cluster of *c*-axes (Table 1, Figures 5 and 6), which is typically regarded as a product of polygonization in shallower ice with compression [18]. Scarcity of small domains with a high misorientation suggests that the numerous air bubbles, in or near grain boundaries, may play an important role in generating fabrics at 50 m (Figure 5b). Between 80–110 m, the analyzed ice samples displayed a steady increase in grain size and fabric strength, and a minor change in grain shape (Figure 5a,b and Figure 7b). At depths between 110 m and 140 m, the ice samples exhibited an abrupt increase in fabric strength. This change corresponds to a more diverse grain size distribution (high standard deviation), and the appearance of lobate grain boundaries (Figure 5), thus indicating recrystallization involves GBM.

Small domains with high misorientation, 120° triple junctions, and straight grain boundaries are distributed at all depths, suggesting the formation of new grains and the occurrence of recovery and subgrain rotation and recovery among the analyzed samples (Figure 5). The evidence suggests, therefore, that recrystallization at depths ≥140 m is dominated by GBM accompanied by recovery and subgrain rotation.

Analysis of low angle boundaries is often useful for delineating the dominant slip systems in ice (e.g., [64,70]). The rotation axes of the analyzed samples, shown in Figure 8, can be classified into two types: one parallel to the *c*-axes that is activated at the depths of 50, 140, and 170 m, and another along the basal plane that has been highlighted at all depths except at 140 m. Assuming that only the basal slip system is active, rotation axes parallel to the *c*-axis can be related to twist walls (screw dislocations) of the basal slip system, while rotation axes along the basal plane can be related to tilt walls (edge dislocations) of the basal slip system. Statistically significant data sets from experiments identify both of these rotation axes [26,71] and subgrain boundary trace analyses, and weighted burgers vector analyses [72] of these same samples (Sheng Fan, personal communication) indicate that the rotation axes in basal plane are clearly related to basal plane dislocations. Other studies have constrained the operation of non-basal dislocations in both experimental [73,74] and natural [70,75] ice and further work is needed to determine the contribution of non-basal dislocations in these samples.

There is clearly a change in misorientation axes between 110 and 140 m, corresponding to the change in CPO. Thus we infer that the changes in deformation kinematics that cause the change in CPOs is probably also responsible for the change in misorientation axes. Whether this in turn relates to a redistribution of slip from tilt walls to twist walls or a changing involvement of non-basal dislocations needs further analysis.

In the finite longitudinal/horizontal shear stress regimes, the operation of the basal slip could explain the formation of a single cluster with weak maxima of *c*-axes at depths ≤110 m and with a girdle at 140–170 m. Under a compressional (and/or a simple shear) stress regime at 50–110 m, slip occurs along the basal plane and the *c*-axes align around grains in high Schmid factor orientations at the expense of grains in the low Schmid factor orientations (Figure 9a). Under an extension regime, at 140–170 m, slip occurs along the basal plane and the *c*-axes disperse to rotate to form a girdle, with fixed *a*-axes along the tensile direction (Figure 9b). These processes can explain the formation of distinctive CPOs under different stress regimes.

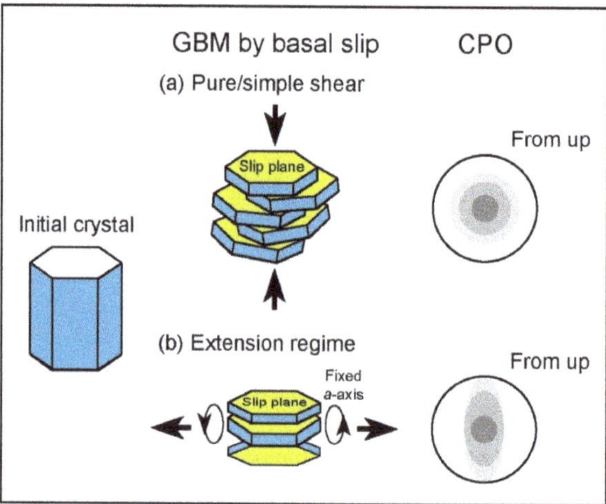

Figure 9. Schematic model to show formation of a single cluster with (**a**) weak maxima around the cluster under a compressional (pure/simple shear) regime, and (**b**) the weak aligned maxima (a girdle) under an extension regime.

4.3. Origin of Extensional Stress Regime and Implications for Ice Sheet Dynamics

Antarctic ice forms by recurrent patterns of snow accumulation, densification, and transformation through firn to ice [38,58,76–78]. The Styx Glacier is a renowned site of snow accumulation; hence, the *c*-axes of ice crystals can be expected to be aligned perpendicular to the surface. The results from the microstructural investigations in this study, however, demonstrate the CPO evolution with depth from a single cluster of *c*-axes to a single cluster with a girdle. The ice in the study area flows and meets a nunatak within ~4 km to the SE. The topography may generate an extensional stress regime that may be responsible for the weak girdle of *c*-axes at depths ≥140 m (Figure 10a). Our GPR data suggest that the bedrock becomes shallower from the sample location towards the nunatak (Figure 3). Additionally, a small bedrock peak corresponds to the core site and may contribute to an extensional stress regime at the sample location (Figure 10b,c).

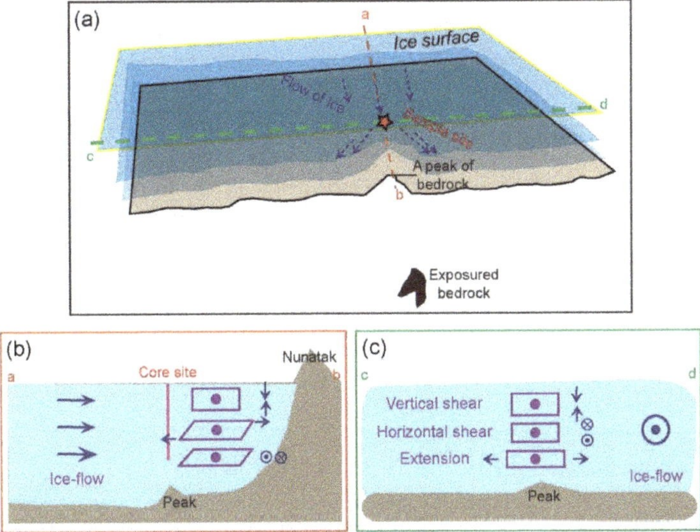

Figure 10. Schematic views of (**a**) sample location with flow of ice, (**b**) a cross section of a–b (marked as the orange line) in (**a**), and (**c**) a cross section (presented as the green line) of c–d in (**a**). The diverse stress regimes at different depths may be the results of the low velocity of ice flow, and a nunatak located ~4 km to the southease (SE) from the study area.

We have compared our results to conventional ice-flow models that regard ice as a viscoelastic material. Conventional numerical models of ice flow [76,79] are characterized by a low velocity at the base due to a high friction with the bedrock, and a logarithmic increase in speed toward the surface. In the study area, the evolution of the CPO from a single cluster of *c*-axes to a single cluster with a girdle, can be attributed to the kinematic control of the underlying bedrock. Although we have concluded that the structure of the bedrock is the most likely cause for the generation of the CPO evolution, other environmental factors, such as the concentration of air bubbles, gas or volcanic tephra due to variations in annual snow accumulation, need to be tested [32–34]. Therefore, we suggest that the microstructural analyses of various ice cores should be explored extensively by sampling from kinematically distinct locations within ice sheets. This process is crucial for a thorough review of the conventional ice flow models. Moreover, new techniques must be devised to facilitate preservation of the azimuthal orientations of cored ice samples.

5. Conclusions

We conducted a microstructural study of five natural ice samples taken at intervals of ~30 m from depths of 50–170 m by a core drilled in the Styx Glacier, located in northern Victoria Land, Antarctica. Our interpretations of the microstructures, based on high-quality EBSD maps combined with GPR, can be summarized as follows:

1. With increasing depth, the analyzed samples show a systematic change from a weak single cluster of *c*-axes at 50 m, to a single cluster with weak multi-maxima between 80–110 m, and to a single cluster with a girdle between 140–170 m. An *a*-axis cluster occurs normal to the *c*-axis girdle. The formation of the girdle that accompanied the strong single cluster is characterized by an abrupt increase in fabric indices, and a decrease in the shape factor; thus, suggesting the extension began to occur at the depth of <140 m.

2. Straight grain boundaries, relatively uniform grain size, triple junction of ~120°, and a weak single cluster of *c*-axes at 50 m (Figures 4 and 6) are typically regarded as products of polygonization under compression, mainly due to bubble extraction. With an increased depth of burial, grain size increased, fabric strengthened, and the irregularity increased in grain shape (Figures 5 and 6). At the depth of 140 m, highly curved grain shapes, abrupt increase of fabric indices, and the occurrence of a girdled *c*-axes ($K < 1$) suggest dynamic recrystallization dominated by GBM.

3. The ice generally flows at ca. 2.3 m yr^{-1} from the northwest (NW) to the SE, dominated under the finite longitudinal and horizontal shear regimes. The presence of a nunatak located ~4 km far from the sample location in the SE direction, and a low surface velocity of ice, allowed the appearance of a girdle of *c*-axes between 140–170 m. The girdle was the result of an extensional stress regime generated at that depths. These results suggest the occurrence of changeable stress regimes as a response to diverse environments, such as an abrupt appearance of a small peak in the bedrock.

Author Contributions: D.K., D.J.P., and Y.H. collaboratively designed the project and wrote the manuscript. D.K., D.J.P., and C.Q. conducted EBSD analysis, and H.H. and H.T.J. provided satellite and GPR data, respectively. All authors contributed to correct the initial version of the manuscript. All authors have read and agreed to the published version of the manuscript.

Funding: This research was funded by a KOPRI project PE20030.

Acknowledgments: This work was originally designed by D.K., Y.H., and Jin-Han Ree (Korea University). Soon Do Hur and Sang-Bum Hong (KOPRI) provided the ice core samples collected during the 2014–2015 austral summer season. Chun-Ki Lee (KOPRI) contributed toward discussions about satellite data. Lisa Craw (University of Tasmania), Kat Lilly, and Pat Langhorne (University of Otago) supported data collection. The authors appreciate the fruitful and constructive reviews by three anonymous reviewers and the careful editorial work by Haemyeong Jung.

References

1. Fahnestock, M.; Abdalati, W.; Joughin, I.; Brozena, J.; Gogineni, P. High Geothermal Heat Flow, Basal Melt, and the Origin of Rapid Ice Flow in Central Greenland. *Science* **2001**, *294*, 2338. [CrossRef] [PubMed]
2. MacGregor, J.A.; Fahnestock, M.A.; Catania, G.A.; Aschwanden, A.; Clow, G.D.; Colgan, W.T.; Gogineni, S.P.; Morlighem, M.; Nowicki, S.M.J.; Paden, J.D.; et al. A synthesis of the basal thermal state of the Greenland Ice Sheet. *J. Geophys. Res. Earth Surf.* **2016**, *121*, 1328–1350. [CrossRef] [PubMed]
3. Rignot, E.; Mouginot, J. Ice flow in Greenland for the International Polar Year 2008–2009. *Geophys. Res. Lett.* **2012**, *39*. [CrossRef]
4. Bell, R.E.; Tinto, K.; Das, I.; Wolovick, M.; Chu, W.; Creyts, T.T.; Frearson, N.; Abdi, A.; Paden, J.D. Deformation, warming and softening of Greenland's ice by refreezing meltwater. *Nat. Geosci.* **2014**, *7*, 497. [CrossRef]
5. Wolovick, M.J.; Creyts, T.T.; Buck, W.R.; Bell, R.E. Traveling slippery patches produce thickness-scale folds in ice sheets. *Geophys. Res. Lett.* **2014**, *41*, 8895–8901. [CrossRef]

6. Hewitt, I.J. Seasonal changes in ice sheet motion due to melt water lubrication. *Earth Planet. Sci. Lett.* **2013**, *371–372*, 16–25. [CrossRef]

7. Stokes, C.R.; Clark, C.D.; Lian, O.B.; Tulaczyk, S. Ice stream sticky spots: A review of their identification and influence beneath contemporary and palaeo-ice streams. *Earth Sci. Rev.* **2007**, *81*, 217–249. [CrossRef]

8. Sergienko, O.V.; Creyts, T.T.; Hindmarsh, R.C.A. Similarity of organized patterns in driving and basal stresses of Antarctic and Greenland ice sheets beneath extensive areas of basal sliding. *Geophys. Res. Lett.* **2014**, *41*, 3925–3932. [CrossRef]

9. Glen, J.W. Experiments on the Deformation of Ice. *J. Glaciol.* **1952**, *2*, 111–114. [CrossRef]

10. Glen, J.W.; Perutz, M.F. The creep of polycrystalline ice. *Proc. R. Soc. Lond. Ser. A Math. Phys. Sci.* **1955**, *228*, 519–538. [CrossRef]

11. Durham, W.B.; Heard, H.C.; Kirby, S.H. Experimental deformation of polycrystalline H2O ice at high pressure and low temperature: Preliminary results. *J. Geophys. Res.* **1983**, *88*. [CrossRef]

12. Weertman, J. Creep of deformation of ice. *Annu. Rev. Earth Planet. Sci.* **1983**, *11*, 215–240. [CrossRef]

13. Budd, W.F.; Jacka, T.H. A review of ice rheology for ice sheet modelling. *Cold Reg. Sci. Technol.* **1989**, *16*, 107–144. [CrossRef]

14. Treverrow, A.; Budd, W.F.; Jacka, T.H.; Warner, R.C. The tertiary creep of polycrystalline ice: Experimental evidence for stress-dependent levels of strain-rate enhancement. *J. Glaciol.* **2012**, *58*, 301–314. [CrossRef]

15. Goldsby, D.L.; Kohlstedt, D.L. Superplastic deformation of ice: Experimental observations. *J. Geophys. Res. Solid Earth* **2001**, *106*, 11017–11030. [CrossRef]

16. Alley, R.B.; Joughin, I. Modeling Ice-Sheet Flow. *Science* **2012**, *336*, 551. [CrossRef]

17. Craw, L.; Qi, C.; Prior, D.J.; Goldsby, D.L.; Kim, D. Mechanics and microstructure of deformed natural anisotropic ice. *J. Struct. Geol.* **2018**, *115*, 152–166. [CrossRef]

18. Alley, R.B. Flow-law hypotheses for ice-sheet modeling. *J. Glaciol.* **1992**, *38*, 245–256. [CrossRef]

19. Faria, S.H.; Weikusat, I.; Azuma, N. The microstructure of polar ice. Part II: State of the art. *J. Struct. Geol.* **2014**, *61*, 21–49. [CrossRef]

20. Hudleston, P.J. Progressive Deformation and Development of Fabric Across Zones of Shear in Glacial Ice. In *Energetics of Geological Processes: Hans Ramberg on His 60th Birthday*; Saxena, S.K., Bhattacharji, S., Annersten, H., Stephansson, O., Eds.; Springer: Berlin/Heidelberg, Germany, 1977; pp. 121–150. [CrossRef]

21. Thorsteinsson, T.; Kipfstuhl, J.; Miller, H. Textures and fabrics in the GRIP ice core. *J. Geophys. Res. Ocean.* **1997**, *102*, 26583–26599. [CrossRef]

22. Lipenkov, V.Y.; Barkov, N.I.; Duval, P.; Pimienta, P. Crystalline Texture of the 2083 m Ice Core at Vostok Station, Antarctica. *J. Glaciol.* **1989**, *35*, 392–398. [CrossRef]

23. Wang, Y.; Thorsteinsson, T.; Kipfstuhl, J.; Miller, H.; Dahl-Jensen, D.; Shoji, H. A vertical girdle fabric in the NorthGRIP deep ice core, North Greenland. *Ann. Glaciol.* **2002**, *35*, 515–520. [CrossRef]

24. Weikusat, I.; Kipfstuhl, S.; Faria, S.H.; Azuma, N.; Miyamoto, A. Subgrain boundaries and related microstructural features in EDML (Antarctica) deep ice core. *J. Glaciol.* **2009**, *55*, 461–472. [CrossRef]

25. Hudleston, P.J. Structures and fabrics in glacial ice: A review. *J. Struct. Geol.* **2015**, *81*, 1–27. [CrossRef]

26. Fan, S.; Hager, T.; Prior, D.J.; Cross, A.J.; Goldsby, D.L.; Qi, C.; Negrini, M.; Wheeler, J. Temperature and strain controls on ice deformation mechanisms: Insights from the microstructures of samples deformed to progressively higher strains at −10, −20 and −30°C. *Cryosphere Discuss* **2020**, *2020*, 1–43. [CrossRef]

27. Piazolo, S.; Wilson, C.J.L.; Luzin, V.; Brouzet, C.; Peternell, M. Dynamics of ice mass deformation: Linking processes to rheology, texture, and microstructure. *Geochem. Geophys. Geosyst.* **2013**, *14*, 4185–4194. [CrossRef]

28. Qi, C.; Goldsby, D.L.; Prior, D.J. The down-stress transition from cluster to cone fabrics in experimentally deformed ice. *Earth Planet. Sci. Lett.* **2017**, *471*, 136–147. [CrossRef]

29. Qi, C.; Prior, D.J.; Craw, L.; Fan, S.; Llorens, M.G.; Griera, A.; Negrini, M.; Bons, P.D.; Goldsby, D.L. Crystallographic preferred orientations of ice deformed in direct-shear experiments at low temperatures. *Cryosphere* **2019**, *13*, 351–371. [CrossRef]

30. Journaux, B.; Chauve, T.; Montagnat, M.; Tommasi, A.; Barou, F.; Mainprice, D.; Gest, L. Recrystallization processes, microstructure and crystallographic preferred orientation evolution in polycrystalline ice during high-temperature simple shear. *Cryosphere* **2019**, *13*, 1495–1511. [CrossRef]

31. Bouchez, J.L.; Lister, G.S.; Nicolas, A. Fabric asymmetry and shear sense in movement zones. *Geol. Rundsch.* **1983**, *72*, 401–419. [CrossRef]

32. Pälli, A.; Kohler, J.C.; Isaksson, E.; Moore, J.C.; Pinglot, J.F.; Pohjola, V.A.; Samuelsson, H. Spatial and temporal variability of snow accumulation using ground-penetrating radar and ice cores on a Svalbard glacier. *J. Glaciol.* **2002**, *48*, 417–424. [CrossRef]

33. Anschütz, H.; Eisen, O.; Oerter, H.; Steinhage, D.; Scheinert, M. Investigating small-scale variations of the recent accumulation rate in coastal Dronning Maud Land, East Antarctica. *Ann. Glaciol.* **2007**, *46*, 14–21. [CrossRef]

34. Bons, P.D.; Jansen, D.; Mundel, F.; Bauer, C.C.; Binder, T.; Eisen, O.; Jessell, M.W.; Llorens, M.-G.; Steinbach, F.; Steinhage, D.; et al. Converging flow and anisotropy cause large-scale folding in Greenland's ice sheet. *Nat. Commun.* **2016**, *7*, 11427. [CrossRef] [PubMed]

35. Udisti, R.; Barbante, C.; Castellano, E.; Vermigli, S.; Traversi, R.; Capodaglio, G.; Piccardi, G. Chemical characterisation of a volcanic event (about AD 1500) at Styx Glacier plateau, northern Victoria Land, Antarctica. *Ann. Glaciol.* **1999**, *29*, 113–120. [CrossRef]

36. Kwak, H.; Kang, J.-H.; Hong, S.-B.; Lee, J.; Chang, C.; Hur, S.D.; Hong, S. A Study on High-Resolution Seasonal Variations of Major Ionic Species in Recent Snow Near the Antarctic Jang Bogo Station. *Ocean. Polar Res.* **2015**, *37*, 127–140. [CrossRef]

37. Van de Velde, K.; Vallelonga, P.; Candelone, J.P.; Rosman, K.J.R.; Gaspari, V.; Cozzi, G.; Barbante, C.; Udisti, R.; Cescon, P.; Boutron, C.F. Pb isotope record over one century in snow from Victoria Land, Antarctica. *Earth Planet. Sci. Lett.* **2005**, *232*, 95–108. [CrossRef]

38. Stenni, B.; Serra, F.; Frezzotti, M.; Maggi, V.; Traversi, R.; Becagli, S.; Udisti, R. Snow accumulation rates in northern Victoria Land, Antarctica, by firn-core analysis. *J. Glaciol* **2000**, *46*, 541–552. [CrossRef]

39. Rignot, E.; Mouginot, J.; Scheuchl, B. *MEaSUREs InSAR-Based Antarctica Ice Velocity Map, Version 2*; NASA National Snow and Ice Data Center Distributed Active Archive Center: Boulder, CO, USA, 2017. [CrossRef]

40. Han, Y.; Jun, S.J.; Miyahara, M.; Lee, H.-G.; Ahn, J.; Chung, J.W.; Hur, S.D.; Hong, S.B. Shallow ice-core drilling on Styx glacier, northern Victoria Land, Antarctica in the 2014-2015 summer. *J. Geol. Soc. Korea* **2015**, *51*. [CrossRef]

41. Kim, S.-J. *Investigation of Climate Change Mechanism by Observation and Simulation of Polar Climate for the Past and Present (Rep. BSPE16010–146–12)*; Korea Polar Research Institute: Incheon, Korea, 2017.

42. Llorens, M.-G.; Griera, A.; Bons, P.D.; Roessiger, J.; Lebensohn, R.; Evans, L.; Weikusat, I. Dynamic recrystallisation of ice aggregates during co-axial viscoplastic deformation: A numerical approach. *J. Glaciol.* **2016**, *62*, 359–377. [CrossRef]

43. Herron, M.M.; Langway, C.C. Firn Densification: An Empirical Model. *J. Glaciol.* **1980**, *25*, 373–385. [CrossRef]

44. Yang, J.-W.; Han, Y.; Orsi, A.J.; Kim, S.-J.; Han, H.; Ryu, Y.; Jang, Y.; Moon, J.; Choi, T.; Hur, S.D.; et al. Surface Temperature in Twentieth Century at the Styx Glacier, Northern Victoria Land, Antarctica, From Borehole Thermometry. *Geophys. Res. Lett.* **2018**, *45*, 9834–9842. [CrossRef]

45. Udisti, R. Multiparametric Approach for Chemical Dating of Snow Layers from Antarctica. *Int. J. Environ. Anal. Chem.* **1996**, *63*, 225–244. [CrossRef]

46. Kovacs, A.; Gow, A.J.; Morey, R.M. The in-situ dielectric constant of polar firn revisited. *Cold Reg. Sci. Technol.* **1995**, *23*, 245–256. [CrossRef]

47. Kim, D.; Wallis, S.; Endo, S.; Ree, J.-H. Seismic properties of lawsonite eclogites from the southern Motagua fault zone, Guatemala. *Tectonophysics* **2016**, *677–678*, 88–98. [CrossRef]

48. Prior, D.J.; Lilly, K.; Seidemann, M.; Vaughan, M.; Becroft, L.; Easingwood, R.; Diebold, S.; Obbard, R.; Daghlian, C.; Baker, I.; et al. Making EBSD on water ice routine. *J. Microsc.* **2015**, *259*, 237–256. [CrossRef]

49. Mainprice, D.; Hielscher, R.; Schaeben, H. Calculating anisotropic physical properties from texture data using the MTEX open-source package. *Geol. Soc. Lond. Spec. Publ.* **2011**, *360*, 175. [CrossRef]

50. Mainprice, D.; Silver, P.G. Interpretation of SKS-waves using samples from the subcontinental lithosphere. *Phys. Earth Planet. Inter.* **1993**, *78*, 257–280. [CrossRef]

51. Bunge, H.J. *Texture Analysis in Materials Science: Mathematical Methods*; Butterworths: London, UK, 1982. [CrossRef]

52. Skemer, P.; Katayama, I.; Jiang, Z.; Karato, S.-I. The misorientation index: Development of a new method for calculating the strength of lattice-preferred orientation. *Tectonophysics* **2005**, *411*, 157–167. [CrossRef]

53. Vollmer, F.W. An application of eigenvalue methods to structural domain analysis. *GSA Bull.* **1990**, *102*, 786–791. [CrossRef]

54. Jung, H.; Karato, S. Water-induced fabric transitions in olivine. *Science* **2001**, *293*, 1460–1463. [CrossRef]

55. Obbard, R.; Baker, I.; Sieg, K. Using electron backscatter diffraction patterns to examine recrystallization in polar ice sheets. *J. Glaciol.* **2006**, *52*, 546–557. [CrossRef]

56. Hansen, L.N.; Zhao, Y.-H.; Zimmerman, M.E.; Kohlstedt, D.L. Protracted fabric evolution in olivine: Implications for the relationship among strain, crystallographic fabric, and seismic anisotropy. *Earth Planet. Sci. Lett.* **2014**, *387*, 157–168. [CrossRef]

57. Kim, D.; Katayama, I.; Wallis, S.; Michibayashi, K.; Miyake, A.; Seto, Y.; Azuma, S. Deformation microstructures of glaucophane and lawsonite in experimentally deformed blueschists: Implications for intermediate-depth intraplate earthquakes. *J. Geophys. Res. Solid Earth* **2015**, *120*, 1229–1242. [CrossRef]

58. Kipfstuhl, S.; Faria, S.H.; Azuma, N.; Freitag, J.; Hamann, I.; Kaufmann, P.; Miller, H.; Weiler, K.; Wilhelms, F. Evidence of dynamic recrystallization in polar firn. *J. Geophys. Res. Solid Earth* **2009**, *114*. [CrossRef]

59. Ligtenberg, S.R.M.; Helsen, M.M.; van den Broeke, M.R. An improved semi-empirical model for the densification of Antarctic firn. *Cryosphere* **2011**, *5*, 809–819. [CrossRef]

60. Kamb, B. Experimental Recrystallization of Ice Under Stress. *Flow Fract. Rocks* **1972**, 211–241. [CrossRef]

61. Gow, A.J.; Williamson, T. Rheological implications of the internal structure and crystal fabrics of the West Antarctic ice sheet as revealed by deep core drilling at Byrd Station. *GSA Bull.* **1976**, *87*, 1665–1677. [CrossRef]

62. Kamb, W.B. Ice petrofabric observations from Blue Glacier, Washington, in relation to theory and experiment. *J. Geophys. Res. (1896–1977)* **1959**, *64*, 1891–1909. [CrossRef]

63. Wilson, C.J.L.; Peternell, M. Ice deformed in compression and simple shear: Control of temperature and initial fabric. *J. Glaciol.* **2012**, *58*, 11–22. [CrossRef]

64. Llorens, M.-G.; Griera, A.; Steinbach, F.; Bons, P.D.; Gomez-Rivas, E.; Jansen, D.; Roessiger, J.; Lebensohn, R.A.; Weikusat, I. Dynamic recrystallization during deformation of polycrystalline ice: Insights from numerical simulations. *Philos. Trans. R. Soc. A Math. Phys. Eng. Sci.* **2017**, *375*, 20150346. [CrossRef]

65. Gow, A.J.; Meese, D. Physical properties, crystalline textures and c-axis fabrics of the Siple Dome (Antarctica) ice core. *J. Glaciol.* **2007**, *53*, 573–584. [CrossRef]

66. Obbard, R.; Baker, I. The microstructure of meteoric ice from Vostok, Antarctica. *J. Glaciol.* **2007**, *53*, 41–62. [CrossRef]

67. Montagnat, M.; Azuma, N.; Dahl-Jensen, D.; Eichler, J.; Fujita, S.; Gillet-Chaulet, F.; Kipfstuhl, S.; Samyn, D.; Svensson, A.; Weikusat, I. Fabric along the NEEM ice core, Greenland, and its comparison with GRIP and NGRIP ice cores. *Cryosphere* **2014**, *8*, 1129–1138. [CrossRef]

68. Esedoğlu, S. Grain size distribution under simultaneous grain boundary migration and grain rotation in two dimensions. *Comput. Mater. Sci.* **2016**, *121*, 209–216. [CrossRef]

69. Doherty, R.D.; Hughes, D.A.; Humphreys, F.J.; Jonas, J.J.; Jensen, D.J.; Kassner, M.E.; King, W.E.; McNelley, T.R.; McQueen, H.J.; Rollett, A.D. Current issues in recrystallization: A review. *Mater. Sci. Eng. A* **1997**, *238*, 219–274. [CrossRef]

70. Weikusat, I.; DA, D.E.W.; Pennock, G.M.; Hayles, M.; Schneijdenberg, C.T.; Drury, M.R. Cryogenic EBSD on ice: Preserving a stable surface in a low pressure SEM. *J. Microsc.* **2011**, *242*, 295–310. [CrossRef]

71. Seidemann, M.; Prior, D.J.; Golding, N.; Durham, W.B.; Lilly, K.; Vaughan, M.J. The role of kink boundaries in the deformation and recrystallisation of polycrystalline ice. *J. Struct. Geol.* **2020**, *136*, 104010. [CrossRef]

72. Wheeler, J.; Mariani, E.; Piazolo, S.; Prior, D.J.; Trimby, P.; Drury, M.R. The weighted Burgers vector: A new quantity for constraining dislocation densities and types using electron backscatter diffraction on 2D sections through crystalline materials. *J. Microsc.* **2009**, *233*, 482–494. [CrossRef]

73. Chauve, T.; Montagnat, M.; Barou, F.; Hidas, K.; Tommasi, A.; Mainprice, D. Investigation of nucleation processes during dynamic recrystallization of ice using cryo-EBSD. *Philos. Trans. Math. Phys. Eng. Sci.* **2017**, *375*. [CrossRef]

74. Wei, Y.; Dempsey, J.P. The motion of non-basal dislocations in ice crystals. *Philos. Mag.* **1994**, *69*, 1–10. [CrossRef]

75. Weikusat, I.; Kuiper, E.-J.N.; Pennock, G.M.; Kipfstuhl, S.; Drury, M.R. EBSD analysis of subgrain boundaries and dislocation slip systems in Antarctic and Greenland ice. *Solid Earth* **2017**, *8*, 883–898. [CrossRef]

76. Kirchner, N.; Hutter, K.; Jakobsson, M.; Gyllencreutz, R. Capabilities and limitations of numerical ice sheet models: A discussion for Earth-scientists and modelers. *Quat. Sci. Rev.* **2011**, *30*, 3691–3704. [CrossRef]

77. Maeno, N.; Ebinuma, T. Pressure sintering of ice and its implication to the densification of snow at polar glaciers and ice sheets. *J. Phys. Chem.* **1983**, *87*, 4103–4110. [CrossRef]

78. Spaulding, N.E.; Meese, D.A.; Baker, I. Advanced microstructural characterization of four East Antarctic firn/ice cores. *J. Glaciol.* **2011**, *57*, 796–810. [CrossRef]
79. Winkelmann, R.; Levermann, A.; Martin, M.A.; Frieler, K. Increased future ice discharge from Antarctica owing to higher snowfall. *Nature* **2012**, *492*, 239. [CrossRef]

Article

Thermo-Structural Evolution of the Val Malenco (Italy) Peridotite: A Petrological, Geochemical and Microstructural Study

Wenlong Liu [1,2], Yi Cao [1], Junfeng Zhang [1,*], Yanfei Zhang [3], Keqing Zong [1] and Zhenmin Jin [1,2]

[1] State Key Laboratory of Geological Processes and Mineral Resources, School of Earth Sciences, China University of Geosciences, Wuhan 430074, China; wenlongliu@cug.edu.cn (W.L.); caoyi0701@126.com (Y.C.); kqzong@hotmail.com (K.Z.); zmjin@cug.edu.cn (Z.J.)

[2] Institute of Geophysics & Geomatics, China University of Geosciences, Wuhan 430074, China

[3] School of Earth Sciences and Engineering, Hohai University, Nanjing 210098, China; yanfzhang@hhu.edu.cn

* Correspondence: jfzhang@cug.edu.cn

Received: 9 August 2020; Accepted: 21 October 2020; Published: 28 October 2020

Abstract: The Val Malenco peridotite massif is one of the largest exposed ultramafic massifs in Alpine orogen. To better constrain its tectonic history, we have performed a comprehensive petro-structural and geochemical study. Our results show that the Val Malenco serpentinized peridotite recorded both pre-Alpine extension and Alpine convergence events. The pre-Alpine extension is recorded by microstructural and geochemical features preserved in clinopyroxene and olivine porphyroblasts, including partial melting and refertilisation, high-temperature (900–1000 °C) deformation and a cooling, and fluid-rock reaction. The following Alpine convergence in a supra-subduction zone setting is documented by subduction-related prograde metamorphism features preserved in the coarse-grained antigorite and olivine grains in the less-strained olivine-rich layers, and later low-temperature (<350 °C) serpentinization in the fine-grained antigorite in the more strained antigorite-rich layers. The strain shadow structure in the more strained antigorite-rich layer composed of dissolving clinopyroxene porphyroblast and the precipitated oriented diopside and olivine suggest dissolution and precipitation creep, while the consistency between the strain shadow structure and alternating less- and more-strained serpentinized domains highlights the increasing role of strain localization induced by the dissolution-precipitation creep with decreasing temperature during exhumation in Alpine convergence events.

Keywords: Val Malenco; serpentinized peridotite; tectonic evolution; deformation; strain localization

---

1. Introduction

Recent years have seen a great number of studies on peridotite massifs [1–8], the peridotite shear zone [9–11], and mantle xenoliths [12–16] aiming at exploring the evolution of the lithospheric mantle and mantle wedge. These studies demonstrate intimate relationships between deformation, syn-kinematic P-T conditions, mineralogy, and chemistry in the upper mantle. Compared with shear zones and mantle xenoliths, the large size of the peridotite massif allows an integration of the deformation structures into the mantle lithosphere [17], because microstructures are easily overprinted in the shear zone during emplacement/exhumation, and information is limited for mantle xenoliths given the heterogeneity of lithospheric mantle. In addition, benefiting from its enormous size, multi-stage tectonic history can be constrained or recovered from the field tectonic relationship and heterogeneous deformation in the massif.

Many integrated structural, petrological, and geochemical studies have greatly improved our understanding of the evolutions of peridotite massifs and their represented upper mantle in the Alps

(Lanzo massif [18,19], Finero Complex [20,21], Cima di Gagnone Massif [7,22]). These peridotite massifs have preserved relics of an earlier tectono-metamorphic evolution related to Jurassic rifting, continental breakup, and development within the Piemonte-Ligurian Ocean which represents part of the Mesozoic Tethys now occurring as ophiolites in the Alps-Apennine orogenic system [23]. However, the tectonic evolution of the Val Malenco ultramafic massif, the largest exposed ultramafic massif in the Alps, is less clear, mainly because of extensive serpentinization in the ocean basin after Jurassic rifting [24] and strong tectono-metamorphic overprinting resulted from the Alpine convergence [25]. Accordingly, previous studies focused on the petrology of the metamorphic events [25–28]. The tectonic evolution of the Val Malenco ultramafic massif is mainly inferred from geochemical studies [29–31], while direct microstructural studies on the peridotite are sparse [3].

Here, we present a comprehensive study on the Val Malenco serpentinized peridotite massif, including microstructural, petrological, and geochemical analyses. The coupled analyses of deformation and metamorphism/metasomatism enabled us to construct an integrated tectonic evolution history of the Val Malenco massif with the emphasis on improving the understanding of deformation processes at a supra-subduction zone setting.

2. Previous Understanding of Tectonic History of the Val Malenco Unit

The Val Malenco unit is part of the Central Alpine mountain belt which represents the oceanic lithosphere of Piemonte-Ligurian Tethys, a small oceanic basin separating the Europe plate and the Adria plate, and pieces of the paleo-Europe and paleo-Africa continental thinned margins during the Mesozoic period [23]. It is overlain by the Margna nappe and underlain by the Suretta nappe, which has a pre-Mesozoic basement and a Mesozoic cover, respectively, suggesting a Mesozoic age for the Val Malenco Unit (Figure 1) [29,32]. The main part of the ultramafic peridotite was uplifted from garnet-facies conditions and equilibrated at the spinel-facies conditions [27]. Afterward, lower crustal rocks were welded into the Val Malenco ultramafic rocks (subcontinental mantle) by gabbroic intrusions at ~270 Ma [29]. During the Jurassic rifting, ultramafic rocks were rapidly uplifted by near-isothermal decompression and exposed in the Piemontese ocean basin near the Adriatic margin, experienced extensive low-temperature serpentinization, and developed an Alpine type ophiolite suite [30]. Accompanying the uplifting, melt/fluid infiltration and melt/fluid reaction modified the ancient subcontinental mantle coeval with the incipient opening of oceanic basins [27,28]. During the Cretaceous convergence of Alpine metamorphism, the emplaced and serpentinized Val Malenco unit was moderately subducted and experienced pervasive metamorphism which was characterized by metamorphosed serpentinite containing a foliated titanian-clinohumite-bearing magnetite-chlorite-diopside-olivine-antigorite mineral assemblage, indicating greenschist-facies metamorphic conditions (460 °C and 0.6 GPa) in the mantle wedge [25,33]. Finally, the exhumation of the massif occurred between 67–73 Ma (the Ar^{39}-Ar^{40} ages of amphibole [30,31]) and 32 Ma (the age of the Bergell tonalite intrusion [34]).

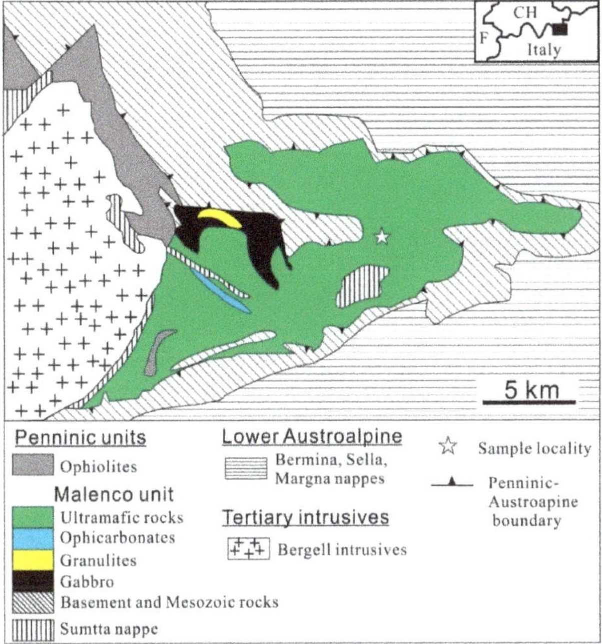

Figure 1. A simplified geological map of the Malenco-Forno unit and sample locality (Franscia, Swiss coordinate: 790.0/129.4). The geological map is modified after Hermann et al. [29]. F: France; CH: Switzerland.

3. Methods

Three serpentinized lherzolite samples were collected from Franscia, Italy (Figure 1). Rock slabs of each sample were cut based on the layering structure and the shape preferred orientations (SPO) of antigorite or olivine before they were doubly polished into 30 μm thick thin sections. The thin sections were treated with vibration polishing using 0.05 μm colloidal silica for more than 4 h before analyses.

Major-element analyses of minerals were performed at the Key Laboratory of Submarine Geosciences, State Oceanic Administration, Hangzhou with a JEOL electron microprobe (Superprobe JXA-8100) and at the China University of Geosciences (Wuhan) using a similar JEOL electron microprobe (Superprobe JXA-8230). Both microprobes are equipped with four wavelength-dispersive spectrometers. The analyses were carried out with a 15 kV accelerating voltage, a 5 μm width beam with a 20 nA beam current, and a counting time of 30 s for peaks and 10 s for backgrounds. All the analytical results were calibrated by natural olivine and diopside standards from the SPI Supplies®.

In-situ trace element analyses were performed using laser ablation inductively coupled plasma-mass spectrometry (LA-ICP-MS) at the State Key Laboratory of Geological Processes and Mineral Resources (GPMR) of the China University of Geosciences in Wuhan. The ion signal intensities were measured using an Agilent 7500a ICP-MS equipped with a 193 nm ArF excimer laser (GeoLas 2005). Helium was used as the carrier gas, and argon was used as the make-up gas. Each analysis included approximately 20–30 s of background acquisition (from a gas blank) and 50 s of data acquisition from the sample. The element concentrations were calibrated against multiple reference materials (BCR-2G, BIR-1G and BHVO-2G), and a summed metal oxide normalization was applied [35].

The crystallographic preferred orientations (CPOs) and orientation maps were acquired using a Quanta 450 Field Emission Gun (FEG) -scanning electron microscope (SEM) equipped with an HKL Nordlys electron backscattered diffraction (EBSD) detector housed at the GPMR. An accelerating voltage of 20 kV, a spot size of 6, and a working distance of 22 mm were used. The analytical conditions for all measurements have been optimized under a low vacuum condition on non-coated samples to obtain high-quality electron backscattered patterns. The maximum accepted angular deviation for measurements was 1.2°, with average values ranging from 0.73 to 0.88°. Orientation maps were obtained in automatic acquisition mode with a step size of 0.3–6 µm. The data were then noise-reduced using a "wildspike" correction and a five-neighbor zero solution extrapolation to fill non-indexed pixels based on neighboring pixels using the CHANNEL5s software (Version 4.3) (see Liu et al. [36] for more details).

The active slip systems in plastic deformed crystals could be determined by analyzing grain orientation data (Reference [37] and references therein). In general, two end-member boundaries are commonly used to describe the relationship between dislocations and low-angle boundaries: tilt boundaries and twist boundaries. A tilt boundary is composed of edge dislocations whose crystallographic rotation axes are mutually perpendicular to slip directions and poles of the slip planes. In this case, both the crystallographic rotation axes and the poles of slip planes lie in the tilt boundary. In contrast, a twist boundary is formed by two or more sets of screw dislocations, whose rotation axes and poles to slip planes are perpendicular to the boundary plane.

4. Petrography

The serpentinized peridotites are characterized by alternating olivine-rich layers and antigorite-rich layers (Figure 2a). The antigorite-rich layers are composed of fine-grained antigorite (>90 vol.%, Atg^F in which the subscript F refers to fine-grained), olivine (5–10 vol.%) (Ol^F) and minor magnetite (Figure 2b). The olivine-rich layers contain olivine (80–85 vol.%), clinopyroxene (5–10 vol.%), antigorite (5–10 vol.%), and magnetite. Chlorite and titanium clinohumite (Ti-Chu) are also observed in the olivine-rich layers (Figure 2a,c).

In the serpentinized peridotite, olivine can be classified into five types based on grain size and occurrence: porphyroblast (Ol^P), coarse-grained (Ol^C), and small-grained (Ol^S) olivine in the olivine-rich layers, fine-grained olivine (Ol^F) in the antigorite-rich layers, and olivine along cracks, the cleavage plane, and grain boundaries (Ol^{cpx}) of clinopyroxene porphyroblasts (Figure 2b–f). The Ol^P grains (0.2–1 mm in diameter), although rare, can be observed in both antigorite-rich and olivine-rich layers. They often show undulatory extinction (Figure 2c). The Ol^C grains (~130 µm in diameter) have straight grain boundaries and well-developed triple junctions (Figure 2d). They often intergrow with coarse-grained antigorite, showing a lepidoblastic-granoblastic texture (Figure 2d). The Ol^S grains (~60 µm in diameter) have curved grain boundaries. Clinopyroxene and olivine porphyroblasts can be observed in small-grained areas (Figure 2e). However, both Ol^C and Ol^S grains are free of undulatory extinction (Figure 2d,e).

Two occurrences of clinopyroxene are observed: porphyroblast (Cpx^P) and diopside neoblast (Di). The porphyroblasts, mainly occurring in the serpentine matrix or small-grained area, often show a dusty core of abundant minute magnetite exsolution lamellae and obvious undulatory extinction (Figure 2g,h). The Ol^{cpx} grains usually surround the Cpx^P grains as well as filling up the fractures (Figure 2f,h). Diopside neoblasts in the olivine-rich layers or together with the Cpx^P grains are tabular-shaped euhedral grains free of intracrystalline deformation features. They show clear chemical zonation (Figure 2i) with grain sizes of 10 to 100 µm (on average 30–40 µm).

Coarse-grained antigorite (Atg^C) intergrows with the Ol^C in the olivine-rich layers. They also have euhedral tabular grain shape and chemical zonation (Figure 2d). In the antigorite-rich layers, fine-grained antigorite (Atg^F) (grain size < 50 µm) shows strong SPOs (Figure 2b).

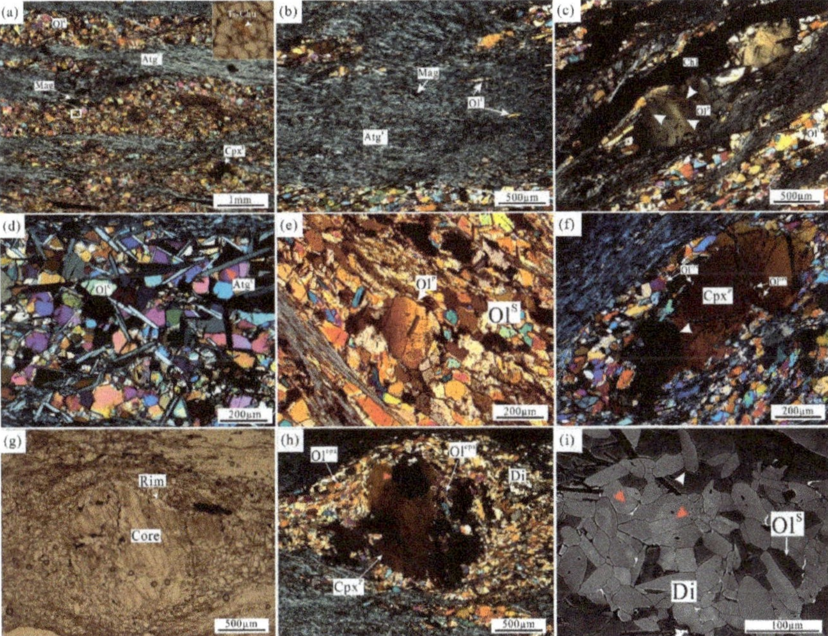

Figure 2. Microstructures of serpentinized peridotite. (**a**) Serpentinized peridotite composed of alternating antigorite-rich layers and olivine-rich layers. The white box is enlarged at the upper right corner to show the occurrence of clinohumite. (**b**) An antigorite-rich layer showing elongated fine-grained olivine (Ol^F) and antigorite (Atg^F). (**c**) An olivine porphyroblast (Ol^P) showing clear undulatory extinction (white arrowheads) and fractures. (**d**) the lepidoblastic-granoblastic texture in an olivine-rich layer, composed of coarse-grained olivine (Ol^C) and antigorite (Atg^C). (**e**) Small-grained olivine (Ol^S) with sinuous grain boundaries. (**f**) Fine-grained olivine surrounding a clinopyroxene porphyroblast and within its fractures (Ol^{cpx}). (**g**) A clinopyroxene porphyroblast with cloudy core and narrow clean rim. (**h**) The same clinopyroxene porphyroblast as in (**g**) showing undulatory extinction and surrounding fine-grained olivine and diopside. (**i**) Lenticular diopside aggregate in an olivine-rich layer showing an embayment structure (white arrowhead) and growth zonations (red arrowheads). Crossed polarized light images (**a**–**f**,**h**); polarized light image (**g**); back-scattered electron image (**i**). Atg: antigorite; Chl: chlorite; Cpx: clinopyroxene; Di: diopside; Mag: Magnetite; Ol: olivine; Ti-Chu: Ti-clinohumite.

5. Deformation Microstructures

5.1. Low-Angle Boundaries and Misorientation Axes of Porphyroblasts

Because the number of porphyroblasts is too few to measure a meaningful CPO, we used crystallographic orientations of low-angle boundaries and misorientation axes within grains to constrain dislocation slip systems activated in olivine and clinopyroxene porphyroblasts.

The clinopyroxene porphyroblasts display continuous crystallographic orientation gradients and low-angle boundaries (misorientation angles between 2° and 10°) (Figure 3a). The rotation of the crystalline lattice could be as large as 20° over a distance of 400 μm. A misorientation profile across a clinopyroxene porphyroblast shows continuous orientation rotation (Figure 3b). The geometry relationship between the trace of low-angle boundaries and cluster of rotation axes suggests that the slip systems responsible for the formation of low-angle boundaries are the (010)[001] and the (100)[001] (Figure 3c), according to the method described in Reddy et al. [37]. The analyses of many clinopyroxene porphyroblasts suggest that the (010)[100] slip system is dominant. Similarly, the olivine porphyroblasts also display continuous crystallographic orientation gradients and well-defined low-angle boundaries (Figure 3d). The low-angle boundaries are commonly subparallel with 15–30 μm spacing. Strong intracrystalline deformation is indicated by orientation distortion of about 20 degrees within the grain (Figure 3e). The misorientation profile across the low-angle boundaries from rim to center shows a progressive change in crystallographic orientation (Figure 3e). The inferred slip systems are the (010)[100], the (021)[100], the (041)[100] and the (001)[100] (Figure 3f). The dominant slip system is the (010)[001] in olivine porphyroblasts.

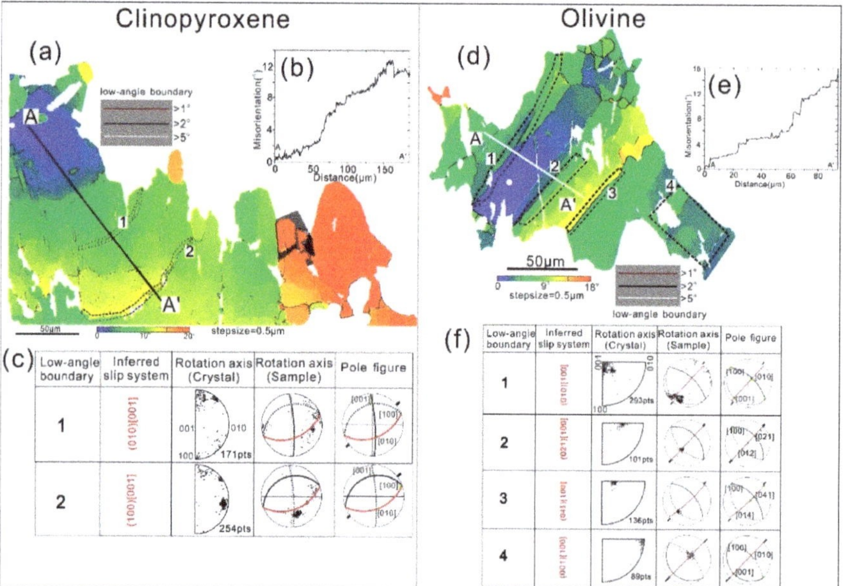

Figure 3. Orientation analyses of clinopyroxene (**a–c**) and olivine (**d–f**) porphyroblasts. (**a,d**) Orientation maps are colored for disorientation angle from a reference point (the white point). Boundaries above 1° are shown. (**b,e**) Misorientation profiles along the AA′ lines in (**a,d**) respectively, suggesting a continuous intracrystalline deformation. (**c,f**) Slip system analyses based on the method from Reddy et al. [37]. Subsets of disorientation axes data are from the dotted areas (#1,2) in (**a**) and (#1–4) in (**d**). The pole figures are lower-hemisphere equal-area projections in (**c,f**). The boundary trace orientation (shown as thick black lines outside the primitive circle), the inferred boundary wall (red lines), and slip plane (dotted lines) have been added assuming a tilt boundary geometry.

5.2. Topotaxial Orientation Analysis

The clinopyroxene porphyroblasts have abundant minute lamellae of magnetite. There is a clear topotaxial relationship between clinopyroxene porphyroblast and magnetite lamellae (Figure 4), which is characterized by $(100)_{Cpx}//(111)_{Mag}$, $(010)_{Cpx}//[\bar{1}10]_{Mag}$, $[001]_{Cpx}//[\bar{1}\bar{1}2]_{Mag}$, $[101]_{Cpx}//[112]_{Mag}$, $(\bar{1}01)_{Cpx}//(\bar{1}\bar{1}1)_{Mag}$. It is noteworthy that the dispersion of the crystallographic axis of magnetite is consistent with the crystallographic axis dispersion of host clinopyroxene, implying that the exsolution took place after intracrystalline deformation.

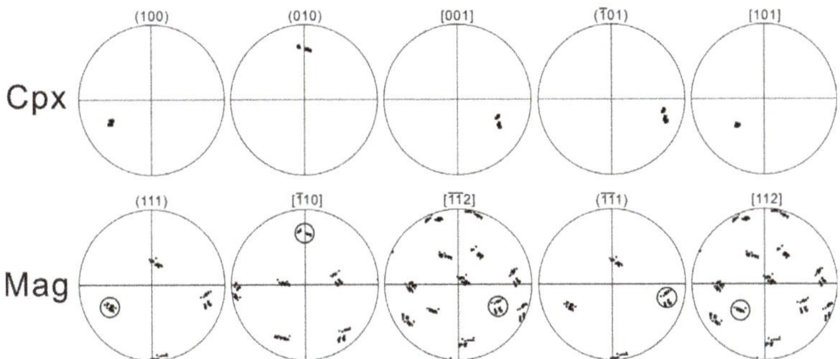

Figure 4. Crystallographic orientations of clinopyroxene porphyroblast and exsolved magnetite. The orientations of magnetite (circled) are consistent with the orientation of the clinopyroxene host. All orientations are presented in the upper hemisphere using a stereographic projection.

5.3. Pressure Shadow Structure

The pressure shadow structures of clinopyroxene porphyroblasts are presented in Figure 2g,h and Figure 5. Figure 5a is a clinopyroxene porphyroblast cut through by olivine veins (Figure 5a). The contact boundary between the olivine vein and the porphyroblast is sinuous. Growth zonation can be observed on both sides of the veins (Figure 5b). The chemical composition profile across the growth zonation displays abrupt variation in major element contents, characterized by the increase of Ca content and the decrease of Al, Cr, Na, and Fe contents (Figure 5e). Fine-grained diopside aggregates are distributed at the lower left and upper right corners of the porphyroblast, while fine-grained olivine aggregates occur at the lower right and upper left corners (Figure 5a). The diopside grains in fine-grained aggregate have brighter cores of higher Ca contents, while the olivine grains in fine-grained aggregate have grayer cores of higher Mg# (~92) (Figure 5c,d).

The orientation map shows that the clinopyroxene porphyroblast is split by olivine veins into three domains containing low-angle boundaries (Figure 6a). Strong intracrystalline deformation is well preserved in clinopyroxene porphyroblast, while the fine-grained diopsides are almost free of intracrystalline deformation, as indicated by the mis2mean misorientation map (Figure 6b). The pole figures of fine-grained diopside aggregate show pronounced CPOs with the [001] and the [100] axes of clinopyroxene clustering at low angles to the lineation and to the foliation normal, respectively (Figure 6c). The orientations of fine-grained diopside are very different from the crystallographic orientation of the clinopyroxene porphyroblast. The orientations of the [100]$_{cpx}$ and [001] clusters indicate a left lateral shear strain around the clinopyroxene porphyroblast.

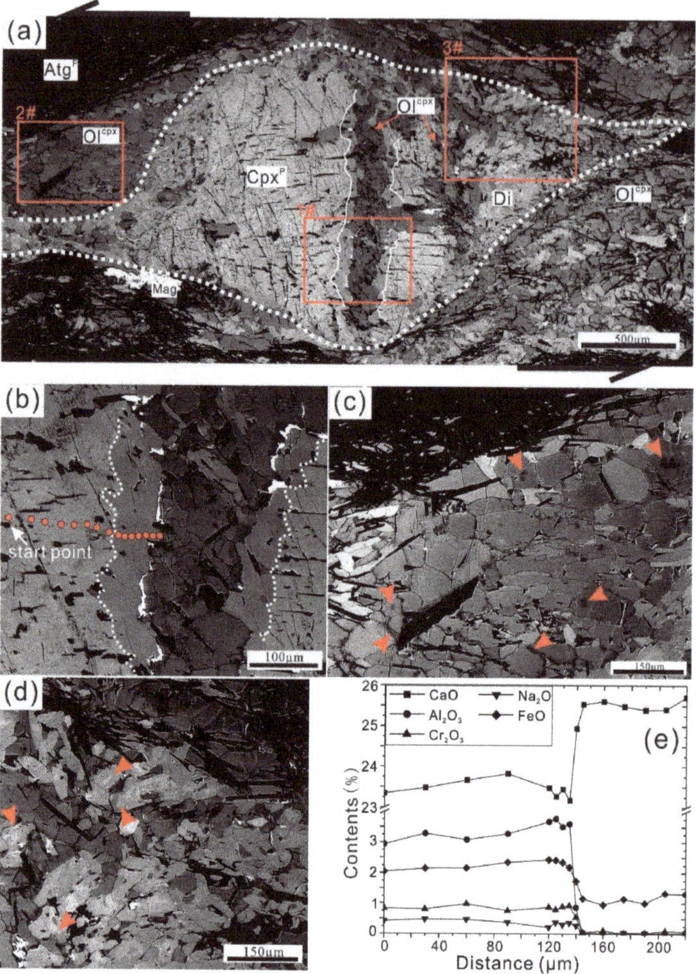

Figure 5. The pressure shadow structure showed by BSE images (**a–d**) and the major element profile (**e**). (**a**) An overview of the pressure shadow structure composed of clinopyroxene porphyroblast, diopside, and olivine. Note the abundance of diopside at the lower-left and upper-right corners, while olivine mainly exists at the lower-right and upper-left corners. The clinopyroxene porphyroblast is dismembered by olivine veins. Red rectangles #1–3 are enlarged and shown in (**b–d**). (**b**) The sinuous contact boundary between the olivine vein and clinopyroxene porphyroblast. The growth zonation is delineated by white dotted curves. The red dots represent the EMPA analytical sites. The corresponding major element profile is shown in (**e**) (see Table S2 for chemical composition data). (**c**) The enlarged olivine growth zonation shows olivine grains with a dark core (Mg# = ~92) (red arrowheads) and bright rim (Mg# = ~86) (see Table S1 for chemical composition data). (**d**) The enlarged diopside growth zonation shows diopside grains with a bright core and dark rim (red arrowheads) (see Table S2 for chemical composition data). (**e**) The major element profile along the red dotted line in (**b**), displaying a sharp change in composition.

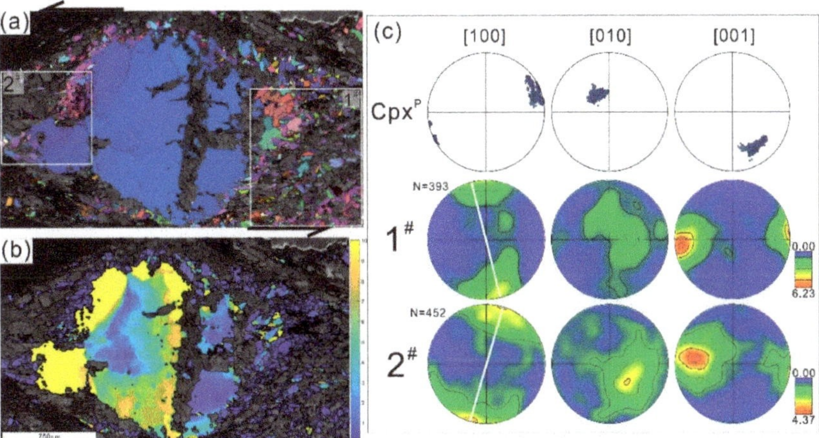

Figure 6. Orientation maps of the clinopyroxene porphyrobalst in Figure 5. (**a**) A Euler map of clinopyroxene overlaid on a band contrast map. (**b**) A mis2mean map of clinopyroxene overlaid on a band contrast map. (**c**) The orientation of the clinopyroxene porphyroblast and pole figures of diopside grains from zone #1 and #2 (**a**). The pole figures are plotted with one point per grain and are presented in the lower hemisphere by using an equal-area projection. A half-width of 20° and a cluster size of 5 are used. The white solid lines in the pole figures delineate the orientations of the $[100]_{cpx}$ clusters.

6. Major and Trace Element Geochemistry in Minerals

6.1. Olivine

Different types of olivine have similar major element compositions (Mg# = 86.2–86.8) (Table 1). The CaO contents are <0.01 wt.% to 0.19 wt.% in Ol^S and Ol^{Cpx} and negligible in Ol^P, Ol^C and Ol^F, respectively. Compared with normal mantle olivine (Mg# = 89–92, NiO = 0.2–0.35 wt.%) [38], olivine in the Val Malenco serpentinized peridotite has lower Mg# and higher NiO contents (0.277–0.46 wt.%) (Figure 7a).

Table 1. Major element compositions of minerals in Val Malenco serpentinized peridotite.

Sample#	V-2				V-3			V-1	V-2				V-3		V-4	V-1	V-2			V-3	
Minerals	Ol^P	Ol^C	Ol^S	Ol^{Cpx}	Ol^F	Ol^C	Ol^S	Cpx^P	Cpx^P	Cpx^P	Di	Di	Cpx^P	Cpx^P	Cpx^P	Atg^F	Atg^C	Atg^C	Atg^F	Atg^C	Atg^C
								core	core	rim	core	rim	core	rim	core	core	core	rim	core	core	rim
SiO_2	40.32	39.91	40	40.23	40.10	40.62	40.41	52.81	53.01	54.16	53.61	55.17	52.24	54.93	51.87	43.41	41.81	42.43	42.18	41.71	43.11
TiO_2	-	0.01	0.02	0.01	-	-	0.02	0.07	0.09	0.03	0.06	0.02	0.06	-	0.12	-	0.02	-	-	0.03	0.01
Al_2O_3	-	0.02	0.01	0.01	0.02	-	0.01	3.37	2.66	0.78	1.83	0.01	3.23	0.03	2.06	1.48	2.41	1.5	1.77	2.46	0.88
FeO	12.95	12.67	12.75	13.16	12.69	12.7	12.87	2.29	2.01	1.63	2.15	1.20	2.20	1.08	2.08	3.62	4.02	3.13	3.53	4.39	3.04
MnO	0.17	0.27	0.24	0.20	0.28	0.26	0.23	0.07	0.04	0.06	0.04	0.03	0.05	0.11	0.03	0.06	0.03	0.01	0.02	0.04	0.01
MgO	46.71	46.72	46.58	45.95	46.75	46.0	46.14	16.52	17.31	17.89	17.60	17.93	16.94	17.79	18.30	38.2	37.22	39.06	37.84	36.79	39.27
CaO	0.01	0.01	0.08	0.05	0.01	-	0.01	23.58	23.79	25.10	23.78	25.66	23.39	25.61	23.67	-	0.01	0.01	0.02	-	0.01
Na_2O	0.01	-	0.01	-	0.01	0.02	0.01	0.23	0.41	0.23	0.21	0.03	0.44	0.06	0.46	0.01	0.02	-	0.01	-	-
Cr_2O_3	0.02	-	0.02	0.01	0.03	0.04	-	0.88	0.82	0.42	0.53	0.09	0.84	0.10	0.92	-	0.64	0.06	0.5	0.59	0.03
NiO	0.35	0.36	0.37	0.37	0.35	0.38	0.44	0.07	0.03	0.03	0.04	0.03	0.07	0.05	0.01	0.11	0.19	0.04	0.11	0.18	0.04
Total	100.54	99.98	100.06	99.99	100.21	99.98	100.13	99.89	100.20	100.45	99.86	100.20	99.45	99.75	99.57	86.89	86.36	86.26	85.99	86.2	86.39
Cation/O	4	4	4	4	4	4	4	3	3	3	3	3	3	3	3	116	116	116	116	116	116
Si	1.00	0.99	1.00	1.00	1.00	1.01	1.01	1.92	1.92	1.96	1.95	2.00	1.91	2.00	1.88	33.68	32.86	33.15	33.17	32.89	33.57
Ti	-	-	-	-	-	-	-	-	-	-	-	-	-	-	-	-	0.01	-	-	0.02	0.01
Cr	-	-	-	-	-	-	-	0.03	0.02	0.01	0.02	-	0.02	-	0.03	-	0.4	0.04	0.31	0.36	0.02
Al	-	-	-	-	-	-	-	0.14	0.11	0.03	0.08	-	0.14	-	0.09	1.35	2.23	1.38	1.64	2.28	0.81
Fe^{3+}	-	0.01	0.01	-	0.01	-	-	-	0.05	0.04	0.02	0.01	0.04	0.01	0.06	-	-	-	-	-	-
Fe^{2+}	0.27	0.25	0.26	0.27	0.26	0.26	0.27	0.07	0.02	0.01	0.04	0.03	0.02	0.03	-	2.35	2.64	2.05	2.32	2.89	1.98
Mn	-	0.01	0.01	-	0.01	0.01	-	-	-	-	-	-	-	-	-	0.04	0.02	0.01	0.01	0.03	-
Mg	1.73	1.73	1.73	1.71	1.73	1.71	1.72	0.90	0.93	0.96	0.95	0.97	0.92	0.96	0.99	44.19	43.6	45.49	44.36	43.24	45.6
Ca	-	-	-	-	-	-	-	0.92	0.92	0.97	0.93	0.99	0.92	1.00	0.92	-	0.01	0.01	0.01	-	0.01
Na	-	-	-	-	-	-	-	0.02	0.03	0.02	0.01	-	0.03	-	0.03	0.02	0.03	-	0.02	-	-
Ni	-	-	-	-	-	-	-	-	-	-	-	-	-	-	-	0.03	0.06	0.01	0.03	0.06	0.01
Total	3.00	3.00	3.00	3.00	3.00	3.00	3.00	4.00	4.01	4.00	4.01	4.00	4.02	4.00	4.00	81.66	81.85	82.14	81.88	81.77	82.01
Mg#	86.5	86.8	86.7	86.2	86.8	86.6	86.5	92.8	93.9	95.2	93.6	96.4	93.2	96.7	94.0	95	94	96	95	94	96

Mg# = $Mg^{2+}/(Mg^{2+} + Fe^{2+}) \times 100$; - below detection limit. Mineral abbreviations are the same as described in the text.

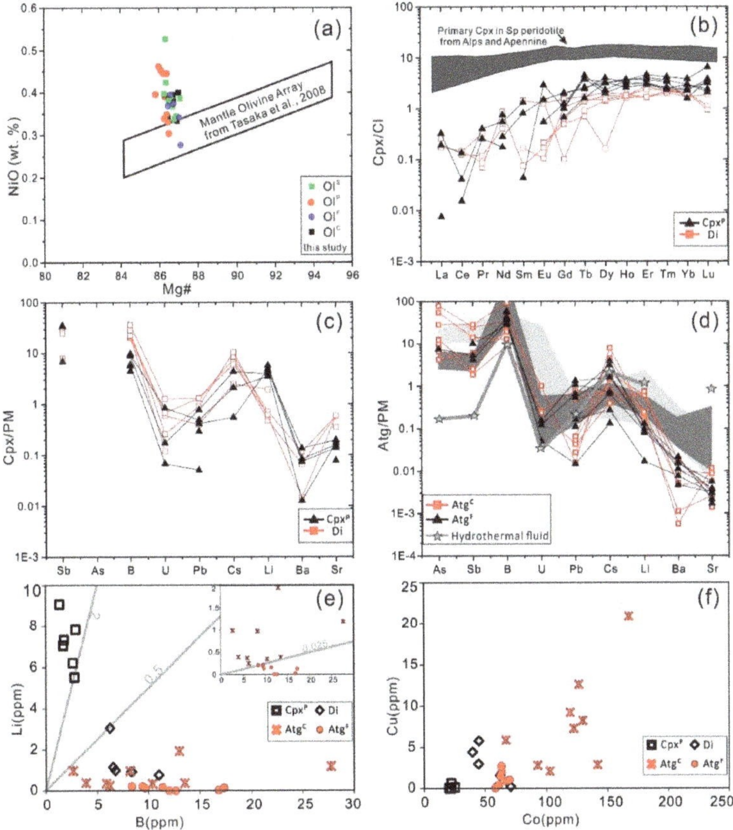

Figure 7. Major and trace element compositions of olivine, clinopyroxene, and antigorite in Val Malenco serpentinized peridotite. (**a**) Mg# (Mg/(Mg+Fe)) vs. NiO (%) in olivine from Val Malenco (this study). The parallelogram represents the values of NiO variation with Mg# for mantle olivine [11]. (**b**) REE in clinopyroxene compared to those in clinopyroxene (dark grey zone) of spinel peridotites from the Alps and Apennines [28]. (**c**) Fluid mobile elements (FMEs) in clinopyroxene. (**d**) FMEs in antigorite. The FMEs in hydrothermal fluids in the ocean [39] are shown by grey stars. FMEs in antigorite from the subduction channel at relatively high temperatures (350–400 °C) (light grey zone) and the mantle wedge at low temperatures (300–350 °C) (dark grey zone) [40] are also shown. (**e**) B and Li contents of clinopyroxene and antigorite. The B vs. Li contents of antigorite are enlarged at the upper-right corner, showing coarse-grained antigorite is enriched in Li (Li/B > 0.025). The grey lines represent the ratios of Li/B. (**f**) Cu and Co contents of clinopyroxene and antigorite. All the composition data are included in Tables S1–S4.

6.2. Clinopyroxene

The major element compositions of clinopyroxene in the Val Malenco serpentinized peridotite are given in Table 1. The cores of clinopyroxene porphyroblasts (Mg# = 92.8–94.0) have a diopside composition, containing 2.06–3.37 wt.% Al_2O_3, 0.82–0.92 wt.% Cr_2O_3, 0.06–0.12 wt.% TiO_2 and 0.23–0.46 wt.% Na_2O. The diopside neoblasts have higher Mg# (93.3–97.0) and lower TiO_2 (<0.01–0.07 wt.%), Al_2O_3 (<0.01–2.3 wt.%), Cr_2O_3 (<0.01–0.59 wt.%), and Na_2O (<0.01–0.37 wt.%) contents.

The rare earth element (REE) compositions of clinopyroxene in the Val Malenco serpentinized peridotite are characterized by depleted light rare earth elements (LREE) (Figure 7b). The REE contents are much lower than those of clinopyroxenes in the primary lithospheric mantle [28].

The clinopyroxenes show similar patterns of fluid mobile elements (FME), characterized by the enrichment of Sb, B, Cs, and Li relative to the primitive mantle (Figure 7c). Compared to diopside neoblasts, clinopyroxene porphyroblasts have higher Li contents and Li/B ratios (Figure 7e), and higher Co and Cu contents (Figure 7f).

6.3. Antigorite

The cores of Atg^C in the olivine-rich layers have lower Mg# (94) and higher Al_2O_3 contents (2.41–2.46 wt.%), compared with the rim and Atg^F in the antigorite-rich layers (Table 1). The antigorite is enriched in FMEs of As, Sb, and B relative to the primitive mantle (Figure 7d). Compared with Atg^F, Atg^C have generally higher Li content and Li/B ratio (Figure 7e), and higher Cu and Co contents.

7. Discussion

7.1. Thermo-Structural Evolution of the Val Malenco Peridotite

7.1.1. Partial Melting and Refertilization

Previous studies suggest that the protolith of the typical Alpine metamorphic, foliated antigorite-olivine-diopside-chlorite-magnetite±Ti-clinohumite ultramafic peridotite, belongs to spinel lherzolite [25], representing the subcontinental mantle [27,28]. Differing from primary clinopyroxene in fresh lherzolite [27,28], the clinopyroxene porphyroblasts in serpentinized peridotite show higher Mg# (92–95) and Cr_2O_3 (0.6–1.1 wt.%) contents, "N-MORB" REE patterns of flat MREE-HREEs, and depleted LREEs (Figure 7b) [4,41]. Thus, the clinopyroxene porphyroblasts represent relict refractory minerals after the partial melting and metasomatism of mantle rocks during Jurassic rifting. The olivine porphyroblasts have low Mg# (86,87) and high NiO (~0.25–0.47 wt.%) contents (Figure 7a), which are distinct from the residue origin of partial melting that was deemed to form olivine grains with higher Mg# (>90) [1,11]. Refertilization is a common process in orogenic lherzolite [4,42] and has been suggested to occur in the Val Malenco ultramafic peridotite during rifting and opening of the Piemonte-Ligurian Ocean during the Jurassic era [27]. Hence, the olivine porphyroblasts are refertilized olivine. The protolith of the Val Malenco serpentinized peridotite may have formed via partial melting of the subcontinental mantle, followed by a refertilization reaction between a refractory harzburgite and ascending basaltic melts [27,29]. The original textures have been erased by later recrystallization and serpentinization (see below).

7.1.2. High-Temperature Deformation and Followed Cooling

Both clinopyroxene and olivine prophyroblasts show obvious intracrystalline deformation and well-organized subparallel low-angle boundaries (Figure 3, Figure A1). Based on the methods described in Reddy et al. [37], the inferred dislocation slip systems are (100)[001], (010)[100] and (010)[001] in clinopyroxene porphyroblasts, and [100]{0kl}, [100](010), [100](001), [001](010), [001](100) in olivine porphyroblasts, respectively (Figure 3). These observations provide solid evidence for the development of multiple independent slip systems in grains deformed in the dislocation creep regime [43].

The temperature dependence of the dislocation slip system has been suggested based on TEM observations of experimentally deformed diopside and natural samples [44,45]. At 800–900 °C, the (100)[001] slip system is the easiest glide system, while the (010)[100], (110)[001] and {110}1/2<110> slip systems can also be activated but require a much larger resolved shear stress [44]. Above 1000 °C, the dominant slip systems are {110}1/2<110>, {110}[001], and (100)[001], corresponding to an increasing critical resolved stress [45]. The observation of the (010)[100] slip system in clinopyroxene porphyroblasts indicates that the clinopyroxene porphyroblasts were deformed by dislocation creep at temperatures lower than 1000 °C

Recent experimental work on olivine deformation has suggested that the deformation of olivine is accommodated by the Peierls mechanism at temperatures <900 °C, and by the power-law dislocation creep at temperatures >900 °C, respectively [46,47]. The activation of multiple slip planes, i.e., the {0kl}, is favored at a temperature of ~1000 °C [48,49]. The [100] and [001] dislocations are found to be dominant at temperatures above 1200 °C and below 1000 °C, respectively [50–52]. A transition from [100]-dominated dislocations at ~1100 °C to [001]-dominated dislocations at ~900 °C was reported in Durinck et al. [53]. Therefore, the inferred [100]{0kl} and [001](010), [001](100) slip systems in olivine porphyroblasts (Figure 3f) indicate a deformation event accommodated by dislocation creep that occurred at 900–1000 °C.

Oriented needles/rods of oxides/silicates in silicate minerals from ultrahigh-pressure (UHP) rocks are indicative of an exsolution or precipitation origin related to cooling [54–56]. Based on the optimization theory of phase boundary, i.e., the temperature of formation is determined by an optimal lattice fit at elevated temperatures estimated from thermal expansion data for the two lattices, the temperature of the magnetite exsolution can be estimated from the angle between different exsolution arrays [56], or crystallographic relationship between magnetite and clinopyroxene [55]. The topotaxial relationship between clinopyroxene porphyroblast and magnetite lamellae, i.e., $(100)_{Cpx}//(111)_{Mag}$, $(010)_{Cpx}//[\bar{1}10]_{Mag}$, $[001]_{Cpx}//[\bar{1}\bar{1}2]_{Mag}$, $[101]_{Cpx}//[112]_{Mag}$ and $(\bar{1}01)_{Cpx}//(\bar{1}\bar{1}1)_{Mag}$ (Figure 4), is the same as those reported in Feinberg et al. [55], indicating an exsolution temperature of ~860 °C [54,55].

7.1.3. Subduction Prograde Metamorphism

The trace of peridotite serpentinization in the Piemontese oceanic basin was fairly overprinted by the Alpine convergence, during which the Val Malenco ultramafic massif was moderately subducted. The OH-Ti-clinohumite in the olivine-rich layers (Figure 2a) formed by the consumption of serpentine [57] is a product of subduction-related prograde metamorphism. A lepidoblastic-granoblastic texture, well preserved in the olivine-rich layers, is characterized by the intergrowth of euhedral antigorite and olivine (Figure 2d), well-developed triple junctions, free of intracrystalline plastic deformation [36,58] and low dislocation density (1.7×10^{10} m^{-2}) (Figure A1) (see Appendix A.1). Meanwhile, the antigorite-inclusion-rich coarse-grained olivine implies that olivine has grown by consuming antigorite (Figure A2) (see Appendices A.2 and A.3). These observations suggest that the lepidoblastic-granoblastic texture is an equilibrium texture [23,25]. The peak metamorphic/equilibrium conditions estimated from Al content of antigorite [36] and mineral assemblages [26] are 450 °C and 0.6 GPa.

The FMEs of the coarse-grained antigorite also provide evidence for subduction prograde metamorphism. The As, Sb, and B are 10 to 100 times more enriched in coarse-grained antigorite compared to those of the primitive mantle and hydrothermal fluids in oceans (Figure 7d). On one hand, the signature of over-enrichment of As, Sb, and B could be inherited from oceanic serpentinite. Compared with clinopyroxene porphyroblasts, antigorite is depleted in Li and Sr, and has comparable Sb and Pb contents. Hence, the enrichment of As and B cannot be inherited from antigorite, especially taking into account the heavy loss of B during the transition from chrysotile/lizardite to antigorite [59,60]. On the other hand, the over-enrichment of As and Sb is found to be the characteristic of high-grade subducted serpentinites [40,61]. Previous works [40,61,62] on FME mobility have demonstrated that the mobility of some elements, such as As, Sb, B, Cs, is associated with low-grade metamorphism (350–400 °C) of metasedimentary rocks which are characterized by high As and Sb concentrations during subduction [63]. Therefore, the enrichment of As, Sb, and B in coarse-grained antigorite could result from the circulation of the sediment-derived fluids during subduction prograde metamorphism at temperatures around 350–400 °C.

7.1.4. Later Serpentinization

In contrast to the lepidoblastic-granoblastic texture preserved in the olivine-rich layer, both Atg^F and Ol^F in the antigorite-rich layers developed strong SPO (Figure 2b) and CPOs [36,58], suggesting a syn-kinematic serpentinization process. The temperatures for the later-stage serpentinization are estimated at 300–370 °C by Al content in the Atg^F and the absence of lizardite [36].

Compared with Atg^C in the olivine-rich layers, Atg^F in the antigorite-rich layers are depleted in almost all the FMEs, while the concentrations of B, Pb, and Ba are comparable. As discussed above, because As and Sb in Atg^C are mainly derived from sediment-derived fluid released at temperatures >350 °C [40,62], the depletion of As and Sb in Atg^F implies that the serpentinization temperature of Atg^F in the antigorite-rich layers is lower than 350 °C. The B/Li ratio is also an indicator of temperature for hydrous metasomatism because boron is released into fluid much faster at lower temperatures, but lithium can remain in the rocks at higher temperatures [64,65]. The higher B/Li ratios in Atg^F than in Atg^C also agree with a decreasing serpentinization temperature (Figure 7e). A sharp decrease of Cu and Co concentrations in vent fluids has been reported as temperature drops below 350 °C [66]. Therefore, the much lower Cu and Co concentrations in Atg^F (Figure 7f) could be explained by their greatly reduced solubility in fluid due to the decrease of serpentinization temperature below 350 °C [66,67].

Clinopyroxene porphyroblasts and diopside neoblasts show similar distributions of REE characterized by a slight enrichment of HREEs and depletion of LREEs, while diopside neoblasts have lower REE concentrations (Figure 7b). The high B/Li ratios of diopside neoblast are comparable to those of the metamorphic olivine and antigorite but differ from those of the clinopyroxene porphyroblasts with low B/Li ratios (Figure 7e), suggesting a similar metamorphic origin for diopside neoblasts and Atg^F at lower temperatures during exhumation [64,65]. The low B/Li ratios in clinopyroxene porphyroblasts are coherent with high-temperature hydrous metasomatism during emplacement.

7.2. Deformation during the Exhumation along Subduction Zone Interface

7.2.1. Dissolution and Precipitation Creep

The consistency between strain shadow structure and antigorite foliation (Figure 2g,h) and the parallelism of the foliation displayed by alternating antigorite-rich and olivine-rich layers (Figure 2a) indicates a syn-kinematic serpentinization process. Clinopyroxene shows a systematic difference in composition between the diopside neoblasts and clinopyroxene porphyroblasts (Table 1, Figure 5e), indicating an adjustment of the composition to new equilibrium conditions during the development of strain shadow structure. The sinuous boundary between the recrystallized diopside and clinopyroxene porphyroblast (Figure 5b) and the growth of both olivine and diopside in the wing of clinopyroxene porphyroblast (Figure 5c,d) indicate a dissolution and precipitation process [68]. Thus, A syn-kinematic dissolution and precipitation process can be applied to describe the dissolution of clinopyroxene porphyroblast and precipitation of diopside and olivine during serpentinization [69,70].

Clinopyroxene porphyroblasts, acting as rigid objects, could cause local perturbations of the stress field and flow pattern during the syn-kinematic low-temperature serpentinization process [71]. Increasing dissolution may occur adjacent to the porphyroblast on the sides of the shortening site (upper right and lower left corners of the porphyroblast in Figure 5a). Meanwhile, new olivine and diopside may nucleate and grow on the sides of the extensional site (the upper-left and lower-right corners of the porphyroblast in Figure 5a). The dissolution of clinopyroxene porphyroblast and local stress-controlled transportation of cations (Ca, Mg, Fe) have been observed at mm to cm scales during the process of serpentinization [69,72]. Local physicochemical gradients can cause precipitation of diopside and olivine. However, the local equilibrium is transient and will evolve during the deformation in a dynamic environment [73], which is evidenced by the growth zonation of both olivine and diopside (Figure 5c,d).

7.2.2. Role of Strain Localization during Exhumation

In contrast to the well-developed SPOs and CPOs for both Atg^F and Ol^F in the antigorite-rich layers and diopside neoblasts in strain shadow structures, both Atg^C and Ol^C in the olivine-rich layers preserve an equilibrium lepidoblastic-granoblastic texture (Figure 2d). The Ol^C displays nearly random fabrics [36,58]. This evidence implies most of the strain was accommodated by the antigorite-rich layers with little in the olivine-rich layers.

It has been suggested that the deformation of olivine in the antigorite-rich layers is accomplished by dissolution creep during serpentinization in our previous study [36]. The pressure-shadow structures indicate also a dissolution and precipitation process (Figures 5 and 6). The viscosity of rocks/minerals deformed by the dissolution-precipitation creep is generally assumed to be lower than the viscosity deformed by dislocation creep, especially at lower temperatures (Reference [74] and references therein). Recent studies have also shown that strain localization can be induced by the dissolution-precipitation creep in ultramafic rocks, especially in the presence of fluid at low temperatures [8,73]. Combined with the temperature decrease from the olivine-rich layers to the antigorite-rich layers, we propose that the differences in microstructure and CPOs of antigorite and olivine from the olivine-rich to the antigorite-rich layers reveal a progressive strain localization coeval with the serpentinization and cooling. Feedback between deformation and permeability probably has resulted in focused fluid flow and the collateral of more effective deformation, thus favoring strain localization and mylonization in the antigorite-rich layers.

Field observations show that dense high-pressure rocks in the subduction zone commonly coexist with a light-weighted and soft antigorite mylonite matrix, suggesting that the buoyant and weak antigorite in the subduction channel can assist with the exhumation of high-pressure rocks [75]. Although the low viscosity of antigorite deformed in the dislocation creep regime is thought to be the main cause [76], antigorite deformed by the dissolution creep is also reported in natural samples [77]. Based on the new results, we suggest that syn-kinematic serpentinization through dissolution-precipitation creep could be an effective mechanism to produce strain localization and exhumation of high-pressure rocks along a subduction interface.

8. Conclusions

The Val Malenco serpentinized peridotite provides constraints to the tectonic evolution and deformation of the mantle wedge corner. Our results show that the serpentinized peridotite has recorded a multi-stage thermo-structural history (Figure 8):

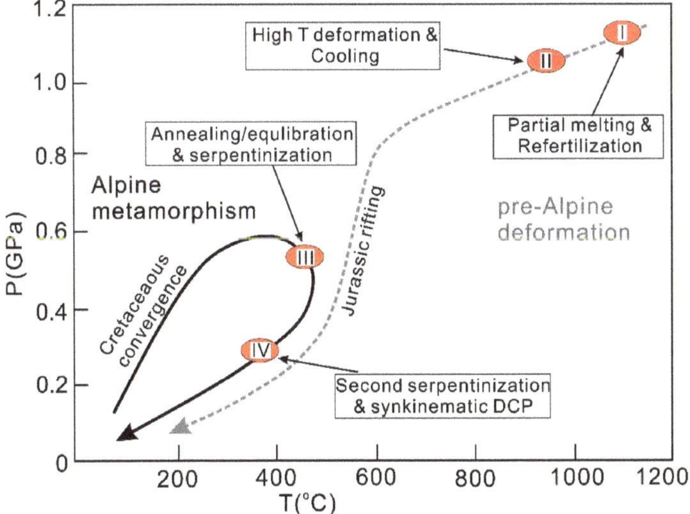

Figure 8. The pressure–temperature paths for the Val Malenco mantle based on previous studies [23,29] and this study. The pre-oceanic exhumation path of the Val Malenco mantle is shown by the grey dotted curve, while the Alpine subduction cycle is shown by the solid curve. The red shaded zones represent different thermo-mechanical stages.

(1) Partial melting, refertilization, and associated high-temperature deformation took place during the Pre-Alpine extension. The clinopyroxene porphyroblasts have high Mg# (92–95) and Cr_2O_3 (0.6–1.1 wt.%) contents, relatively low REE contents, depleted LREE, and a flat MREE-HREE pattern. The olivine porphyroblasts show low Mg# (86–87) and high NiO content (0.3–0.46 wt.%). These geochemical features imply that the partial melting of the subcontinental mantle and following refertilization occurred locally, possibly related to the rifting and opening of the Piemonte-Ligurian Ocean during the Jurassic era. Both clinopyroxene and olivine porphyroblasts developed nearly parallel low-angle boundaries. The (100)[001], (010)[100] and (010)[001] slip systems are responsible for the development of clinopyroxene low-angle boundaries, while the [100]{0kl}, [100](010), [100](001), [001](010), [001](100) slip systems account for the development of olivine low-angle boundaries. All these slip systems are indicative of a high-temperature (900–1000 °C) dislocation creep in olivine and clinopyroxene porphyroblasts during the early stage of rifting.

(2) The following Alpine convergence in a supra-subduction zone setting is documented by subduction prograde metamorphism and low-temperature serpentinization. The subduction prograde metamorphism is represented by the lepidoblastic-granoblastic structure. The enrichment of As, Sb, and B in coarse-grained antigorite could result from the circulation of the sediment-derived fluids during subduction at temperatures of 350–400 °C. The later serpentinization is responsible for the formation of the antigorite-rich layers with pronounced SPOs and CPOs in antigorite and olivine, revealing a subsequent syn-kinematic serpentinization process. Compared with coarse-grained antigorite, fine-grained antigorite is more depleted in highly fluid mobile elements (e.g., Sb and As), and has a higher B/Li ratio and lower Cu and Co concentrations, suggesting a lower serpentinization temperature (<350 °C).

The deformation of the mantle wedge at low temperatures (300–350 °C) is manifested by the concurrent development of antigorite-rich layers and pressure shadow structures. Dissolution-precipitation creep was responsible for the development of pressure shadow structures and antigorite-rich layers. Most of the strain is accommodated by the antigorite-rich layers with strongly aligned antigorite, while little is imposed in the olivine-rich layers with the well-preserved lepidoblastic-granoblastic structure. We suggest that the dissolution-precipitation creep accompanying serpentinization may result in strain localization and exhumation of high-pressure rocks along the subduction interface.

Supplementary Materials: The following are available online at http://www.mdpi.com/2075-163X/10/11/962/s1, Table S1: Major element compositions of olivine, Table S2: Major element compositions of clinopyroxene, Table S3: REE compositions of clinopyroxene, Table S4: REE compositions of antigorite.

Author Contributions: Conceptualization, W.L., J.Z.; methodology and investigation, W.L., Y.C., K.Z., Y.Z.; writing: W.L., J.Z., Y.C.; supervision and project administration, J.Z., Y.Z. and Z.J. All authors have read and agreed to the published version of the manuscript.

Funding: This work was supported by funds from the National Natural Science Foundation of China (Grant No. 41902221, 41425012), the National Programme on Global Change and Air-Sea Interaction (GASI-GEOGE-02), and the MOST Special Fund for the State Key Laboratory of GPMR at CUG-Wuhan.

Acknowledgments: We would like to acknowledge Haijin Xu, Xiang Zhou, Feng Shi, and Tao Luo for their help with discussions of geochemical interpretations and/or LA-ICP-MS analyses. Harry W. Green II and Jöerg Hermann are deeply appreciated for kindly providing some of the studied samples.

Conflicts of Interest: The authors declare no conflict of interest.

Appendix A

Appendix A.1. Dislocation Structure

Dislocation microstructures after oxygen decoration of the sample are shown in Figure A1. The dislocations, which are shown as white dots and lines, are distributed heterogeneously among different olivine grains. Low angle boundaries defined by arrays of dislocations are only observed in Ol^P

(Figure A1a,b). The dislocation density is low (5.6×10^{10} m^{-2}) between low angle boundaries as most of the free dislocations tend to accumulate into the low angle boundaries (Figure A1b). The dislocation density of Ol^C is also low (1.7×10^{10} m^{-2}). Most of the Ol^C are free of dislocations (Figure A1c). The Ol^S grains have higher heterogeneously distributed dislocation density (1.7×10^{12} m^{-2}) (Figure A1d). Free dislocations are found in Ol^{Cpx} with low dislocation density (1×10^{10} m^{-2}) (Figure A1e). As for Ol^F in the antigorite-rich layers, some grains are dislocation free, while others have dislocation densities up to 2.2×10^{11} m^{-2} (Figure A1f).

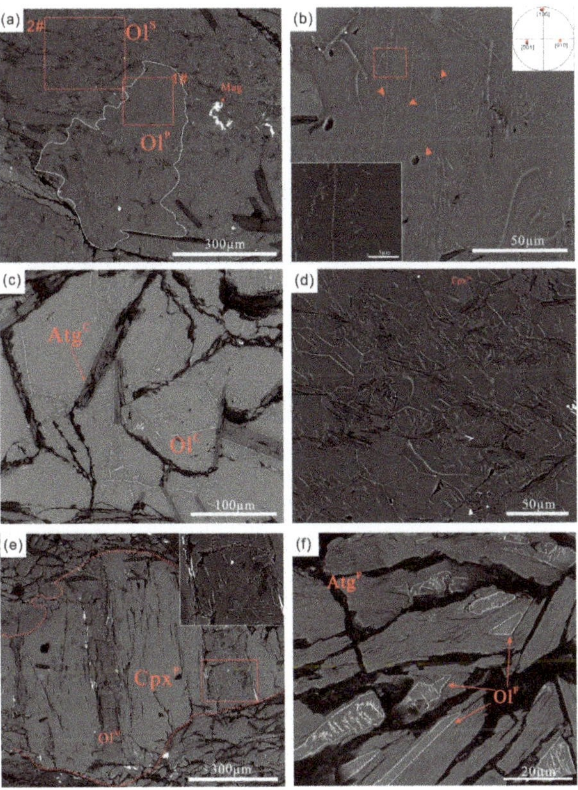

Figure A1. Back-scattered electron (BSE) images showing the dislocation microstructure of different types of olivine after oxygen decoration at 900 °C for 1 h. The dislocations are shown as white dots and lines. (**a**) Overview of an olivine porphyroblast and small-grained olivine. Red rectangles #1 and #2 are enlarged and shown in (**b,d**), respectively. (**b**) An enlarged view of olivine porphyroblast in (**a**), showing most dislocations are concentrated into low-angle boundaries. The dislocation density is low between boundaries. (**c**) The dislocation structure of coarse-grained olivine, showing that the cores of olivine grains are almost dislocation free. (**d**) An enlarged view of small-grained olivine in (**a**), showing heterogeneous dislocation distribution. (**e**) The dislocation structure of olivine along clinopyroxene porphyrobalst fractures. (**f**) The dislocation structure of fine-grained olivine in the antigorite-rich layers.

Appendix A.2. FTIR Measurements

The water contents and representative FTIR spectra for each mineral are shown in Figure A2. All the spectra are normalized to a thickness of 1 mm. Nearly half of the olivine grains show no detectable OH absorption bands in the 3000–4000 cm^{-1} range, while the others show IR absorption peaks at wavenumbers ranging from 3668 cm^{-1} to 3691 cm^{-1} (Figure A2a), which can be attributed to micro-inclusions of serpentine and/or talc in olivine [78].

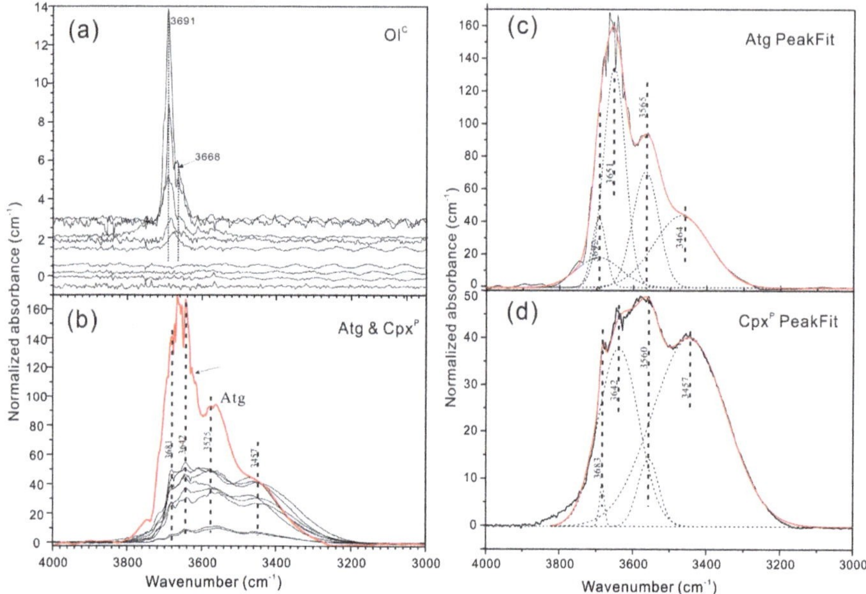

Figure A2. The unpolarized FTIR spectra of coarse-grained olivine and clinopyroxene porphyroblast. (**a**) The coarse-grained olivine. The observed IR absorption peaks correspond to antigorite IR absorption peaks, as shown in (**b,c**). (**b**) Clinopyroxene porphyroblast (black) and antigorite (red). The clinopyroxene spectra are characterized by absorption peaks at 3681, 3642, 3575 and 3457 cm^{-1}, which are all correlated to the absorption peaks of antigorite spectra. (**c,d**) Atigorte and clinopyroxene porphyroblast spectra analyzed by PeakFit.

Because fractures are well-developed in clinopyroxene porphyroblasts and filled up by antigorite, the PeakFit technique was used to remove the influence of overlapped antigorite IR absorption bands. Both clinopyroxene porphyroblasts and antigorite in the antigorite-rich layers were analyzed (Figure A2b). Peak positions, integrated absorbances, band maxima, and the area-weighted average of band positions were determined by applying a Gaussian distribution function to all component bands [79] (Figure A2c,d). Water contents were calculated by subtracting the integrated absorbances area of antigorite from clinopyroxene at 3457 cm^{-1}. The results represent the minimum water contents in clinopyroxene. The water contents of clinopyroxene porphyroblasts vary between 340 and 600 ppm.

The IR absorption spectra of antigorite were collected for comparison. Antigorite shows characteristic IR absorption bands at 3692, 3651, 3565 and 3464 cm^{-1} (Figure A2b).

Appendix A.3. Discussion

The subduction prograde metamorphism is also recorded by the inclusions in coarse-grained olivine. The water contents in these olivines can be categorized into two groups: "wet" grains with antigorite inclusions and dry grains (Figure A2a). The antigorite inclusion could be formed through OH migrating or replacement of the high-pressure hydrous phase inclusion during decompression [79,80], serpentinization along the cracks at low temperature and pressure, and olivine growth at the expense of antigorite. The observed hydrous IR peaks around 3673~3689 cm^{-1} (corresponding to serpentine) suggest that deep origin could not be the case because some other IR peaks should also be present [80,81]. Although brittle fractures are common in porphyroblasts, they are rare in coarse-grained olivine

(Figure 2d). Thus, the "wet" olivine grains indicate olivine growth via the consumption of antigorite. The dry coarse-grained olivine may imply that the hydrogen solubility in olivine is negligible at low temperatures, as shown by the hydrogen solubility dependence on temperature [79,82].

References

1. Cao, Y.; Song, S.; Su, L.; Jung, H.; Niu, Y. Highly refractory peridotites in Songshugou, Qinling orogen: Insights into partial melting and melt/fluid-rock reactions in forearc mantle. *Lithos* **2016**, *252*, 234–254. [CrossRef]
2. Cao, Y.; Jung, H.; Song, S. Seismic anisotropies of the Songshugou peridotites (Qinling orogen, central China) and their seismic implications. *Tectonophysics* **2018**, *722*, 432–446. [CrossRef]
3. Jung, H. Deformation fabrics of olivine in Val Malenco peridotite found in Italy and implications for the seismic anisotropy in the upper mantle. *Lithos* **2009**, *109*, 341–349. [CrossRef]
4. Le Roux, V.; Bodinier, J.L.; Tommasi, A.; Alard, O.; Dautria, J.M.; Vauchez, A.; Riches, A.J.V. The Lherz spinel lherzolite: Refertilized rather than pristine mantle. *Earth Planet. Sci. Lett.* **2007**, *259*, 599–612. [CrossRef]
5. Park, M.; Jung, H. Microstructural evolution of the Yugu peridotites in the Gyeonggi Massif, Korea: Implications for olivine fabric transition in mantle shear zones. *Tectonophysics* **2017**, *709*, 55–68. [CrossRef]
6. Rogkala, A.; Petrounias, P.; Tsikouras, B.; Giannakopoulou, P.P.; Hatzipanagiotou, K. Mineralogical Evidence for Partial Melting and Melt-Rock Interaction Processes in the Mantle Peridotites of Edessa Ophiolite (North Greece). *Minerals* **2019**, *9*, 120. [CrossRef]
7. Skemer, P.; Katayama, I.; Karato, S.I. Deformation fabrics of the Cima di Gagnone peridotite massif, Central Alps, Switzerland: Evidence of deformation at low temperatures in the presence of water. *Contrib. Mineral. Petrol.* **2006**, *152*, 43–51. [CrossRef]
8. Tommasi, A.; Langone, A.; Padrón-Navarta, J.A.; Zanetti, A.; Vauchez, A. Hydrous melts weaken the mantle, crystallization of pargasite and phlogopite does not: Insights from a petrostructural study of the Finero peridotites, southern Alps. *Earth Planet. Sci. Lett.* **2017**, *477*, 59–72. [CrossRef]
9. Michibayashi, K.; Oohara, T. Olivine fabric evolution in a hydrated ductile shear zone at the Moho Transition Zone, Oman Ophiolite. *Earth Planet. Sci. Lett.* **2013**, *377*, 299–310. [CrossRef]
10. Prigent, C.; Guillot, S.; Agard, P.; Ildefonse, B. Fluid-Assisted Deformation and Strain Localization in the Cooling Mantle Wedge of a Young Subduction Zone (Semail Ophiolite). *J. Geophys. Res. Solid Earth* **2018**, *123*, 7529–7549. [CrossRef]
11. Tasaka, M.; Michibayashi, K.; Mainprice, D. B-type olivine fabrics developed in the fore-arc side of the mantle wedge along a subducting slab. *Earth Planet. Sci. Lett.* **2008**, *272*, 747–757. [CrossRef]
12. Arai, S.; Ishimaru, S. Insights into petrological characteristics of the lithosphere of mantle wedge beneath arcs through peridotite xenoliths: A review. *J. Petrol.* **2008**, *49*, 665–695. [CrossRef]
13. Chin, E.J.; Soustelle, V.; Hirth, G.; Saal, A.E.; Kruckenberg, S.C.; Eiler, J.M. Microstructural and geochemical constraints on the evolution of deep arc lithosphere. *Geochem. Geophys. Geosyst.* **2016**, *17*, 2497–2521. [CrossRef]
14. Liu, S.; Tommasi, A.; Vauchez, A.; Mazzucchelli, M. Deformation, annealing, melt-rock interaction, and seismic properties of an old domain of the equatorial Atlantic lithospheric mantle. *Tectonic* **2019**, *38*, 1–25. [CrossRef]
15. Payot, B.D.; Arai, S.; Yoshikawa, M.; Tamura, A.; Okuno, M.; Rivera, D.J. Mantle Evolution from Ocean to Arc: The Record in Spinel Peridotite Xenoliths inMt. Pinatubo, Philippines. *Minerals* **2018**, *8*, 515. [CrossRef]
16. Tommasi, A.; Vauchez, A.; Ionov, D.A. Deformation, static recrystallization, and reactive melt transport in shallow subcontinental mantle xenoliths (Tok Cenozoic volcanic field, SE Siberia). *Earth Planet. Sci. Lett.* **2008**, *272*, 65–77. [CrossRef]
17. Strating, E.H.H.; Rampone, E.; Piccardo, G.B.; Drury, M.R.; Vissers, R.L.M. Subsolidus emplacement of mantle peridotites during incipient oceanic rifting and opening of the mesozoic tethys (voltri massif, NW Italy). *J. Petrol.* **1993**, *34*, 901–927. [CrossRef]
18. Kaczmarek, M.A.; Tommasi, A. Anatomy of an extensional shear zone in the mantle, Lanzo massif, Italy. *Geochem. Geophys. Geosyst.* **2012**, *12*, 1–24. [CrossRef]
19. Bodinier, J.L.; Godard, M. Orogenic, Ophiolitic, and Abyssal Peridotites. In *Treatise on Geochemistry*, 2nd ed.; Holland, H.D., Turekian, K.K., Eds.; Elsevier: Amsterdam, The Netherlands, 2014; Volume 3.

20. Matysiak, A.K.; Trepmann, C.A. The deformation record of olivine in mylonitic peridotites from the Finero Complex, Ivrea Zone: Separate deformation cycles during exhumation. *Tectonics* **2015**, *34*, 2514–2533. [CrossRef]

21. Zanetti, A.; Mazzucchelli, M.; Rivalenti, G.; Vannucci, R. The Finero phlogopite-peridotite massif: An example of subduction-related metasomatism. *Contrib. Mineral. Petrol.* **1999**, *134*, 107–122. [CrossRef]

22. Scambelluri, M.; Pettke, T.; Cannaò, E. Fluid-related inclusions in Alpine high-pressure peridotite reveal trace element recycling during subduction-zone dehydration of serpentinized mantle (Cima di Gagnone, Swiss Alps). *Earth Planet. Sci. Lett.* **2015**, *429*, 45–59. [CrossRef]

23. Trommsdorff, V.; Hermann, J.; Müntener, O.; Pfiffner, M.; Risold, A.C. Geodynamic cycles of subcontinental lithosphere in the Central Alps and the Arami enigma. *J. Geodyn.* **2000**, *30*, 77–92. [CrossRef]

24. Puschnig, A.R. Metasomatic alterations at mafic-ultramafic contacts in Valmalenco (Rhetic Alps, N-Italy). *Schweiz. Mineral. Petrogr. Mitt.* **2002**, *82*, 515–536.

25. Trommsdorff, V.; Evans, B.W. Alpine metamorphism of peridotitic rocks. *Schweiz. Mineral. Petrogr. Mitt.* **1974**, *54*, 333–354.

26. Trommsdorff, V.; Evans, B.W. Progressive metamorphism of antigorite schist in the Bergell tonalite aureole (Italy). *Am. J. Sci.* **1972**, *272*, 423–437. [CrossRef]

27. Müntener, O.; Pettke, T.; Desmurs, L.; Meier, M.; Schaltegger, U. Refertilization of mantle peridotite in embryonic ocean basins: Trace element and Nd isotopic evidence and implications for crust-mantle relationships. *Earth Planet. Sci. Lett.* **2004**, *22*, 293–308. [CrossRef]

28. Müntener, O.; Manatschal, G.; Desmurs, L.; Pettke, T. Plagioclase peridotites in ocean-continent transitions: Refertilized mantle domains generated by melt stagnation in the shallow mantle lithosphere. *J. Petrol.* **2010**, *51*, 255–294. [CrossRef]

29. Hermann, J.; Müntener, O.; Trommsdorff, V.; Hansmann, W.; Piccardo, B. Fossil crust-to-mantle transition, Val Malenco (Italian Alps). *J. Geophys. Res.* **1997**, *102*, 20123–20132. [CrossRef]

30. Müntener, O.; Hermann, J.; Trommsdorff, V. Cooling History and Exhumation of Lower-Crustal Granulite and Upper Mantle (Malenco, Eastern Central Alps). *J. Petrol.* **2000**, *41*, 175–200. [CrossRef]

31. Villa, I.M.; Hermann, J.; Müntener, O.; Trommsdorff, V. 39Ar-40Ar dating of multiply zoned amphibole generations (Malenco, Italian Alps). *Contrib. Mineral. Petrol.* **2000**, *140*, 363–381. [CrossRef]

32. Peretti, A.; Dubessy, J.; Mullis, J.; Frost, B.R.; Trommsdorff, V. Highly reducing conditions during Alpine metamorphism of the Malenco peridotite (Sondrio, northern Italy) indicated by mineral paragenesis and H2 in fluid inclusions. *Contrib. Mineral. Petrol.* **1992**, *112*, 329–340. [CrossRef]

33. Mellini, M.; Trommsdorff, V.; Compagnoni, R. Antigorite polysomatism: Behaviour during progressive metamorphism. *Contrib. Mineral. Petrol.* **1987**, *97*, 147–155. [CrossRef]

34. Von Blackenburg, F. Combined high-precision chronometry and geochemical tracing using accessory minerals: Applied to the Central-Alpine Bergell intrusion (central Europe). *Chem. Geol.* **1992**, *100*, 19–40. [CrossRef]

35. Liu, Y.; Hu, Z.; Gao, S.; Günther, D.; Xu, J.; Gao, C.; Chen, H. In situ analysis of major and trace elements of anhydrous minerals by LA-ICP-MS without applying an internal standard. *Chem. Geol.* **2008**, *257*, 34–43. [CrossRef]

36. Liu, W.; Zhang, J.; Barou, F. B-type olivine fabric induced by low temperature dissolution creep during serpentinization and deformation in mantle wedge. *Tectonophysics* **2018**, *722*, 1–10. [CrossRef]

37. Reddy, S.M.; Timms, N.E.; Pantleon, W.; Trimby, P. Quantitative characterization of plastic deformation of zircon and geological implications. *Contrib. Mineral. Petrol.* **2007**, *153*, 625–645. [CrossRef]

38. De Hoog, J.C.M.; Gall, L.; Cornell, D.H. Trace-element geochemistry of mantle olivine and application to mantle petrogenesis and geothermobarometry. *Chem. Geol.* **2010**, *270*, 196–215. [CrossRef]

39. Schmidt, K.; Koschinsky, A.; Garbe-Schönberg, D.; de Carvalho, L.; Seifert, R. Geochemistry of hydrothermal fluids from the ultramafic-hosted Logatchev hydrothermal field, 15° N on the Mid-Atlantic Ridge: Temporal and spatial investigation. *Chem. Geol.* **2007**, *242*, 1–21. [CrossRef]

40. Deschamps, F.; Godard, M.; Guillot, S.; Chauvel, C.; Andreani, M.; Hattori, K.; Wunder, B.; France, L. Behavior of fluid-mobile elements in serpentines from abyssal to subduction environments: Examples from Cuba and Dominican Republic. *Chem. Geol.* **2012**, *312*, 93–117. [CrossRef]

41. McDonough, W.F.; Frey, F.A. Rare earth elements in upper mantle rocks. In *Geochemistry and Mineralogy of Rare Earth Elements*; Lipin, B.R., McKay, G.A., Eds.; Walter de Gruyter GmbH & Co KG.: Berlin, Germany, 1989; pp. 99–139.

42. Varas-Reus, M.; Garrido, C.; Marchesi, C.; Bodinier, J.; Frets, E.; Bosch, D.; Tommasi, A.; Hidas, K.; Targuisti, K. Refertilization Processes in the Subcontinental Lithospheric Mantle: The Record of the Beni Bousera Orogenic Peridotite (Rif Belt, Northern Morocco). *J. Petrol.* **2016**, *57*, 2251–2270. [CrossRef]

43. Parks, D.; Ahzi, S. Polycrystalline plastic deformation and texture evolution for crystals lacking five independent slip systems. *J. Mech. Phys. Solids* **1990**, *38*, 701–724. [CrossRef]

44. Ingrin, J.; Doukhan, N.; Doukhan, J.-C. Dislocation glide systems in diopside single crystals deformed at 800–900 °C. *Eur. J. Mineral.* **1992**, *4*, 1291–1302. [CrossRef]

45. Raterron, P.; Doukhan, N.; Jaoul, O.; Doukhan, J.C. High temperature deformation of diopside IV: Predominance of {110} glide above 1000 °C. *Phys. Earth Planet. Inter.* **1994**, *82*, 209–222. [CrossRef]

46. Demouchy, S.; Schneider, S.E.; Mackwell, S.J.; Zimmerman, M.E.; Kohlstedt, D.L. Experimental deformation of olivine single crystals at lithospheric temperatures. *Geophys. Res. Lett.* **2009**, *36*, 1–5. [CrossRef]

47. Long, H.; Weidner, D.J.; Li, L.; Chen, J.; Wang, L. Deformation of olivine at subduction zone conditions determined from in situ measurements with synchrotron radiation. *Phys. Earth Planet. Inter.* **2011**, *186*, 23–35. [CrossRef]

48. Carter, N.L.; Avé Lallemant, H.G. High temperature flow of dunite and peridotite. *Geol. Soc. Am. Bull.* **1970**, *81*, 2181–2202. [CrossRef]

49. Durham, W.B.; Goetze, C.; Blake, B. Plastic flow of oriented single crystals of olivine: 2. Observations and interpretations of the dislocation structures. *J. Geophys. Res.* **1977**, *82*, 5755–5770. [CrossRef]

50. Bai, Q.; Kohlstedt, D.L. High-temperature creep of olivine single crystals, 2. dislocation structures. *Tectonophysics* **1992**, *206*, 1–29. [CrossRef]

51. Goetze, C. The Mechanisms of Creep in Olivine. Philosophical Transactions of the Royal Society A: Mathematical. *Phys. Eng. Sci.* **1978**, *288*, 99–119.

52. Katayama, L.; Karato, S.I. Effect of temperature on the B- to C-type olivine fabric transition and implication for flow pattern in subduction zones. *Phys. Earth Planet. Inter.* **2006**, *157*, 33–45. [CrossRef]

53. Durinck, J.; Devincre, B.; Kubin, L.; Cordier, P. Modeling the plastic deformation of olivine by dislocation dynamics simulations. *Am. Mineral.* **2007**, *92*, 1346–1357. [CrossRef]

54. Hwang, S.L.; Yui, T.F.; Chu, H.T.; Shen, P.; Iizuka, Y.; Yang, H.Y.; Yang, J.S.; Xu, Z. Hematite and magnetite precipitates in olivine from the Sulu peridotite: A result of dehydrogenation-oxidation reaction of mantle olivine? *Am. Mineral.* **2008**, *93*, 1051–1060. [CrossRef]

55. Feinberg, J.M.; Wenk, H.R.; Renne, P.R.; Scott, G.R. Epitaxial relationships of clinopyroxene-hosted magnetite determined using electron backscatter diffraction (EBSD) technique. *Am. Mineral.* **2004**, *89*, 462–466. [CrossRef]

56. Renne, P.R.; Scott, G.R.; Glen, J.M.G.; Feinberg, J.M. Oriented inclusions of magnetite in clinopyroxene: Source of stable remanent magnetization in gabbros of the Messum Complex, Namibia. *Geochem. Geophys. Geosyst.* **2002**, *3*, 1–11. [CrossRef]

57. López Sánchez-Vizcaíno, V.; Trommsdorff, V.; Gómez-Pugnaire, M.T.; Garrido, C.J.; Müntener, O.; Connolly, J.A.D. Petrology of titanian clinohumite and olivine at the high-pressure breakdown of antigorite serpentinite to chlorite harzburgite (Almirez Massif, S. Spain). *Contrib. Mineral. Petrol.* **2005**, *149*, 627–646. [CrossRef]

58. Liu, W.; Zhang, J.; Cao, Y.; Jin, Z. Geneses of two contrasting antigorite crystal preferred orientations (CPOs) and their implications for seismic anisotropy in the forearc mantle. *J. Geophys. Res. Solid Earth* **2020**, *125*, e2020JB019354. [CrossRef]

59. Kodolányi, J.; Pettke, T.; Spandler, C.; Kamber, B.S.; Ling, K.G. Geochemistry of ocean floor and fore-arc serpentinites: Constraints on the ultramafic input to subduction zones. *J. Petrol.* **2012**, *53*, 235–270. [CrossRef]

60. Vils, F.; Müntener, O.; Kalt, A.; Ludwig, T. Implications of the serpentine phase transition on the behaviour of beryllium and lithium-boron of subducted ultramafic rocks. *Geochim. Cosmochim. Acta* **2011**, *75*, 1249–1271. [CrossRef]

61. Deschamps, F.; Godard, M.; Guillot, S.; Hattori, K. Geochemistry of subduction zone serpentinites: A review. *Lithos* **2013**, *178*, 96–127. [CrossRef]

62. Bebout, G.E.; Bebout, A.E.; Graham, C.M. Cycling of B, Li, and LILE (K, Cs, Rb, Ba, Sr) into subduction zones: SIMS evidence from micas in high-P/T metasedimentary rocks. *Chem. Geol.* **2007**, *239*, 284–304. [CrossRef]

63. Jochum, K.P.; Verma, S.P. Extreme enrichment of Sb, Tl and other trace elements in altered MORB. *Chem. Geol.* **1996**, *130*, 289–299. [CrossRef]

64. Marschall, H.R.; Altherr, R.; Ludwig, T.; Kalt, A.; Gméling, K.; Kasztovszky, Z. Partitioning and budget of Li, Be and B in high-pressure metamorphic rocks. *Geochim. Cosmochim. Acta* **2006**, *70*, 4750–4769. [CrossRef]

65. Marschall, H.R.; Altherr, R.; Rüpke, L. Squeezing out the slab—Modelling the release of Li, Be and B during progressive high-pressure metamorphism. *Chem. Geol.* **2007**, *239*, 323–335. [CrossRef]

66. Metz, S.; Trefry, J.H. Chemical and mineralogical influences on concentrations of trace metals in hydrothermal fluids. *Geochim. Cosmochim. Acta* **2000**, *64*, 2267–2279. [CrossRef]

67. Hannington, M.D.; Jonasson, I.R.; Herzig, P.M.; Petersen, S. Physical and chemical processes of seafloor mineralization at mid-ocean ridges. In *Seafloor Hydrothermal Systems: Physical, Chemical, Biological, and Geological Interactions*; Humphris, S.E., Zierenberg, R.A., Mullineaux, L.S., Thomson, R.E., Eds.; American Geophysical Union: Washington, DC, USA, 1995; Volume 91, pp. 115–157.

68. Ruiz-Agudo, E.; Putnis, C.V.; Putnis, A. Coupled Dissolution and Precipitation at Mineral—Fluid Interfaces. *Chem. Geol.* **2014**, *383*, 132–146. [CrossRef]

69. Austrheim, H.; Prestvik, T. Rodingitization and hydration of the oceanic lithosphere as developed in the Leka ophiolite, north-central Norway. *Lithos* **2008**, *104*, 177–198. [CrossRef]

70. Lamadrid, H.M.; Rimstidt, J.D.; Schwarzenbach, E.M.; Klein, F.; Ulrich, S.; Dolocan, A.; Bodnar, R.J. Effect of water activity on rates of serpentinization of olivine. *Nat. Commun.* **2017**, *8*, 1–9. [CrossRef]

71. Passchier, C.W.; Trouw, R.A.J. *Microtectonics*; Springer: Berlin/Heidelberg, Germany, 2005; pp. 159–187.

72. Iyer, K.; Austrheim, H.; John, T.; Jamtveit, B. Serpentinization of the oceanic lithosphere and some geochemical consequences: Constraints from the Leka Ophiolite Complex, Norway. *Chem. Geol.* **2008**, *249*, 66–90. [CrossRef]

73. Hidas, K.; Tommasi, A.; Garrido, C.J.; Padrón-Navarta, J.A.; Mainprice, D.; Vauchez, A.; Barou, F.; Marchesi, C. Fluid-assisted strain localization in peridotites during emplacement of the shallow subcontinental lithospheric mantle. *Lithos* **2016**, *262*, 636–650. [CrossRef]

74. Wassmann, S.; Stöckhert, B. Rheology of the plate interface—Dissolution precipitation creep in high pressure metamorphic rocks. *Tectonophysics* **2013**, *608*, 1–29. [CrossRef]

75. Guillot, S.; Hattori, K.; Agard, P.; Schwartz, S.; Vidal, O. Exhumation Processes in Oceanic and Continental Subduction Contexts: A Review. In *Subduction Zone Geodynamics*; Lallemand, S., Funiciello, F., Eds.; Springer: Berlin/Heidelberg, Germany, 2009; pp. 175–204.

76. Hilairet, N.; Reynard, B. Stability and dynamics of serpentinite layer in subduction zone. *Tectonophysics* **2009**, *465*, 24–29. [CrossRef]

77. Wassmann, S.; Stöckhert, B.; Trepmann, C.A. Dissolution precipitation creep versus crystalline plasticity in high-pressure metamorphic serpentinites. *Geol. Soc. Lond. Spec. Publ.* **2011**, *360*, 129–149. [CrossRef]

78. Beran, A.; Libowitzky, E. Water in natural mantle minerals II: Olivine, garnet and accessory minerals, Water in nominally anhydrous minerals. *Rev. Mineral. Geochem.* **2006**, *62*, 169–191. [CrossRef]

79. Litasov, K.D.; Ohtani, E.; Kagi, H.; Jacobsen, S.D.; Ghosh, S. Temperature dependence and mechanism of hydrogen incorporation in olivine at 12.5–14.0 GPa. *Geophys. Res. Lett.* **2007**, *34*, 3–7. [CrossRef]

80. Khisina, N.R.; Wirth, R.; Andrut, M.; Ukhanov, A.V. Extrinsic and intrinsic mode of hydrogen occurrence in natural olivines: FTIR and TEM investigation. *Phys. Chem. Miner.* **2001**, *28*, 291–301.

81. Koch-Müller, M.; Matsyuk, S.S.; Rhede, D.; Wirth, R.; Khisina, N. Hydroxyl in mantle olivine xenocrysts from the Udachnaya kimberlite pipe. *Phys. Chem. Miner.* **2006**, *33*, 276–287. [CrossRef]

82. Bali, E.; Bolfan-Casanova, N.; Koga, K.T. Pressure and temperature dependence of H solubility in forsterite: An implication to water activity in the Earth interior. *Earth Planet. Sci. Lett.* **2008**, *268*, 354–363. [CrossRef]

Publisher's Note: MDPI stays neutral with regard to jurisdictional claims in published maps and institutional affiliations.

Article

Deformation Microstructures of Phyllite in Gunsan, Korea, and Implications for Seismic Anisotropy in Continental Crust

Seokyoung Han and Haemyeong Jung *

Tectonophysics Laboratory, School of Earth and Environmental Sciences, Seoul National University, Seoul 08826, Korea; hs04111@snu.ac.kr
* Correspondence: hjung@snu.ac.kr; Tel.: +82-2-880-6733

Abstract: Muscovite is a major constituent mineral in the continental crust that exhibits very strong seismic anisotropy. Muscovite alignment in rocks can significantly affect the magnitude and symmetry of seismic anisotropy. In this study, deformation microstructures of muscovite-quartz phyllites from the Geumseongri Formation in Gunsan, Korea, were studied to investigate the relationship between muscovite and chlorite fabrics in strongly deformed rocks and the seismic anisotropy observed in the continental crust. The [001] axes of muscovite and chlorite were strongly aligned subnormal to the foliation, while the [100] and [010] axes were aligned subparallel to the foliation. The distribution of quartz c-axes indicates activation of the basal<a>, rhomb<a> and prism<a> slip systems. For albite, most samples showed (001) or (010) poles aligned subnormal to the foliation. The calculated seismic anisotropies based on the lattice preferred orientation and modal compositions were in the range of 9.0–21.7% for the P-wave anisotropy and 9.6–24.2% for the maximum S-wave anisotropy. Our results indicate that the modal composition and alignment of muscovite and chlorite significantly affect the magnitude and symmetry of seismic anisotropy. It was found that the coexistence of muscovite and chlorite contributes to seismic anisotropy constructively when their [001] axes are aligned in the same direction.

Keywords: phyllite; lattice preferred orientation; seismic anisotropy; deformation microstructures; muscovite; chlorite

Citation: Han, S.; Jung, H. Deformation Microstructures of Phyllite in Gunsan, Korea, and Implications for Seismic Anisotropy in Continental Crust. *Minerals* **2021**, *11*, 294. https://doi.org/10.3390/min11030294

Academic Editor: Paul Bons

Received: 31 December 2020
Accepted: 6 March 2021
Published: 11 March 2021

Publisher's Note: MDPI stays neutral with regard to jurisdictional claims in published maps and institutional affiliations.

1. Introduction

Seismic anisotropy originating in the interior of the earth provides important information for understanding tectonic processes, deep earth structures and geodynamics [1–8]. In continental crust, strong seismic anisotropies have been observed in large-scale tectonic structures such as mountain belts in orogenic systems [9–11], strike-slip faults or shear zones near plate boundaries [3,4], and the overriding upper crust in subduction zones [12,13]. In many cases, the fast S-wave polarization direction in continental crust is parallel to the major tectonic boundary formed by compressive regime [4,11,13]. These seismic patterns are usually attributed to the fault or fluid-filled cracks [14–16] and the orientation of anisotropic fabrics and structures [2,12,17].

Many studies using the receiver function technique [9,18], acoustic wave velocity measurements in laboratory settings [15,16,19], and fabric analysis on naturally [5,19–23] and experimentally [24,25] deformed rock samples have suggested that the layering and lattice preferred orientation (LPO) of anisotropic minerals is one of the important factors controlling the seismic anisotropy in the middle crust below the depth of microcrack closure (P ≈ 150–250 MPa [15,16,18,26]). Mica and amphibole groups are major constituent minerals in continental crust, which are elastically anisotropic [27–29]. In particular, phyllosilicate minerals show very strong anisotropy and it has been suggested that these minerals play an important role in affecting seismic anisotropies observed in various tectonic settings [5,17,22,30–34].

Owing to its abundance in the crust, the deformation mechanism and LPO of quartz have long been investigated. Quartz has been analyzed to provide structural data to construct the tectonic history of regional geology [35–40] and to understand its deformation mechanism [41–47]. The LPO data of albite or plagioclase have also been extensively studied [48–59]. Because quartz and albite crystals do not develop strong LPO in rocks, seismic anisotropy seems to be weakened when seismic waves pass through quartz- and albite-rich rocks [15,60,61]. Their diluting effect on seismic anisotropy must be compared with other phyllosilicate minerals based on their modal composition and LPOs to elucidate the relationship between the observed seismic anisotropy in the crust and the elastic properties of deformed rocks.

The phyllite from the Geumseongri Formation in Gunsan, Korea, has typical greenschist-facies mineral assemblages including muscovite, chlorite, biotite, quartz, and albite. The metamorphic condition and constituent minerals of this phyllite are representative of the middle crust where rocks were plastically deformed and subsequent LPO was formed. Electron backscatter diffraction (EBSD) analysis using fast and high-resolution mapping techniques is necessary because seismic properties should be calculated based on representative LPOs of minerals that reflect the exact volume proportions and deformation structures of natural rock samples. In this paper, we present data on deformation microstructures and LPOs of minerals revealed by EBSD, and the seismic properties of strongly deformed phyllite collected from the Geumseongri Formation in Gunsan, Korea, to understand the deformation mechanism of minerals and the causes of seismic anisotropy in the middle crust of highly deformed tectonic boundaries.

2. Geological Setting and Outcrop Description

The Korean Peninsula is composed of three Precambrian crystalline basements, the Nangrim, Gyeonggi and Yeongnam massifs from north to south (Figure 1a). Regional metamorphism and magmatism occurred around 1.9–1.8 Ga in these Paleoproterozoic massifs [62,63]. Two metamorphic belts, the Imjingang Belt and the Ogcheon Belt, comprising intensely deformed and metamorphosed sedimentary and volcanic rocks, separate these terranes. Permo-Triassic high-grade metamorphism strongly affected the Gyeonggi massif and two metamorphic belts, which have often been considered the possible eastward extension of the Qinling-Dabie-Sulu HP/UHP metamorphic belt in China. However, the Permo-Triassic tectonic model of the Korean Peninsula remains controversial [64–73]. All terranes were extensively intruded by Mesozoic plutonic rocks.

The Ogcheon Belt consists of Neoproterozoic to Paleozoic sedimentary and volcanic rocks which have been metamorphosed and strongly deformed in the Permian to Triassic. This fold-thrust belt separates the Gyeonggi and Yeongnam massifs, delineating the major tectonic boundary. The Ogcheon Belt is subdivided into two tectonic provinces: non-metamorphosed or slightly deformed early to late Paleozoic Taebaeksan Basin and strongly deformed and metamorphosed Neoproterozoic to Paleozoic Ogcheon Metamorphic Belt [68].

Most petrological and structural studies have focused on the northeastern part of the Ogcheon Metamorphic Belt. Cluzel et al. [64,65] divided the Ogcheon Metamorphic Belt into five main tectonic units bounded by thrusts: Iwharyeong, Poeun, Turungsan, Chungju, and Pibanryeong units. Although there has been some controversy regarding deformation stages and the exact timing of metamorphism, it has been suggested that bimodal volcanisms and sedimentation in the rift setting occurred from the Neoproterozoic [67,74–77] to late Paleozoic [78,79], followed by the subsequent Permian to Triassic tectonic event that extensively affected the Ogcheon Metamorphic Belt, forming the southeastward stacking of the nappes [64,65,67,72,79,80].

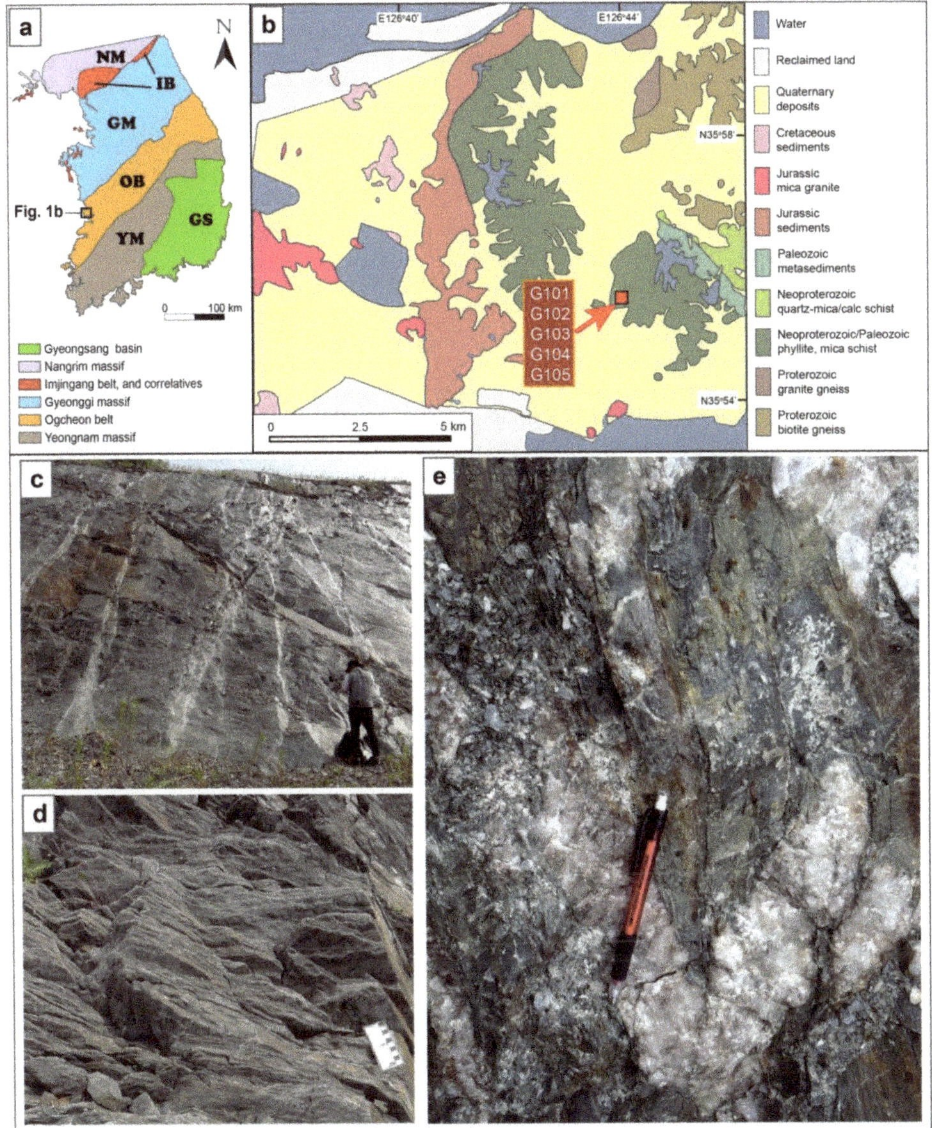

Figure 1. (**a**) Simplified tectonic sketch map showing the major tectonic units in the southern part of the Korean Peninsula. NM: Nangrim massif; GM: Gyeonggi massif; YM: Yeongnam massif; IB: Imjingang belt and correlatives; OB: Ogcheon belt; GS: Gyeongsang basin (reprinted with permission from ref. [81]. Copyright 2020 Elsevier). (**b**) Geological map of the Gunsan area and sample location (reprinted with permission from ref. [82] and ref. [83]. Copyright 2020 Korea Institute of Geoscience and Mineral Resources). (**c**) Field photograph showing phyllite outcrop of the Geumseongri Formation. (**d**) Close-up view of a strongly deformed phyllite showing well-developed foliation. (**e**) Folded and boudinaged quartz veins discordantly cutting phyllite foliation.

The P-T conditions of metamorphism of the NE Ogcheon Metamorphic Belt have been calculated in the range of 4.2–8.2 kbar and 490–540 °C in the Poeun unit and 5.4–9.4 kbar and 520–630 °C in the Pibanryeong unit based on garnet-biotite geothermometer and garnet-plagioclase-biotite-quartz or muscovite geobarometer [67,84]. A decrease in metamorphic grade towards the southwest along the strike of the Ogcheon Metamorphic Belt was suggested based on the change in the mineral assemblage of the Poeun unit [85]. Another study suggested the lower metamorphic condition of the Poeun unit (3.6–4.4 kbar and 350–450 °C) close to the study area in the SE Ogcheon Metamorphic Belt, based on the mineral assemblage and chemistry of muscovite [72,86].

The studied samples were obtained from the Geumseongri Formation in Gunsan, Korea (Figure 1a,b). There have not been many petrological and structural studies on the southwestern edge of the Ogcheon Metamorphic Belt. The lithologies in the study area consist of Paleoproterozic gneiss, Neoproterozoic to Paleozoic phyllite, schist and metasediments, Jurassic sedimentary rock, and granites (Figure 1b). The depositional age of the Geumseongri Formation is controversial. Two geological maps of adjacent regions [82,83] proposed different depositional ages of the protolith of the Geumseongri Formation as Neoproterozoic and Paleozoic, respectively. The detrital zircon age distribution of the Geumsengri Formation showed the youngest age of 806±19 Ma, which is the maximum age of sedimentation [82]. The ages of metamorphism and deformation in the Geumseongri Formation have not yet been reported.

At the sample location, dark-grey phyllite outcrops with shiny mica-rich foliation are observed (Figure 1c,d). Kink bands and chevron folds are abundant. Folded and boudinaged 10–50 cm thick quartz veins, with some including fragments of phyllite, discordantly cut the phyllite foliation (Figure 1e). The axial planes and fold limbs of quartz veins are typically subparallel to the foliation of phyllite, indicating the contemporary ductile deformation of the quartz vein and phyllite. Outcrop-scale folds and faults affected the strikes of phyllites, which are WSW–ENE to WNW–ESE with varying dips.

3. Methods

Five samples of phyllite were selected from the Geumseongri Formation in Gunsan, Korea. The foliation and lineation of each sample were determined by observing the compositional layering and stretching lineation of muscovite and quartz. For samples showing no clear lineation in the hand specimen, we analyzed grain shapes of digitized lines from elongated minerals on the foliation to ensure that lineation was determined by the orientation of maximum elongation, following the method of Panozzo [87]. Based on the decided foliation and lineation, standard 30 μm thick–thin sections of the XZ plane were made, where X is parallel to the lineation and Z is perpendicular to the foliation of each sample. Thin sections were polished using 1 μm powder and colloidal silica (0.06 μm) to remove surface damage.

The LPOs of the minerals were measured using a scanning electron microscope equipped with an EBSD system. EBSD data were collected using a JEOL JSM-7100F field emission scanning electron microscope (FE-SEM, JEOL, Tokyo, Japan) equipped with a Symmetry detector (Oxford Instruments, Abingdon, UK) installed at the School of Earth and Environmental Sciences, Seoul National University, Korea. The samples were analyzed with an accelerating voltage of 20 kV at a 25.0 mm working distance on a stage tilted 70°. The Kikuchi patterns were automatically obtained and indexed using AZtec software (Version 4.3, Oxford Instrument, Abingdon, UK) with a step size of 5 μm for four samples (Table 1). The step size was determined to be approximately 1/10 of the average diameter of the major minerals forming each sample, which clearly demonstrated the shape of the grains. We analyzed one additional sample with a step size of 0.31 μm to observe the internal microstructures of the deformed grains (Table 1). The rates of zero solutions in raw data ranged from 4.7% to 18.3%, which were usually found along grain boundaries and fractures. We conducted zero solution correction as follows: (1) wild spikes were eliminated first; (2) zero solutions were corrected if each was surrounded by at least six consistent pixels; and (3) wild spikes were eliminated again. By performing minimal correction in (2), EBSD data were conserved, and grain shape was not distorted by correction. To avoid oversampling from large grains, pole figures were plotted from one point per grain using HKL Channel 5. Mean aspect ratios of grains were calculated for quartz and albite composed of more than 10 pixels (60 pixels for G105) (Table 1).

Table 1. Mineral modal composition, mean aspect ratio of grains in the phyllite samples, and step size of electron backscatter diffraction (EBSD) analysis.

Sample No.	Modal Composition (%) [1]						Mean Aspect Ratio		Step Size (μm)
	Ms	Chl	Qtz	Ab	Cal	Accessory	Qtz	Ab	
G101	34.7	7.6	38.6	13.4	0.9	4.8	1.91	2.44	5
G102	31.1	11.3	33.0	15.0	0.9	8.7	1.93	2.11	5
G103	23.4	8.8	35.7	22.8	4.6	4.7	1.87	2.25	5
G104	22.1	13.8	47.6	14.0	0.8	1.7	2.12	3.39	5
G105	4.6	5.7	67.6	12.1	2.3	7.7	2.04	1.87	0.31

[1] Modal composition was estimated based on the EBSD map data. Ms: muscovite, Qtz: quartz, Chl: chlorite, Ab: albite, Cal: calcite

To understand the relationship between the fabric strength of the constituent minerals and seismic anisotropy, the misorientation index (M-index [88]) was calculated for minerals in each sample (Table 2) using the MTEX toolbox for Matlab, version 5.4.0 [89], using one point per grain data. We used the Bootstrap method to estimate the uncertainties of M-index [90]. First, a group of grains was drawn from the original data with replacement. Based on the number of grains in the original data, about 1/6–1/15 of grains (for example, 200 grains for quartz from each sample) were drawn for each group [91]. Second, this process was replicated 1000 times for each mineral. Third, the M-index was calculated for each random group of grains. Finally, with the M-indices of groups, the mean and ±95% confidence intervals were calculated (Table 2) except for G105, which had too few grains.

Table 2. Fabric strength of minerals in phyllite samples.

Sample No.	M-Index							
	Ms		Chl		Qtz		Ab	
	Mean	CI [1]	Mean	CI	Mean	CI	Mean	CI
G101	0.3172	±0.01170	0.2510	±0.04770	0.0387	±0.00890	0.0627	±0.01395
G102	0.1693	±0.02335	0.1711	±0.04980	0.0326	±0.00670	0.0632	±0.01460
G103	0.2194	±0.02090	0.1578	±0.04575	0.0537	±0.01055	0.0639	±0.01295
G104	0.3175	±0.02775	0.2928	±0.06080	0.0555	±0.00570	0.0821	±0.01860
G105	0.21	-	0.293	-	0.09	-	0.379	-

[1] CI: ± 95% confidence interval. Ms: muscovite, Qtz: quartz, Chl: chlorite, Ab: albite.

The seismic velocity and anisotropy of each mineral in the samples were calculated using the single-crystal elastic constant, all LPOs, and crystal density. We used the Voigt–Reuss–Hill averaging scheme and Fortran program created by Mainprice [92]. Single crystal elastic constants were used for quartz [93], plagioclase [94], muscovite [29], and chlorite [28]. Using volume proportions measured by EBSD analysis, the seismic velocity and anisotropy of whole rocks were calculated. To investigate the relationship between the seismic properties of platy minerals, quartz and albite, the P-wave anisotropy (AVp) and the maximum S-wave anisotropy (max. AVs) of muscovite + quartz + albite, chlorite + quartz + albite and muscovite + chlorite + quartz + albite for sample G102 were calculated.

4. Results

4.1. Sample Description and Microstructures

The phyllite samples are mainly composed of quartz (33–68%), albite (12–23%), muscovite (5–35%), chlorite (6–14%), with minor biotite, calcite, and ilmenite (Table 1). Hand specimens are fine-grained and exhibit shiny mica-rich foliation. All samples were strongly foliated and compositionally layered, showing alternating Q-domain (quartz and albite) and M-domain (muscovite and chlorite) in photomicrographs (Figure 2a). The grain sizes of quartz and albite vary in the range of 0.02 to 0.5 mm, and they have a subangular and elliptical grain shape elongated parallel to the foliation of each sample.

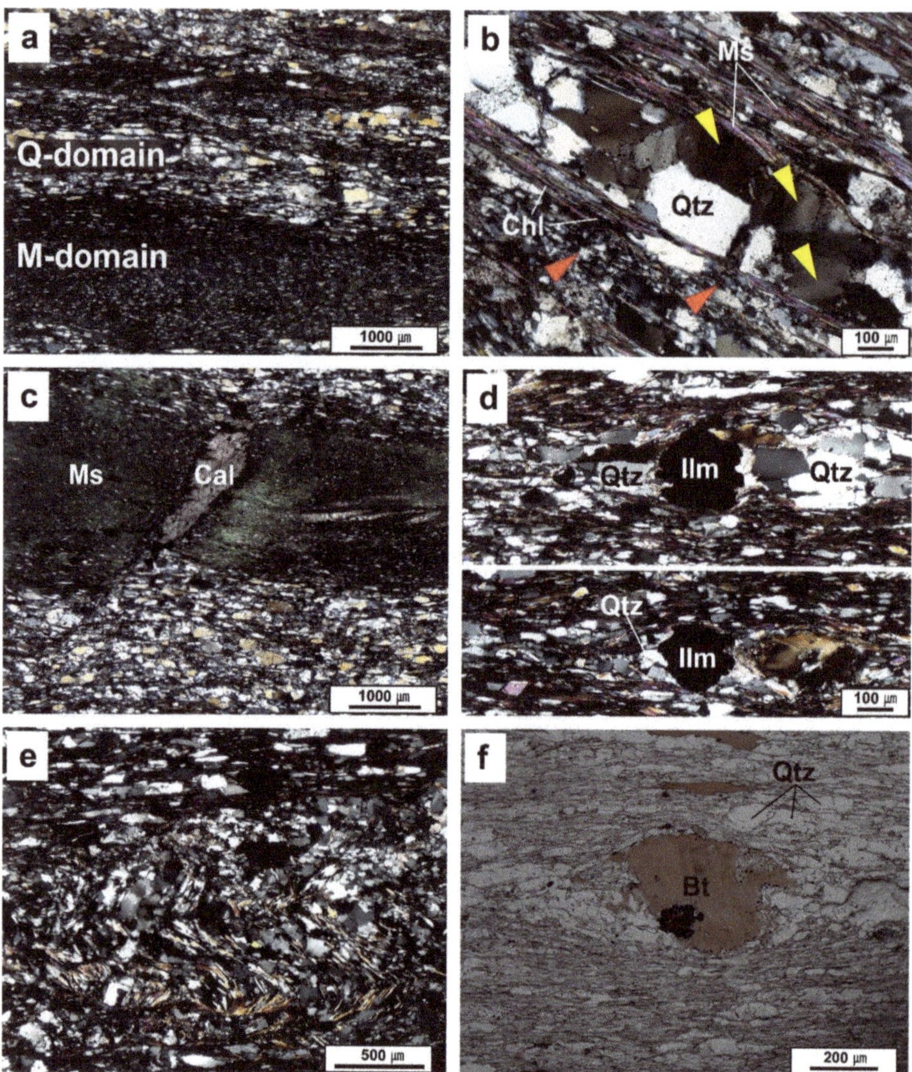

Figure 2. Optical photomicrographs of phyllite samples from the Geumseongri Formation. (**a**) Phyllite showing well-developed foliation with compositional layering of alternating Q-domain mainly composed of quartz and albite, and M-domain of mica (sample G103). (**b**) Quartz ribbon displaying undulose extinction and subgrain boundaries (sample G104). Yellow arrows represent undulose extinction of quartz. Red arrows represent recrystallized small grains of quartz. Qtz: quartz, Ms: muscovite. (**c**) Microcrack in M-domain filled with a calcite crystal (sample G103). Cal: calcite. (**d**) Pressure shadow of quartz around ilmenite porphyroclasts. Strain cap is partly observed (sample G102). Ilm: ilmenite. (**e**) Small-scale microfolds showing limbs and axial planes parallel to the foliation (G101). (**f**) Sheared biotite grain showing dextral shear sense (G102). Bt: biotite.

Mica grains usually fill intergranular spaces and/or are strongly aligned parallel to the foliation, forming relatively thick mica bands (Figure 2a,c).The old, relatively large quartz grains exhibit undulose extinction (Figure 2b) and subgrain boundaries indicating intracrystalline deformation, which can be observed in the inverse pole figure map created using a fine step size (Figures 3b and 4a). Quartz grains commonly form ribbons consisting of deformed quartz grains with some recrystallized small grains (Figure 2b).

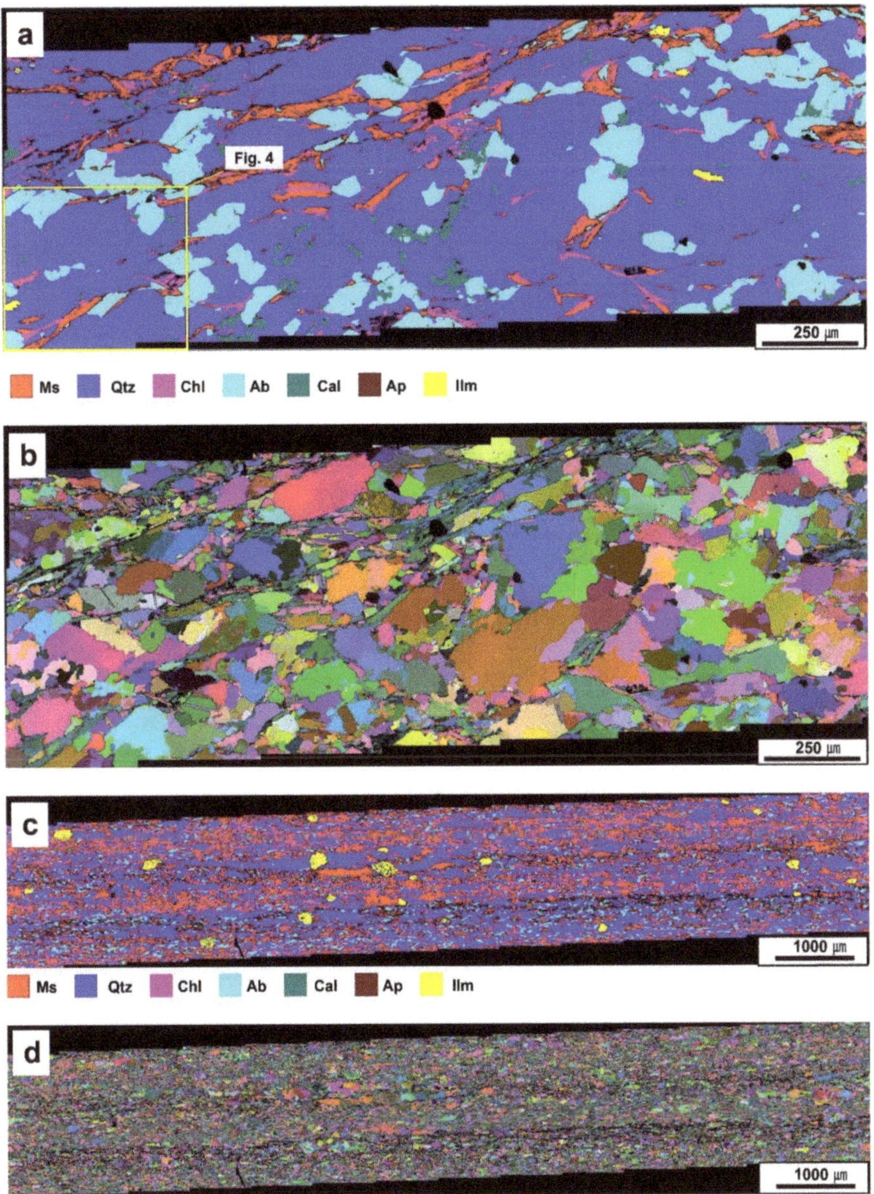

Figure 3. EBSD phase maps (**a**,**c**), and Euler maps (**b**,**d**) of phyllite samples. (**a**,**b**) Sample G105 analyzed with step size of 0.31 μm. (**c**,**d**) Sample G102 analyzed with step size of 5 μm. Magnified EBSD map of the yellow box in Figure 3a is shown in Figure 4. Ms: muscovite, Qtz: quartz, Chl: chlorite, Ab: albite, Cal: calcite, Ap: apatite, Ilm: ilmenite.

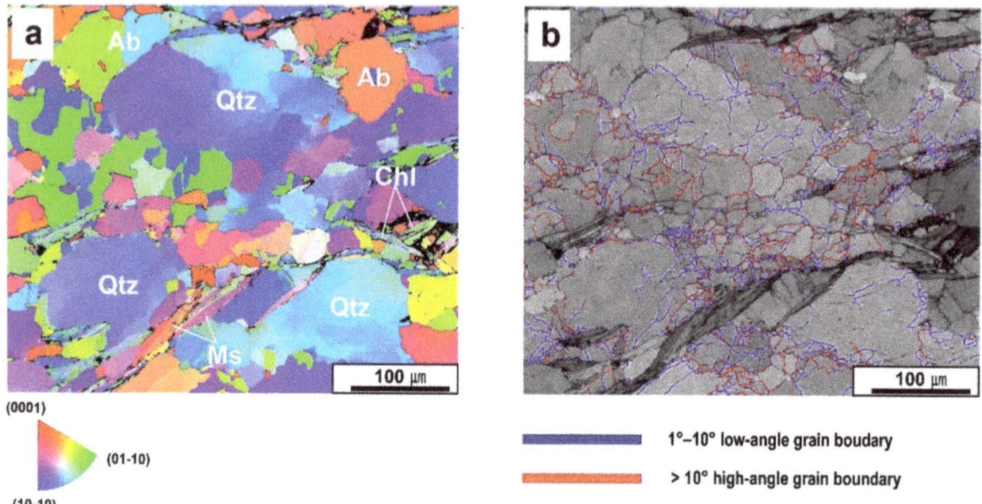

Figure 4. Magnified EBSD maps of sample G105 (Yellow box in Figure 3a). (**a**) Inverse pole figure showing crystallographic axes aligned parallel to the lineation. Quartz grains have subgrain boundaries, indicating intracrystalline plastic deformation, while albite rarely show microstructures of ductile deformation. (**b**) Low-angle (1°–10°) and high-angle (<10°) grain boundary map of quartz on the band contrast map. Note that grain boundaries of albite were excluded.

The pressure shadow of quartz grains around ilmenite porphyroclasts was observed frequently, indicating that the dissolution–precipitation was active (Figure 2d). Small-scale microfolds showing limbs and axial planes parallel to the foliation of the sample are observed, indicating overlapped deformation stages (Figure 2e). Asymmetrically deformed and rotated porphyritic biotite grains are occasionally observed, showing concordant shear direction in each sample (Figure 2f). However, penetrative non-coaxial deformation structures such as S-C fabric are not well developed (Figures 2 and 3). Albite is difficult to distinguish from quartz in an optical microscope with the exception of rare grains exhibiting twinning.

4.2. LPO and Fabric Strength of Minerals

4.2.1. LPO of Muscovite and Chlorite

The LPOs of muscovite and chlorite are shown in Figure 5. In general, [001] axes of both muscovite and chlorite show strong point maxima normal to foliation while (110) poles, [100] axes and (010) poles are aligned in a girdle parallel to the foliation (Figure 5). The fabric strength of each mineral is shown as an M-index in Table 2. The M-index of muscovite was the strongest among minerals, ranging from 0.169 to 0.317, while that of quartz was the weakest. Chlorite also showed a strong fabric strength in the range of M = 0.158–0.293.

4.2.2. LPO of Quartz

The LPOs of quartz are generally weak, with the c-axes aligned around the pole of foliation. In samples G102 and G104, the c-axes are also aligned subnormal to the lineation. The G102 sample shows a typical pattern of quartz LPO forming a crossed girdle. Sample G104 showed some c-axes aligned subparallel to the lineation. The fabric strength of quartz is weak in the range of M = 0.033–0.090 (Table 2). No significant difference was observed in the M-index of quartz among the samples, excluding sample G105. Note that the LPO and fabric strength of quartz from sample G105 was overestimated because the number of analyzed grains was too small.

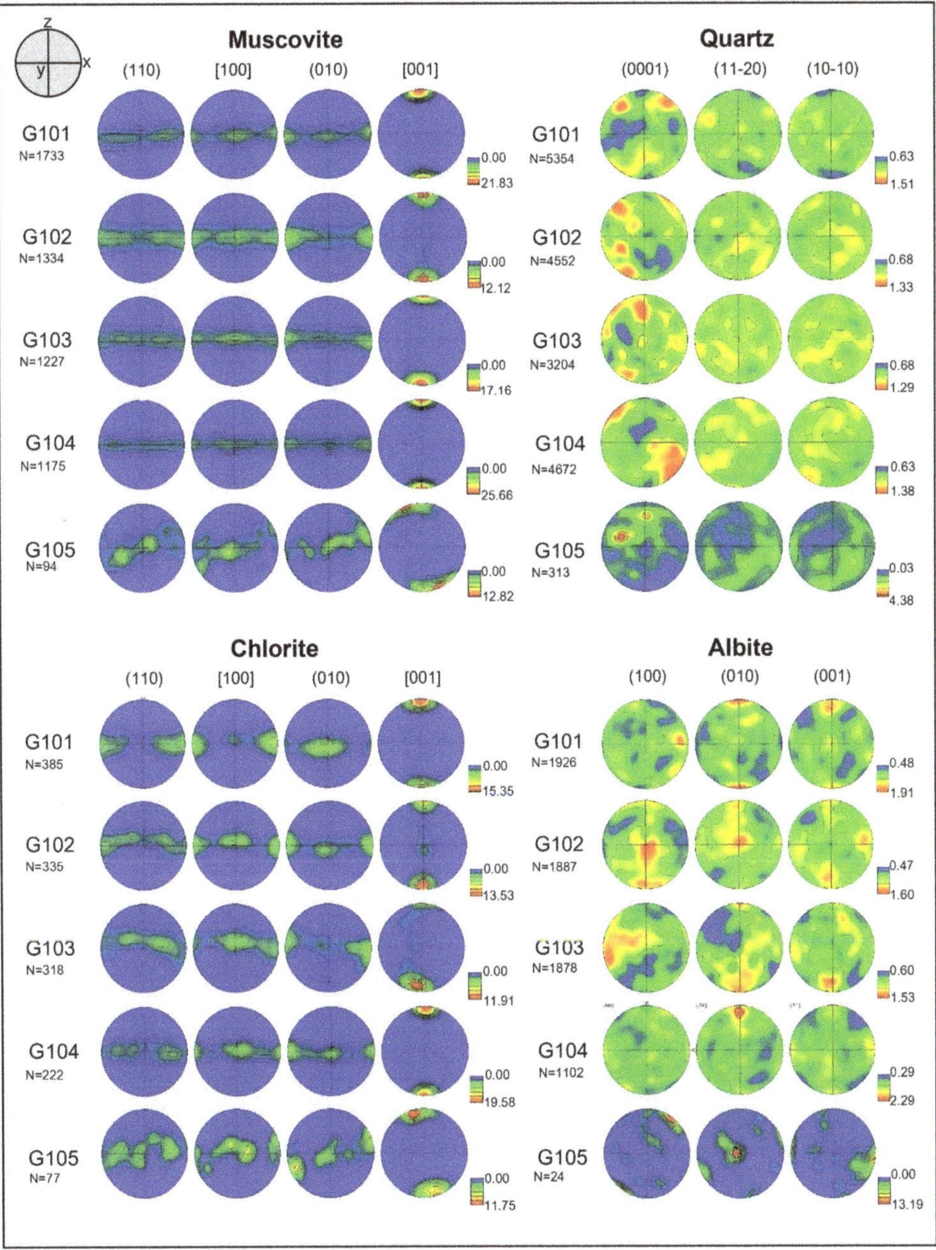

Figure 5. Pole figures of the quartz, muscovite, albite, and chlorite presented in the lower hemisphere using equal-area projection. A half-scatter width of 20° was used. The color coding indicates the data point density. Multiples of uniform distribution are denoted as the numbers in the legend. E–W direction of each figure corresponds to stretching lineation (X), and N–S direction (Z) is perpendicular to the foliation. N = number of data points.

4.2.3. LPO of Albite

The most common pattern of albite LPOs was characterized by the (001) pole aligned subnormal to the foliation (samples G101, G102, G103, and G104). In two samples (G102, G104), (001) poles are also partly aligned subparallel to lineation (X), forming two conjugate maxima in the XZ plane. One sample (G103) showed the (001) pole aligned subparallel to the lineation. Two samples (G101 and G104) showed that the (010) poles strongly aligned subnormal to the foliation. The fabric strength of albite is in the range of M = 0.063–0.082, and it is weaker than other minerals, excluding that in sample G105 (Table 2). Note that the LPO and fabric strength of albite from sample G105 were overestimated because too few grains were analyzed. In Figure S1, additional LPOs of the albite axes measured in this study are displayed as pole figures. The <001> axes of two samples (G102 and G104) and the <100> axis of one sample (G101) were aligned subparallel to the lineation (Figure S1).

4.3. Seismic Anisotropies of Phyllite Samples

The seismic properties of muscovite, chlorite, quartz, albite, and whole rocks are shown in Figures 6–10 and Table 3. The seismic anisotropies of the P-wave (AVp) were in the range of 43.4–51.1% for muscovite, 21.4–29.4% for chlorite, 2.8–7.1% for quartz, and 1.2–15.6% for albite. The maximum anisotropy of the S-wave (max. AVs) was in the range of 41.9–59.1% for muscovite, 34.9–56.8% for chlorite, 3.5–9.8% for quartz, and 1.3–24.8% for albite. Muscovite showed the strongest P- and S-wave anisotropy while albite showed the weakest P- and S-waves anisotropy. The abnormally high AVp and maximum AVs of albite in sample G105 resulted from the exaggerated LPO data due to the limited number of albite grains analyzed. Therefore, the realistic AVp and maximum AVs for albite in this study are in the ranges of 1.2–2.1% and 1.3–1.7%, respectively. Integrating all seismic anisotropies of minerals based on the modal composition of each sample, the calculated AVp and maximum AVs of whole rocks were 9.0–21.7% and 9.6–24.2%, respectively. The patterns of contours and the polarization direction of the fast S-wave of whole rocks were quite similar to those of muscovite and chlorite, except for sample G105, which has ~80% quartz and albite (Figure 10).

Table 3. Mineral and whole-rock Seismic velocity and anisotropy in phyllite samples.

Sample No.	Muscovite		Chlorite		Quartz		Albite		Whole Rock	
	AVp (%)	Max. AVs (%)	AVp (%)	Max. AVs (%)	AVp (%)	Max. AVs (%)	AVp (%)	Max. AVs (%)	AVp (%)	Max. AVs (%)
G101	51.1	58.1	26.7	51.1	4.5	5.7	1.2	1.5	21.7	24.2
G102	43.4	41.9	25.3	40.2	2.8	3.5	2.1	1.3	18.9	19.7
G103	44.4	47.7	21.4	34.9	3.9	4.7	1.3	1.7	14.3	15.4
G104	51.1	59.1	29.4	56.8	4.3	5.7	1.3	1.6	17.2	20.3
G105	43.9	42.8	26.5	47.4	7.1	9.8	15.6	24.8	9	9.6

To understand the relationship between various mineral assemblages in phyllite and seismic anisotropies, the AVp and maximum AVs of sample G102 were calculated based on different groups of minerals (Figure 11). The calculated results showed that the anisotropies of the P-wave were 18.9% for muscovite + chlorite + quartz + albite (Figure 11a), 17.7% for muscovite + quartz + albite (Figure 11b), and 6.2% for chlorite + quartz + albite (Figure 11c). The maximum anisotropies of the S-wave were 8.2% for chlorite + quartz + albite, 16.5% for muscovite + quartz + albite, and 19.7% for muscovite + chlorite + quartz + albite. The results also elucidated the constructive role of coexisting muscovite and chlorite on seismic anisotropy.

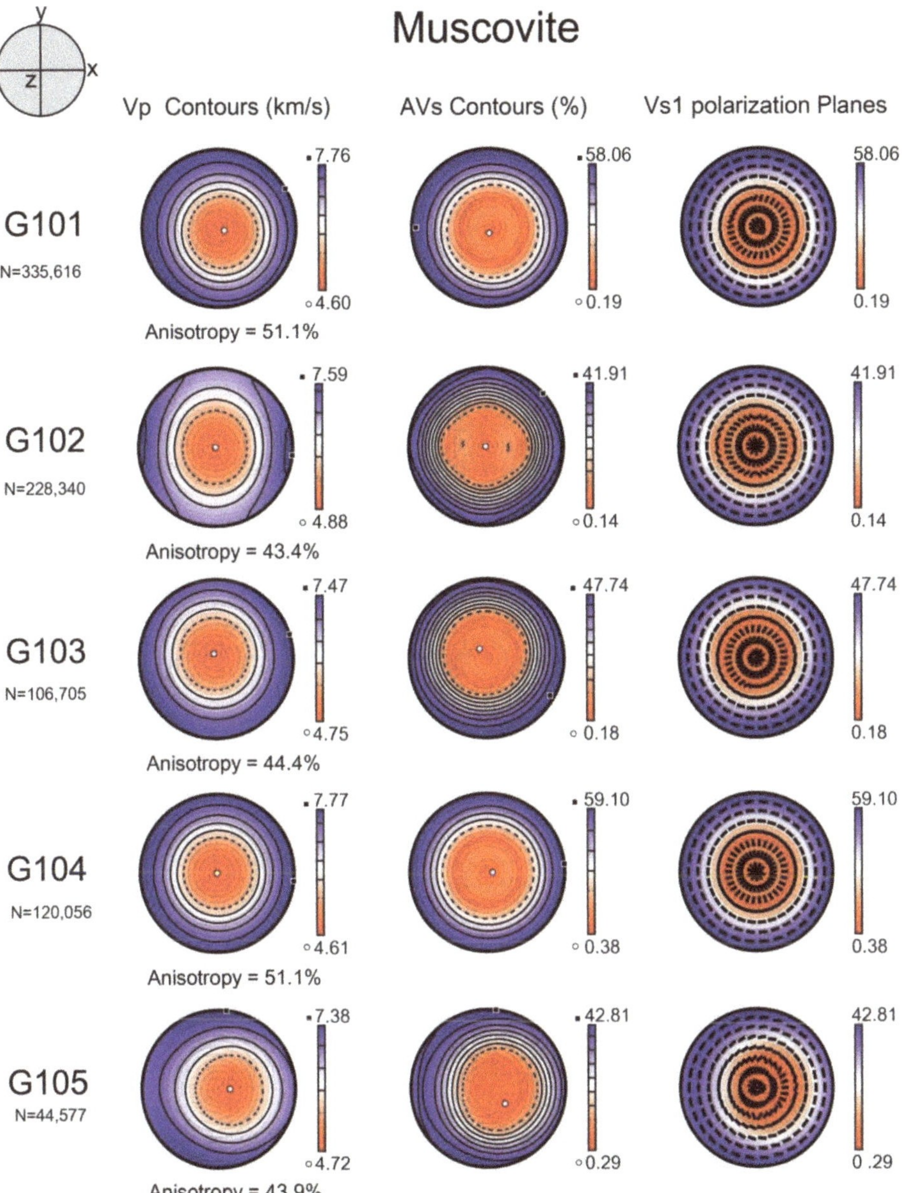

Figure 6. Seismic properties of muscovite calculated from lattice preferred orientation (LPO) and elastic constant. The P-wave velocity (Vp), amplitude of the shear-wave anisotropy (AVs), and polarization direction of the faster shear wave (Vs1) are plotted in the lower hemisphere using an equal-area projection. E–W direction of each figure corresponds to stretching lineation (X), and Z is normal to the foliation. N = number of data points.

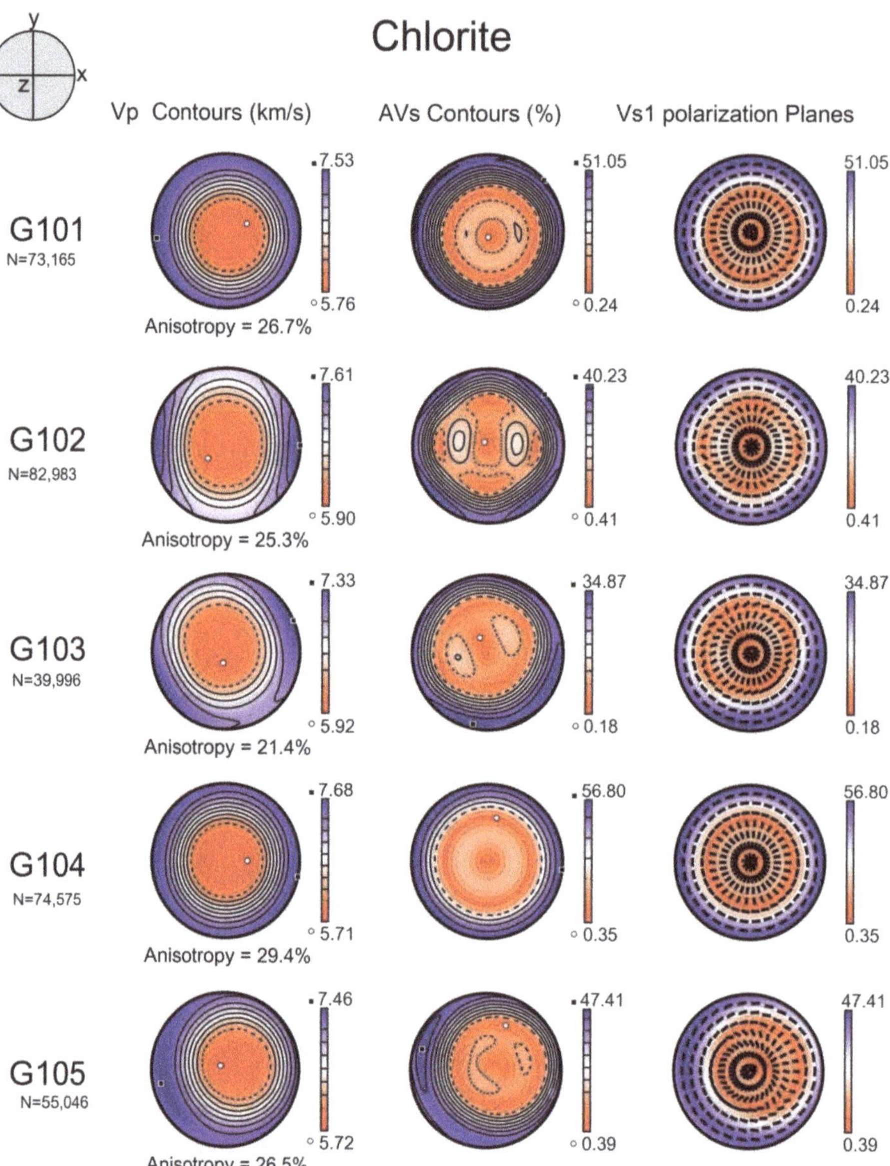

Figure 7. Seismic properties of chlorite calculated from LPO and elastic constant. The P-wave velocity (Vp), amplitude of the shear-wave anisotropy (AVs), and polarization direction of the faster shear wave (Vs1) are plotted in the lower hemisphere using an equal-area projection. E–W direction of each figure corresponds to stretching lineation (X), and Z is normal to the foliation. N = number of data points.

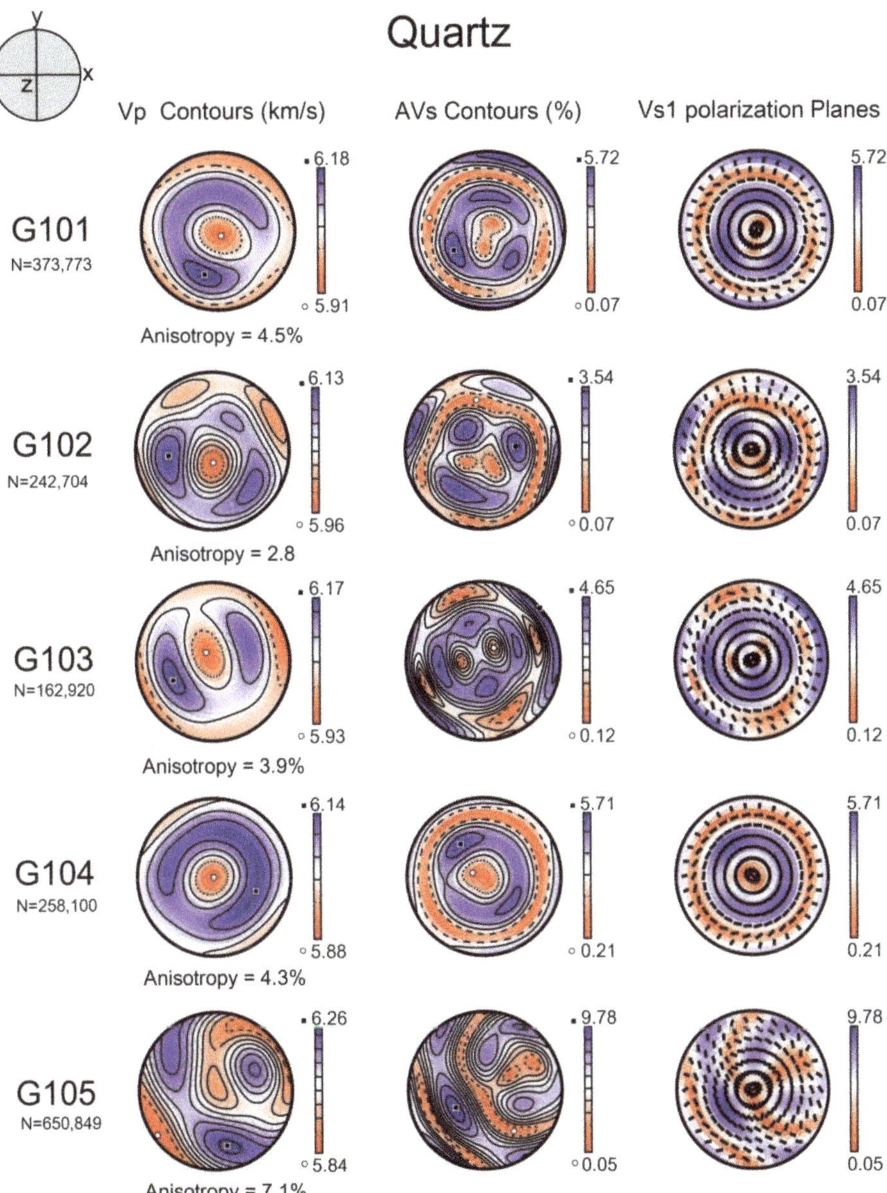

Figure 8. Seismic properties of quartz calculated from LPO and elastic constant. The P-wave velocity (Vp), amplitude of the shear-wave anisotropy (AVs), and polarization direction of the faster shear wave (Vs1) are plotted in the lower hemisphere using an equal-area projection. E–W direction of each figure corresponds to stretching lineation (X), and Z is normal to the foliation. N = number of data points.

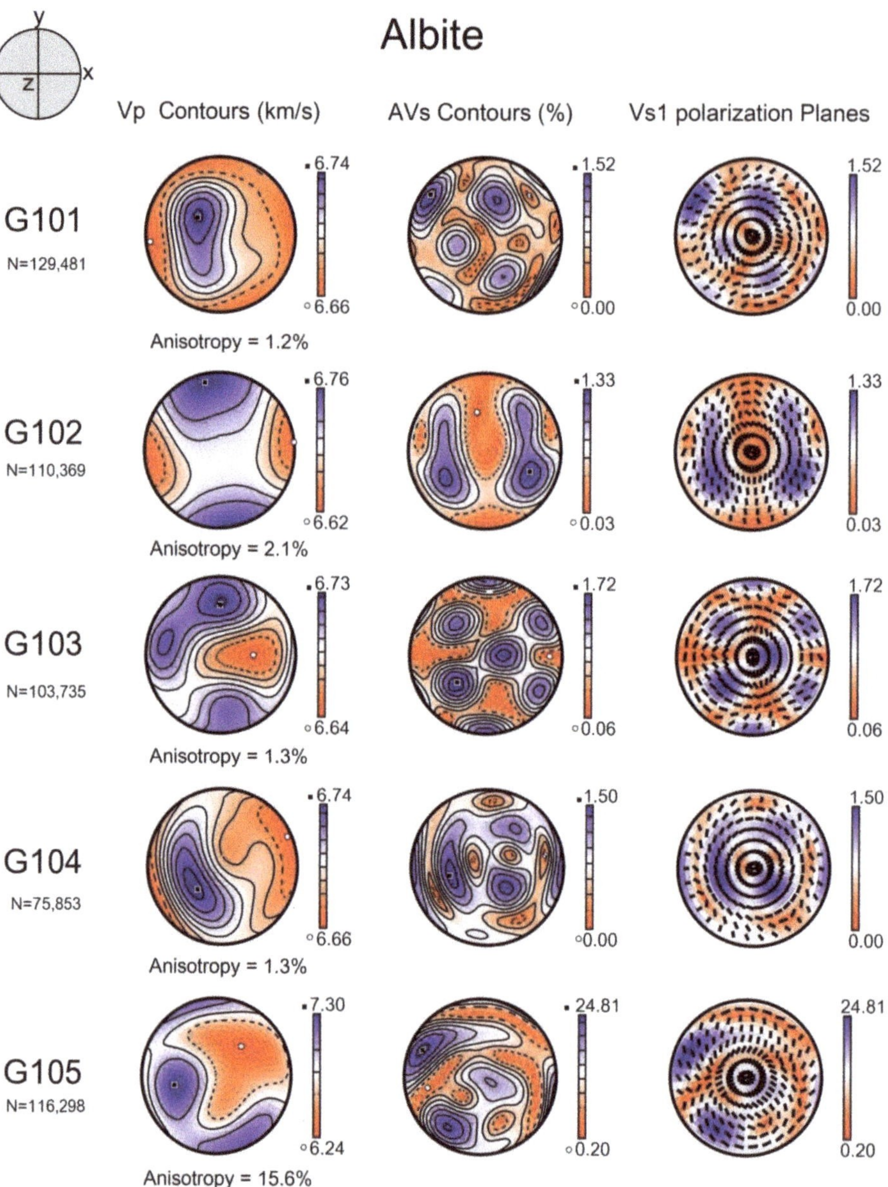

Figure 9. Seismic properties of albite calculated from LPO and elastic constant. The P-wave velocity (Vp), amplitude of the shear-wave anisotropy (AVs), and polarization direction of the faster shear wave (Vs1) are plotted in the lower hemisphere using an equal-area projection. E–W direction of each figure corresponds to stretching lineation (X), and Z is normal to the foliation. N = number of data points.

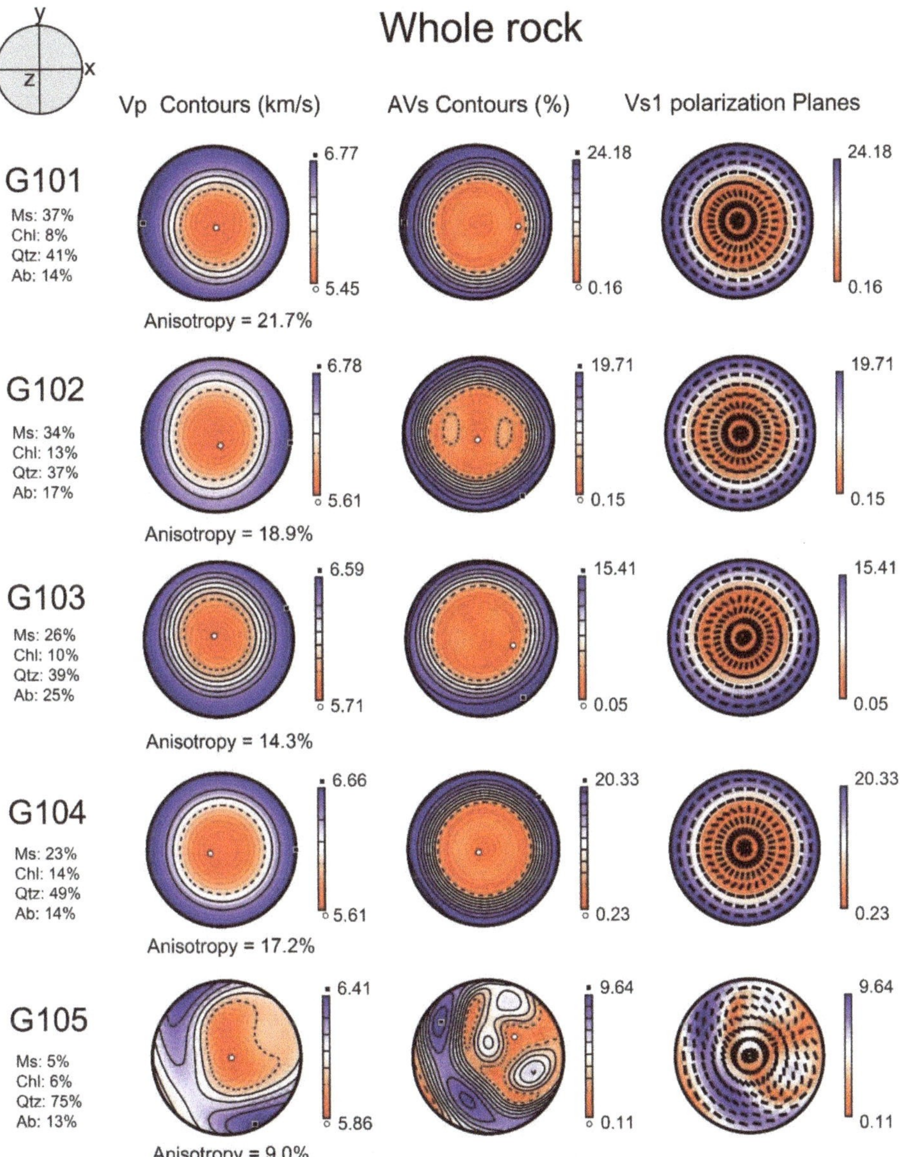

Figure 10. Whole rock seismic properties calculated from LPOs, volume proportions and elastic constant of each minerals. The P-wave velocity (Vp), amplitude of the shear-wave anisotropy (AVs), and polarization direction of the faster shear wave (Vs1) are plotted in the lower hemisphere using an equal-area projection. E–W direction of each figure corresponds to stretching lineation (X), and Z is normal to the foliation.

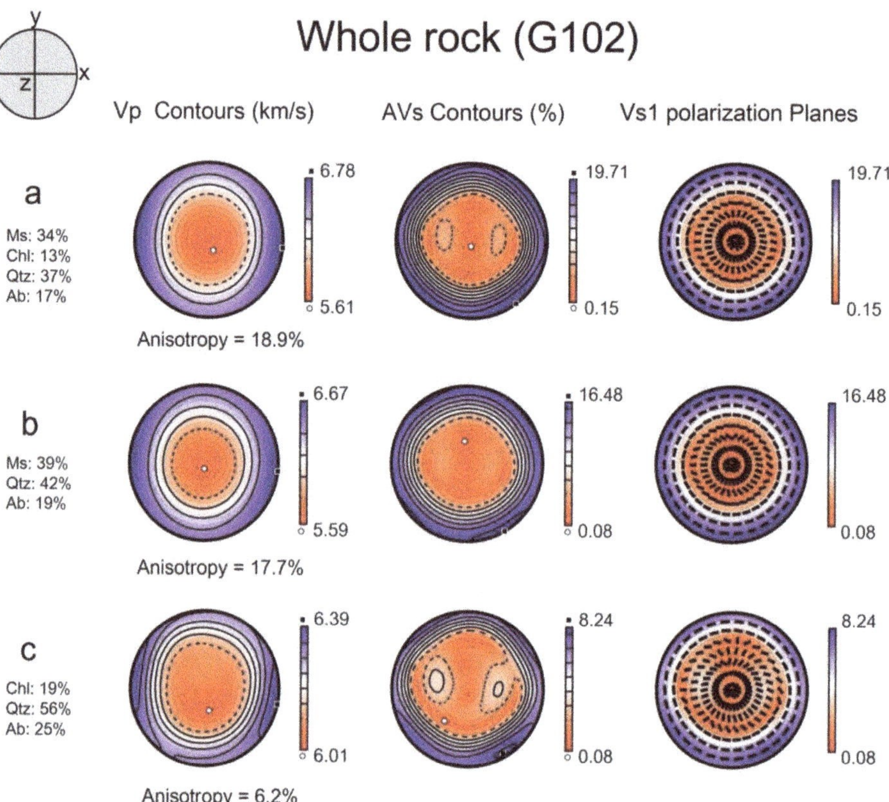

Figure 11. Whole rock seismic property of sample G102 calculated from LPOs, volume proportions and elastic constant of each mineral. (**a**) Seismic anisotropy of muscovite + chlorite + quartz + albite. (**b**) Seismic anisotropy of muscovite + quartz + albite. (**c**) Seismic anisotropy of chlorite + quartz + albite. The P-wave velocity (Vp), amplitude of the shear-wave anisotropy (AVs), and polarization direction of the faster shear wave (Vs1) are plotted in the lower hemisphere using an equal-area projection. E–W direction of each figure corresponds to stretching lineation (X), and Z is normal to the foliation.

5. Discussion

5.1. LPO Development of Minerals

The [001] axes of both muscovite and chlorite show strong point maxima normal to foliation while (110) poles, [100] axes and (010) poles are aligned in a girdle parallel to the foliation (Figure 5). The LPOs of muscovite and chlorite in this study are well known fabric patterns of phyllosilicates reported from naturally [34,95–100] and experimentally deformed rocks [101]. The platy shape of muscovite and chlorite is a primary factor in the rotation of these minerals parallel to foliation during deformation [102]. Recrystallization has also been suggested as the main deformation mechanism of muscovite [34]. The strong alignment of the muscovite [001] axis in this study is interpreted as the easy glide on the (001) planes [95,97,99,103]. Experimental studies have reported <100>(001) and <110>(001) slip systems of muscovite [103]. Muscovite bands in M-domains clearly showed undulose extinction of muscovite and chlorite (Figures 3a and 4a) indicating the operation of dislocation creep in muscovite and chlorite. Therefore, it is suggested that strong LPOs of muscovite and chlorite were formed due to dislocation creep on the basal plane and crystal rotation parallel to the foliation.

The LPO of quartz depends on many factors, including the deformation mechanism, active slip system, strain rate, temperature, kinematic framework, and water contents [38,42,43]. However, in naturally deformed samples, it is very difficult to measure the strain rate of samples in thin sections.

Therefore, we focused on the role of temperature in this study. Folds of quartz veins in the outcrop (Figure 1e) and prevalent subgrain boundaries of quartz (Figures 3b and 4b) indicate that the deformation temperature was high enough for the ductile deformation of quartz. In the quartz LPO data, one of the main features observed was the c-axes maximum at a small angle to the direction normal to the foliation (Figure 5). It is correlated to the basal<a> slip system, which is active under relatively low temperature conditions [38,102]. In sample G102, the cross girdle of the quartz c-axis indicates multiple slip systems activated including basal<a>, rhomb<a>, and prism<a>. An early study of naturally deformed quartz reported that an abrupt change from these multiple slip systems to a single slip system (prism<a>) occurs at approximately 500 °C, which was observed in metasedimentary rocks with similar mineral assemblages to those of our phyllites in the Tonale fault zone [37]. This is also supported by the observation of subangular albite grains with rare intracrystalline deformation structures in this study (Figures 3a and 4a), indicating that albite was mainly deformed in brittle manner. In addition, the mineral assemblage of phyllites seems to correspond to the Poeun unit in the area of the decreased metamorphic condition [85,86], which might indicate a lower metamorphic grade than the NE Ogcheon Metamorphic Belt [84]. Therefore, it is suggested that the deformation temperature of phyllite was below 490–540 °C and possibly in the range of 350–450 °C [86]. This argument reflects our finding of the suggested maximum deformation temperature (about 500 °C) of quartz based on the measured LPO. Additional petrological analysis should be performed to confirm the exact metamorphic conditions and deformation temperature of the phyllite.

In sample G104, some quartz c-axes are aligned subparallel to the lineation, which might indicate the increased prism<c> slip. However, we did not observe any evidence of high-temperature deformation of quartz, such as grain boundary migration recrystallization, chessboard extinction pattern, or pinning structure. Several studies have reported that water can change the relative activity of different slip systems [41,104], and decreasing the strain rate can increase the activity of prism<c> slip [38,41]. Anisotropic growth promoted by dissolution–precipitation creep could affect the quartz c-axes aligned subparallel to the lineation [105,106]. In this study, pressure shadows of quartz around ilmenite porphyroclasts were frequently observed (Figure 3d), indicating the dissolution–precipitation process of quartz. We suggest that the main deformation mechanism of quartz is represented by dislocation creep on the basal<a>, prism<a>, and rhomb<a> slip system, and some alignment of quartz c-axes subparallel to the lineation might have been affected by the dissolution–precipitation creep where the fluid was abundant.

Notably, the LPO of plagioclase in this study showed that the (010) and (001) poles simultaneously showed maxima subnormal to the foliation (Figure 5), which has been documented independently from the study of magmatic processes on anorthosite [51] and the deformation mechanism of amphibolite and mafic schist [52]. In the case of anorthosite, the LPO of plagioclase was interpreted as the interchange between the poles of (010) and (001) of rod-shaped prismatic plagioclase crystals in melt flow [51]. The LPO of plagioclase in the amphibolite and mafic schist was explained by changes in dislocation slip systems influenced by strain and temperature [52]. Our albite LPO data may indicate that multiple slip planes of dislocation glide have been activated. The LPO of albite (G104) indicates an activation of the (010)[001] slip system, which has been reported as the dominant slip system in plagioclase (Figure 5 and Figure S1; [57–59]). However, the phyllites in this study are thought to have been deformed in greenschist facies where the temperature is lower than 500 °C. Subangular albite grains (Figures 3a and 4a) and a relatively small number of internal structures such as subgrain boundaries (Figure 4a) indicate that crystal plastic deformation was only a minor component. In addition, the low symmetries of feldspar and the large unit cell may make it difficult to have multiple activated slip systems particularly in low temperature conditions [54,107]. The suitable interpretation in this study is based on the physical properties of albite with two good (001) and (010) cleavage planes [107]. These planes formed via brittle deformation can be related to the shape preferred orientation (SPO) of albite with a mean aspect ratio of about two in this study (Table 1). Fracturing has been pointed out as the active deformation mechanism of plagioclase even at high temperatures [53,57,59,107,108]. Particularly, the LPO of albite in two samples (G102 and G104) showed their (001) poles (and partly, (010) poles) aligned subnormal to the foliation and (001) poles aligned subparallel to the lineation, indicating interchange of two poles during rotation of albite grains. Therefore, it is suggested that the predominant deformation mechanism of albite in the samples is represented by rigid body rotation with alignment of (010) and (001) cleavage planes parallel to the foliation, combined with a minor dislocation creep with a (010)[001] slip system. This interpretation is concordant with the quartz LPO showing multiple slip systems that are active at relatively low temperature conditions below 500 °C [37,102]. It is also possible that anisotropic growth by dissolution–precipitation creep might have affected the LPO of albite [53,105].

5.2. Implications for Seismic Anisotropy in Continental Crust

Because the mineral assemblages of the phyllite samples analyzed in this study correspond to typical greenschist-facies minerals of a pelitic protolith (biotite zone [67]), it is suggested that the calculated seismic properties reflect a part of the observed seismic anisotropies in the highly deformed middle crustal zone. Tectonic processes during compressive or extensional regimes in the middle and lower crust result in ductilely deformed planar structures aligned subnormal to compressive or extensional stress, inducing strong seismic anisotropies when the rocks are rich in mica and/or amphibole [5,18,20,22,60,109]. It is particularly important below the depth of microcrack closure (approximately 200–250 MPa [15,16,18]), where the plastically deformed shear fabrics and aligned minerals play an important role in controlling the seismic properties.

If the seismic waves pass through the strongly foliated phyllites rich in muscovite, the maximum P-wave velocity (Vp) would be observed where the seismic wave propagated subparallel to the foliation, according to the results of this study (Figure 10). In particular, the maximum Vp was observed in the direction subparallel to lineation. The S-wave anisotropy (AVs) of the samples were also dramatically high in the direction subparallel to the foliation (Figure 10). These seismic patterns are closely related to the hexagonal symmetry with the slow symmetry axis of phyllosilicates [19]. The overall results, including very high AVp and maximum AVs measured from phyllites (Table 3), imply that phyllosilicate-rich deformed rocks comprising fold-thrust belts in continental collision belts, subduction zones or strike-slip shear zones may significantly affect the observed crustal seismic anisotropies parallel to the tectonic boundaries [4,9,11–13,109,110] where the foliation of rocks is steeply aligned owing to tectonic processes.

The fabric strength of rock-forming minerals has been suggested as an important factor for controlling seismic anisotropies in nature [22,97,111,112]. Figure 12 shows the relationship between the seismic anisotropy of whole rock samples, volume proportion and fabric strength of muscovite and chlorite. The general trend of seismic anisotropy shows that AVp and maximum AVs increase with increasing muscovite and chlorite proportions (Figure 12a,b). However, the volume proportion of phyllosilicates is not the sole factor controlling the magnitudes of seismic anisotropy in this study. The proportion of Ms + Chl in G102 was larger than that in G104. However, the seismic anisotropy of G102 is similar to that of G104 because of the stronger fabric of G104. The role of fabric strength was also clarified when the seismic anisotropies of G102 were compared with G101, yielding an increase of approximately 14% and 22% for AVp and maximum AVs, respectively, owing to fabric strength. These results show that the fabric strength of muscovite and chlorite, as well as their proportion can affect the seismic properties of phyllite.

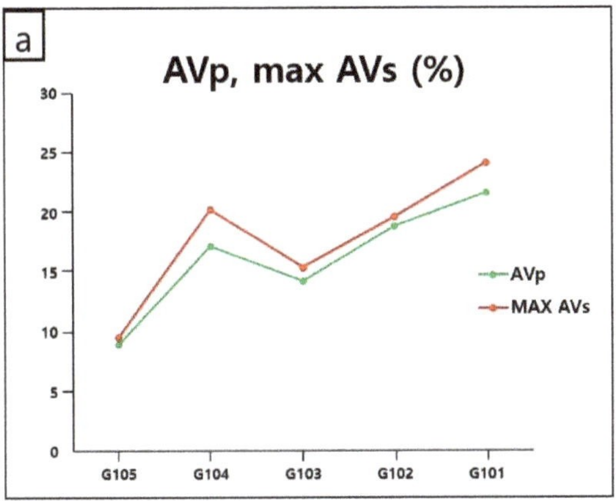

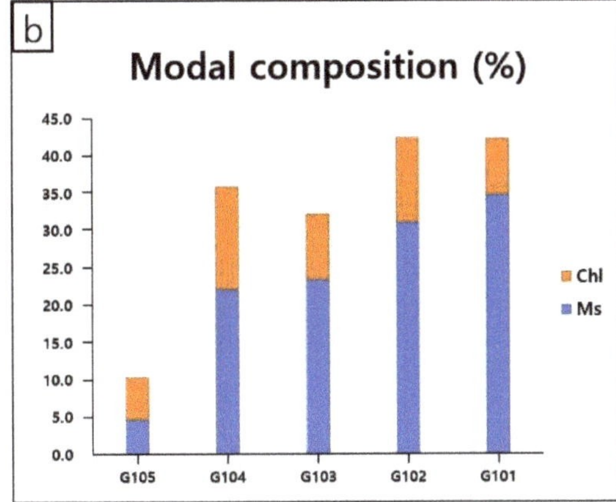

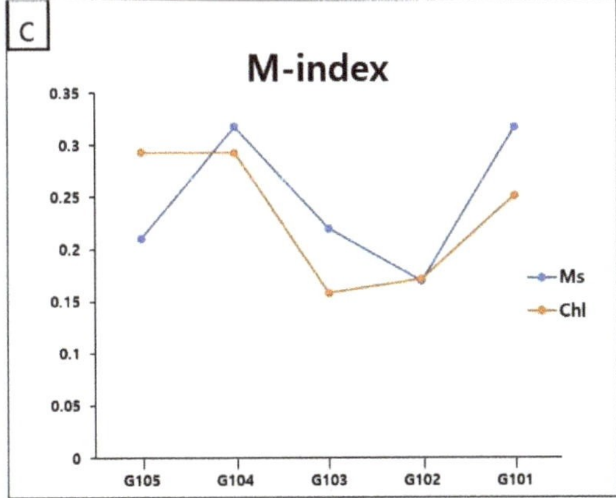

Figure 12. Whole rock seismic anisotropies depending on modal compositions and fabric strengths of phyllosilicates. The order of samples follows the increasing volume proportion of muscovite. (**a**) anisotropy of P-wave (AVp) and maximum anisotropy of S-wave (max. AVs) for each sample. (**b**) Modal composition of muscovite and chlorite for each sample. (**c**) M-index indicating fabric strengths for each mineral. Ms: muscovite; Chl: chlorite.

Additional seismic properties were calculated to understand the role of the coexistence of muscovite and chlorite (Figure 11). Whole rock seismic anisotropies of sample G102 with the assemblages of Qtz + Ab + Ms + Chl, Qtz + Ab + Ms, and Qtz + Ab + Chl were calculated. The calculated data showed that both AVp and maximum AVs increased with increasing proportion of muscovite and chlorite (Figure 11a). The strongest anisotropy was calculated in the assemblage of Qtz + Ab + Ms + Chl, indicating the constructive role of muscovite and chlorite.

This contribution to the whole rock seismic anisotropies is due to their similar seismic patterns. Both minerals showed the maximum and minimum Vp and AVs passing in the subparallel and subnormal directions to the foliation, respectively (Figures 6 and 7). The coincident seismic patterns related to their [001] axes were aligned subnormal to the foliation (Figure 5) and induced a constructive effect on the whole rock seismic anisotropy (Figures 10 and 11). This pattern can be applied to rocks including other phyllosilicate minerals such as serpentine, biotite, phengite, and talc, which have their [001] axes subnormal to the foliation and similar seismic properties [20,22,95,97,99,100,113].

6. Conclusions

The deformation microstructures and seismic anisotropies of greenschist-facies phyllites from the Geumseongri Formation in Gunsan, Korea, were studied to understand the relationship between LPOs, deformation conditions, mineral assemblages, and seismic patterns of deformed rocks in the middle crust. EBSD map data combined with observations of deformation microstructures revealed that the dominant deformation mechanism of quartz was dislocation creep with multiple slip systems and a minor dissolution–precipitation creep. It is also suggested that albite was mainly deformed by rigid body rotation, combined with a minor dislocation creep. The LPOs of muscovite and chlorite were formed by dislocation creep and crystal rotation parallel to the foliation. Based on our results and previous literature, we suggest that the phyllites in this study were deformed under typical greenschist-facies conditions.

Calculated seismic anisotropies based on the LPOs and volume proportions of constituent minerals indicated that muscovite and chlorite in phyllites play a significant role in controlling the seismic pattern and magnitude of seismic anisotropy. The whole rock seismic properties showed typical hexagonal symmetry with slow-axis symmetry. It was also revealed that fabric strength is a primary factor controlling the strength of seismic anisotropy, especially for muscovite and chlorite in phyllites. The increase in fabric strength of muscovite and chlorite induced significant increase of seismic anisotropy in this study. The calculation of seismic anisotropy based on various mineral assemblages showed that the coexistence of muscovite and chlorite contributes to seismic anisotropy constructively when their [001] axes are aligned in the same direction. Finally, it is suggested that the strong LPOs of muscovite and chlorites in phyllosilicates in greenschist-facies rocks play an important role in inducing large seismic anisotropy observed in crustal tectonic boundaries such as the continental collision zone or strike-slip shear zone.

Supplementary Materials: The following are available online at https://www.mdpi.com/2075-163 X/11/3/294/s1, Figure S1: Pole figures of the albite axes presented in the lower hemisphere using equal-area projection.

Author Contributions: Conceptualization, H.J.; methodology, S.H. and H.J.; software, S.H.; validation, S.H. and H.J.; formal analysis, S.H.; investigation, S.H.; resources, H.J.; data curation, S.H.; writing—original draft preparation, S.H.; writing—review and editing, S.H. and H.J.; visualization, S.H.; supervision, H.J.; project administration, H.J.; funding acquisition, H.J. All authors have read and agreed to the published version of the manuscript.

Funding: This research was funded by the Korea Meteorological Administration Research Development Program (KMI2019-00110) to H.J.

Institutional Review Board Statement: Not applicable.

Informed Consent Statement: Not applicable.

Data Availability Statement: Not applicable.

Acknowledgments: The authors are grateful to the anonymous reviewers whose suggestions and comments have notably improved the manuscript.

Conflicts of Interest: The authors declare no conflict of interest.

References

1. Kern, H.; Wenk, H.R. Fabric-related velocity anisotropy and shear wave splitting in rocks from the Santa Rosa mylonite zone, California. *J. Geophys. Res.* **1990**, *95*. [CrossRef]
2. Mainprice, D.; Ildefonse, B. Seismic Anisotropy of Subduction Zone Minerals–Contribution of Hydrous Phases. In *Subduction Zone Geodynamics*; Lallemand, S., Funiciello, F., Eds.; Springer: Berlin/Heidelberg, Germany, 2009; pp. 63–84.
3. Ozacar, A.A.; Zandt, G. Crustal structure and seismic anisotropy near the San Andreas Fault at Parkfield, California. *Geophys. J. Int.* **2009**, *178*, 1098–1104. [CrossRef]
4. Bonnin, M.; Barruol, G.; Bokelmann, G.H.R. Upper mantle deformation beneath the North American–Pacific plate boundary in California from SKS splitting. *J. Geophys. Res.* **2010**, *115*, 1–17. [CrossRef]
5. Lloyd, G.E.; Butler, R.W.H.; Casey, M.; Tatham, D.J.; Mainprice, D. Constraints on the seismic properties of the middle and lower continental crust. In *Deformation Mechanisms, Rheology and Tectonics: Microstructures, Mechanics and Anisotropy*; Prior, D.J., Rutter, E.H., Tatham, D.J., Eds.; Geological Society: London, UK, 2011; Volume 360, pp. 7–32.
6. Long, M.D.; Silver, P.G. Shear wave splitting and mantle anisotropy: Measurements, interpretations, and new directions. *Surv. Geophys.* **2009**, *30*, 407–461. [CrossRef]
7. Long, M.D.; Wirth, E.A. Mantle flow in subduction systems: The mantle wedge flow field and implications for wedge processes. *J. Geophys. Res. Solid Earth* **2013**, *118*, 583–606. [CrossRef]
8. Castellanos, J.; Perry-Houts, J.; Clayton, R.; Kim, Y.; Stanciu, A.C.; Niday, B.; Humphreys, E. Seismic anisotropy reveals crustal flow driven by mantle vertical loading in the Pacific NW. *Sci. Adv.* **2020**, *6*, eabb0476. [CrossRef] [PubMed]
9. Ozacar, A.A.; Zandt, G. Crustal seismic anisotropy in central Tibet: Implications for deformational style and flow in the crust. *Geophys. Res. Lett.* **2004**, *31*, 1–4. [CrossRef]
10. Kong, F.; Wu, J.; Liu, K.H.; Gao, S.S. Crustal anisotropy and ductile flow beneath the eastern Tibetan Plateau and adjacent areas. *Earth Planet. Sci. Lett.* **2016**, *442*, 72–79. [CrossRef]
11. Agius, M.R.; Lebedev, S. Complex, multilayered azimuthal anisotropy beneath Tibet: Evidence for co-existing channel flow and pure-shear crustal thickening. *Geophys. J. Int.* **2017**, *210*, 1823–1844. [CrossRef]
12. Brocher, T.M.; Christensen, N.I. Seismic anisotropy due to preferred mineral orientation observed in shallow crustal rocks in southern Alaska. *Geology* **1990**, *18*, 737–740. [CrossRef]
13. Huang, Z.; Zhao, D.; Wang, L. Shear wave anisotropy in the crust, mantle wedge, and subducting Pacific slab under northeast Japan. *Geochem. Geophys. Geosyst.* **2011**, *12*, 1–17. [CrossRef]
14. Crampin, S. Geological and industrial implications of extensive-dilatancy anisotropy. *Nature* **1987**, *328*, 491–496. [CrossRef]
15. Ji, S.; Shao, T.; Michibayashi, K.; Long, C.; Wang, Q.; Kondo, Y.; Zhao, W.; Wang, H.; Salisbury, M.H. A new calibration of seismic velocities, anisotropy, fabrics, and elastic moduli of amphibole-rich rocks. *J. Geophys. Res. Solid Earth* **2013**, *118*, 4699–4728. [CrossRef]
16. Hefny, M.; Zappone, A.; Makhloufi, Y.; de Haller, A.; Moscariello, A. A laboratory approach for the calibration of seismic data in the western part of the Swiss Molasse Basin: The case history of well Humilly-2 (France) in the Geneva area. *Swiss J. Geosci.* **2020**, *113*, 1–30. [CrossRef]
17. Almqvist, B.S.G.; Mainprice, D. Seismic properties and anisotropy of the continental crust: Predictions based on mineral texture and rock microstructure. *Rev. Geophys.* **2017**, *55*, 367–433. [CrossRef]
18. Bianchi, I.; Bokelmann, G. Probing crustal anisotropy by receiver functions at the deep continental drilling site KTB in Southern Germany. *Geophys. Prospect.* **2019**, *67*, 2450–2464. [CrossRef]
19. Brownlee, S.J.; Schulte-Pelkum, V.; Raju, A.; Mahan, K.; Condit, C.; Orlandini, O.F. Characteristics of deep crustal seismic anisotropy from a compilation of rock elasticity tensors and their expression in receiver functions. *Tectonics* **2017**, *36*, 1835–1857. [CrossRef]
20. Lloyd, G.E.; Butler, R.W.H.; Casey, M.; Mainprice, D. Mica, deformation fabrics and the seismic properties of the continental crust. *Earth Planet. Sci. Lett.* **2009**, *288*, 320–328. [CrossRef]
21. Brownlee, S.J.; Hacker, B.R.; Salisbury, M.; Seward, G.; Little, T.A.; Baldwin, S.L.; Abers, G.A. Predicted velocity and density structure of the exhuming Papua New Guinea ultrahigh-pressure terrane. *J. Geophys. Res.* **2011**, *116*, B08206. [CrossRef]
22. Ji, S.; Shao, T.; Michibayashi, K.; Oya, S.; Satsukawa, T.; Wang, Q.; Zhao, W.; Salisbury, M.H. Magnitude and symmetry of seismic anisotropy in mica- and amphibole-bearing metamorphic rocks and implications for tectonic interpretation of seismic data from the southeast Tibetan Plateau. *J. Geophys. Res. Solid Earth* **2016**, 3782–3803. [CrossRef]
23. Wenk, H.R.; Vasin, R.N.; Kern, H.; Matthies, S.; Vogel, S.C.; Ivankina, T.I. Revisiting elastic anisotropy of biotite gneiss from the Outokumpu scientific drill hole based on new texture measurements and texture-based velocity calculations. *Tectonophysics* **2012**, *570–571*, 123–134. [CrossRef]

24. Kim, J.; Jung, H. New Crystal Preferred Orientation of Amphibole Experimentally Found in Simple Shear. *Geophys. Res. Lett.* **2019**, *46*, 1–10. [CrossRef]

25. Ko, B.; Jung, H. Crystal preferred orientation of an amphibole experimentally deformed by simple shear. *Nat. Commun.* **2015**, *6*, 6586. [CrossRef]

26. Jung, H. Crystal preferred orientations of olivine, orthopyroxene, serpentine, chlorite, and amphibole, and implications for seismic anisotropy in subduction zones: A review. *Geosci. J.* **2017**, *21*, 985–1011. [CrossRef]

27. Brown, J.M.; Abramson, E.H. Elasticity of calcium and calcium-sodium amphiboles. *Phys. Earth Planet. Inter.* **2016**, *261*, 161–171. [CrossRef]

28. Alexandrov, K.S.; Ryzhova, T.V. *The Elastic Properties of Rock-Forming Minerals. II: Layered Silicates*; Geophysics Series; Bulletin USSR Academy of Science: Moscow, Russia, 1961; Volume 9.

29. Vaughan, M.T.; Guggenheim, S. Elasticity of Muscovite and Its Relationship to Crystal Structure. *J. Geophys. Res.* **1986**, *91*, 4657–4664. [CrossRef]

30. Weiss, T.; Siegesmund, S.; Rabbel, W.; Bohlen, T.; Pohl, M. Seismic velocities and anisotropy of the lower continental crust: A review. *Pure Appl. Geophys.* **1999**, *156*, 97–122. [CrossRef]

31. Shapiro, N.M.; Ritzwoller, M.H.; Molnar, P.; Levin, V. Thinning and flow of Tibetan crust constrained by seismic anisotropy. *Science* **2004**, *305*, 233–236. [CrossRef]

32. Mahan, K. Retrograde mica in deep crustal granulites: Implications for crustal seismic anisotropy. *Geophys. Res. Lett.* **2006**, *33*, 1–6. [CrossRef]

33. Meissner, R.; Rabbel, W.; Kern, H. Seismic lamination and anisotropy of the lower continental crust. *Tectonophysics* **2006**, *416*, 81–99. [CrossRef]

34. Wenk, H.R.; Yu, R.; Cárdenes, V.; Lopez-Sanchez, M.A.; Sintubin, M. Fabric and anisotropy of slates: From classical studies to new results. *J. Struct. Geol.* **2020**, *138*. [CrossRef]

35. Law, R.D. Crystallographic fabrics: A selective review of their applications to research in structural geology. In *Deformation Mechanisms, Rheology and Tectonics*; Knipe, R.J., Rutter, E.H., Eds.; Geological Society Special Publication: London, UK, 1990; Volume 54, pp. 335–352.

36. Little, T.A.; Hacker, B.R. Microstructures and quartz lattice-preferred orientations in the eclogite-bearing migmatitic gneisses of the D'Entrecasteaux Islands, Papua New Guinea. *Geochem. Geophys. Geosyst.* **2013**, *14*, 2030–2062. [CrossRef]

37. Stipp, M.; Stünitz, H.; Heilbronner, R.; Schmid, S.M. The eastern Tonale fault zone: A "natural laboratory" for crystal plastic deformation of quartz over a temperature range from 250 to 700 °C. *J. Struct. Geol.* **2002**, *24*, 1861–1884. [CrossRef]

38. Law, R.D. Deformation thermometry based on quartz c-axis fabrics and recrystallization microstructures: A review. *J. Struct. Geol.* **2014**, *66*, 129–161. [CrossRef]

39. Cross, A.J.; Kidder, S.; Prior, D.J. Using microstructures and TitaniQ thermobarometry of quartz sheared around garnet porphyroclasts to evaluate microstructural evolution and constrain an Alpine Fault Zone geotherm. *J. Struct. Geol.* **2015**, *75*, 17–31. [CrossRef]

40. Savignano, E.; Reddy, S.M.; Bridges, J.; Mazzoli, S. Quartz fabric variations across the greenschist facies shear zone separating the zermatt-saas and combin ophiolitic zones, upper val gressoney, western alps. *Ofioliti* **2016**, *41*, 85–98. [CrossRef]

41. Tullis, J.; Christie, J.M.; Griggs, D.T. Microstructures and preferred orientations of experimentally deformed quartzites. In *Bulletin of the Geological Society of America*; Geological Society of America: Boulder, CO, USA, 1973; Volume 84, pp. 297–314.

42. Lister, G.S.; Hobbs, B.E. The simulation of fabric development during plastic deformation and its application to quartzite: The influence of deformation history. *J. Struct. Geol.* **1980**, *2*, 355–370. [CrossRef]

43. Schmid, S.M.; Casey, M. Complete fabric analysis of some commonly observed quartz C-axis patterns. *Geophys. Monogr.* **1986**, *36*, 263–286. [CrossRef]

44. Hirth, G.; Tullis, J. Dislocation creep regimes in quartz aggregates. *J. Struct. Geol.* **1992**, *14*, 145–159. [CrossRef]

45. Heilbronner, R.; Tullis, J. Evolution of c axis pole figures and grain size during dynamic recrystallization: Results from experimentally sheared quartzite. *J. Geophys. Res.* **2006**, *111*, 1–19. [CrossRef]

46. Faleiros, F.M.; Moraes, R.; Pavan, M.; Campanha, G.A.C. A new empirical calibration of the quartz c-axis fabric opening-angle deformation thermometer. *Tectonophysics* **2016**, *671*, 173–182. [CrossRef]

47. Tokle, L.; Hirth, G.; Behr, W.M. Flow laws and fabric transitions in wet quartzite. *Earth Planet. Sci. Lett.* **2019**, *505*, 152–161. [CrossRef]

48. Wenk, H.R.; Bunge, H.J.; Jansen, E.; Pannetier, J. Preferred orientation of plagioclase-neutron diffraction and U-stage data. *Tectonophysics* **1986**, *126*, 271–284. [CrossRef]

49. Prior, D.J.; Wheeler, J. Feldspar fabrics in a greenschist facies albite-rich mylonite from electron backscatter diffraction. *Tectonophysics* **1999**, *303*, 29–49. [CrossRef]

50. Jiang, Z.; Prior, D.J.; Wheeler, J. Albite crystallographic preferred orientation and grain misorientation distribution in a low-grade mylonite: Implications for granular flow. *J. Struct. Geol.* **2000**, *22*, 1663–1674. [CrossRef]

51. Morales, L.F.G.; Boudier, F.; Nicolas, A. Microstructures and crystallographic preferred orientation of anorthosites from Oman ophiolite and the dynamics of melt lenses. *Tectonics* **2011**, *30*, 1–21. [CrossRef]

52. Díaz-Azpiroz, M.; Lloyd, G.E.; Fernández, C. Deformation mechanisms of plagioclase and seismic anisotropy of the Acebuches metabasites (SW Iberian massif). In *Deformation Mechanisms, Rheology and Tectonics: Microstructures, Mechanics and Anisotropy*; Prior, D.J., Rutter, E.H., Tatham, D.J., Eds.; Geological Society Special Publications: London, UK, 2011; pp. 79–95. ISBN 9781862393387.

53. Menegon, L.; Stünitz, H.; Nasipuri, P.; Heilbronner, R.; Svahnberg, H. Transition from fracturing to viscous flow in granulite facies perthitic feldspar (Lofoten, Norway). *J. Struct. Geol.* **2013**, *48*, 95–112. [CrossRef]

54. Eberlei, T.; Habler, G.; Grasemann, B.; Abart, R. Upper-greenschist facies intragrain deformation of albite in mylonitic meta-pegmatite and the influence of crystallographic anisotropy on microstructure formation. *J. Struct. Geol.* **2014**, *69*, 47–58. [CrossRef]

55. Ji, S.; Shao, T.; Salisbury, M.H.; Sun, S.; Michibayashi, K.; Zhao, W.; Long, C.; Liang, F.; Satsukawa, T. Plagioclase preferred orientation and induced seismic anisotropy in mafic igneous rocks. *J. Geophys. Res. Solid Earth* **2014**, *119*, 8064–8088. [CrossRef]

56. Miranda, E.A.; Hirth, G.; John, B.E. Microstructural evidence for the transition from dislocation creep to dislocation-accommodated grain boundary sliding in naturally deformed plagioclase. *J. Struct. Geol.* **2016**, *92*, 30–45. [CrossRef]

57. Stünitz, H.; Fitz Gerald, J.D.; Tullis, J. Dislocation generation, slip systems, and dynamic recrystallization in experimentally deformed plagioclase single crystals. *Tectonophysics* **2003**, *372*, 215–233. [CrossRef]

58. Olsen, T.S.; Kohlstedt, D.L. Analysis of dislocations in some naturally deformed plagioclase feldspars. *Phys. Chem. Miner.* **1984**, *11*, 153–160. [CrossRef]

59. Marshall, D.B.; McLaren, A.C. Deformation mechanisms in experimentally deformed plagioclase feldspars. *Phys. Chem. Miner.* **1977**, *1*, 351–370. [CrossRef]

60. Erdman, M.E.; Hacker, B.R.; Zandt, G.; Seward, G. Seismic anisotropy of the crust: Electron-backscatter diffraction measurements from the Basin and Range. *Geophys. J. Int.* **2013**, *195*, 1211–1229. [CrossRef]

61. Ward, D.; Mahan, K.; Schulte-Pelkum, V. Roles of quartz and mica in seismic anisotropy of mylonites. *Geophys. J. Int.* **2012**, *190*, 1123–1134. [CrossRef]

62. Lee, S.R.; Cho, K. Precambrian Crustal Evolution of the Korean Peninsula. *J. Petrol. Soc. Korea* **2012**, *21*, 89–112. [CrossRef]

63. Cho, M.; Kim, T.; Yang, S.; Yi, K. Paleoproterozoic to Triassic crustal evolution of the Gyeonggi Massif, Korea: Tectonic correlation with the North China craton. In *Linkages and Feedbacks in Orogenic Systems*; Law, R.D., Thigpen, J.R., Merschat, A.J., Stowell, H., Eds.; Geological Society of America Memoir: Boulder, CO, USA, 2017; Volume 213, pp. 165–197.

64. Cluzel, D.; Cadet, J.P.; Lapierre, H. Geodynamics of the Ogcheon Belt (South Korea). *Tectonophysics* **1990**, *183*, 41–56. [CrossRef]

65. Cluzel, D.; Jolivet, L.; Cadet, J. Early Middle Paleozoic Intraplate Orogeny in the Ogcheon Belt (South Korea): A new insight on the Paleozoic buildup of east Asia. *Tectonics* **1991**, *10*, 1130–1151. [CrossRef]

66. Ree, J.H.; Cho, M.; Kwon, S.T.; Nakamura, E. Possible eastward extension of Chinese collision belt in South Korea: The Imjingang belt. *Geology* **1996**, *24*, 1071–1074. [CrossRef]

67. Cho, M.; Kim, H. Metamorphic Evolution of the Ogcheon Belt, Korea: A Review and New Age Constraints. *Int. Geol. Rev.* **2005**, *47*, 41–57. [CrossRef]

68. Chough, S.K.; Kwon, S.-T.; Ree, J.H.; Choi, D.K. Tectonic and sedimentary evolution of the Korean peninsula: A review and new view. *Earth Sci. Rev.* **2000**, *52*, 175–235. [CrossRef]

69. Kwon, S.; Sajeev, K.; Mitra, G.; Park, Y.; Kim, S.W.; Ryu, I.C. Evidence for Permo-Triassic collision in Far East Asia: The Korean collisional orogen. *Earth Planet. Sci. Lett.* **2009**, *279*, 340–349. [CrossRef]

70. Cho, M.; Lee, Y.; Kim, T.; Cheong, W.; Kim, Y.; Lee, S.R. Tectonic evolution of Precambrian basement massifs and an adjoining fold-and-thrust belt (Gyeonggi Marginal Belt), Korea: An overview. *Geosci. J.* **2017**, *21*, 845–865. [CrossRef]

71. Lee, Y., II; Lee, J., II. Paleozoic sedimentation and tectonics in Korea: A review. *Isl. Arc* **2003**, *12*, 162–179. [CrossRef]

72. Oh, C.W.; Kim, S.W.; Ryu, I.C.; Okada, T.; Hyodo, H.; Itaya, T. Tectono-metamorphic evolution of the Okcheon Metamorphic Belt, South Korea: Tectonic implications in East Asia. *Isl. Arc* **2004**, *13*, 387–402. [CrossRef]

73. Oh, C.W.; Choi, S.G.; Seo, J.; Rajesh, V.J.; Lee, J.H.; Zhai, M.; Peng, P. Neoproterozoic tectonic evolution of the Hongseong area, southwestern Gyeonggi Massif, South Korea; implication for the tectonic evolution of Northeast Asia. *Gondwana Res.* **2009**, *16*, 272–284. [CrossRef]

74. Cho, M.; Kim, T.; Kim, H. SHRIMP U-Pb Zircon Age of a Felsic Meta-tuff in the Ogcheon Metamorphic Belt, Korea: Neoproterozoic (ca. 750 Ma) Volcanism. *J. Petrol. Soc. Korea* **2004**, *13*, 119–125.

75. Choi, D.K.; Woo, J.; Park, T. The Okcheon Supergroup in the Lake Chungju area, Korea: Neoproterozoic volcanic and glaciogenic sedimentary successions in a rift basin. *Geosci. J.* **2012**, *16*, 229–252. [CrossRef]

76. Lee, K.S.; Chang, H.W.; Park, K.H. Neoproterozoic bimodal volcanism in the central Ogcheon belt, Korea: Age and tectonic implication. *Precambrian Res.* **1998**, *89*, 47–57. [CrossRef]

77. Kim, S.W.; Kee, W.S.; Santosh, M.; Cho, D.L.; Hong, P.S.; Ko, K.; Lee, B.C.; Byun, U.H.; Jang, Y. Tracing the Precambrian tectonic history of East Asia from Neoproterozoic sedimentation and magmatism in the Korean Peninsula. *Earth Sci. Rev.* **2020**, *209*, 103311. [CrossRef]

78. Lim, S.; Chun, H.Y.; Kim, Y.B.; Kim, B.C.; Cho, D. Geologic ages, stratigraphy and geological structures of the metasedimentary strata in Bibong Yeonmu area, NW Okcheon belt, Korea. *J. Geol. Soc. Korea* **2005**, *41*, 335–368.

79. Kang, J.-H.; Hayasaka, Y.; Ryoo, C.-R. Tectonic evolution of the Central Ogcheon Belt, Korea. *J. Petrol. Soc. Korea* **2012**, *21*, 129–150. [CrossRef]

80. Cho, M.; Min, K.; Kim, H. Geology of the 2018 Winter Olympic site, Pyeongchang, Korea. *Int. Geol. Rev.* **2018**, *60*, 267–287. [CrossRef]

81. De Jong, K.; Han, S.; Ruffet, G. Fast cooling following a Late Triassic metamorphic and magmatic pulse: Implications for the tectonic evolution of the Korean collision belt. *Tectonophysics* **2015**, *662*, 271–290. [CrossRef]
82. Kim, H.; Kihm, Y.H.; Kee, W. *Geological Report of the Iri Sheet (Scale 1:50,000)*; Korea Institute of Geoscience and Mineral Resources: Daejeon, Korea, 2012.
83. Choi, P.; Hwang, J.H. *Geological Report of the Gunsan, Buan, Banchukdo, Jangjado Sheets (Scale 1:50,000)*; Korea Institute of Geoscience and Mineral Resources: Daejeon, Korea, 2013.
84. Kim, H.; Cho, M.; Koh, H.J. Tectonometamorphic evolution of the central Ogcheon belt in the Jeungpyeong-Deokpyeong area. *J. Geol. Soc. Korea* **1995**, *31*, 299–314.
85. Kim, H.; Cho, M. Polymetamorphism of Ogcheon Supergroup in the Miwon area, central Ogcheon metamorphic belt, South Korea. *Geosci. J.* **1999**, *3*, 151–162. [CrossRef]
86. Kim, S.W.; Itaya, T.; Hyodo, H.; Matsuda, T. Metamorphic K-Feldspar in Low-grade Metasediments from the Ogcheon Metamorphic Belt in South Korea. *Gondwana Res.* **2002**, *5*, 849–855. [CrossRef]
87. Panozzo, R.H. Two-dimensional analysis of shape-fabric using projections of digitized lines in a plane. *Tectonophysics* **1983**, *95*, 279–294. [CrossRef]
88. Skemer, P.; Katayama, I.; Jiang, Z.; Karato, S.I. The misorientation index: Development of a new method for calculating the strength of lattice-preferred orientation. *Tectonophysics* **2005**, *411*, 157–167. [CrossRef]
89. Mainprice, D.; Hielscher, R.; Schaeben, H. Calculating anisotropic physical properties from texture data using the MTEX open-source package. In *Deformation Mechanisms, Rheology and Tectonics: Microstructures, Mechanics and Anisotropy*; Prior, D.J., Rutter, E.H., Tatham, D.J., Eds.; Geological Society Special Publications: London, UK, 2011; pp. 175–192.
90. Efron, B.; Tibshirani, R.J. *An Introduction to the Bootstrap*; Chapman & Hall/CRC: Boca Raton, FL, USA, 1994; ISBN 9781000064988.
91. Morales, L.F.G.; Mainprice, D.; Kern, H. Olivine-antigorite orientation relationships: Microstructures, phase boundary misorientations and the effect of cracks in the seismic properties of serpentinites. *Tectonophysics* **2018**, *724–725*, 93–115. [CrossRef]
92. Mainprice, D. A Fortran Program to Calculate Seismic Anisotropy from the Lattice Preferred Orientation of Minerals. *Comput. Geosci.* **1990**, *16*, 385–393. [CrossRef]
93. McSkimin, H.J.; Andreatch, P.; Thurston, R.N. Elastic moduli of quartz versus hydrostatic pressure at 25° and −195.8 °C. *J. Appl. Phys.* **1965**, *36*, 1624–1632. [CrossRef]
94. Ryzhova, T.V. Elastic properties of plagioclases. *Akad SSSR Izv Ser Geofiz* **1964**, *7*, 1049–1051.
95. Wenk, H.R.; Kanitpanyacharoen, W.; Voltolini, M. Preferred orientation of phyllosilicates: Comparison of fault gouge, shale and schist. *J. Struct. Geol.* **2010**, *32*, 478–489. [CrossRef]
96. Kim, D.; Jung, H. Deformation microstructures of olivine and chlorite in chlorite peridotites from Almklovdalen in the Western Gneiss Region, southwest Norway, and implications for seismic anisotropy. *Int. Geol. Rev.* **2015**, *57*, 650–668. [CrossRef]
97. Park, M.; Jung, H. Relationships Between Eclogite-Facies Mineral Assemblages, Deformation Microstructures, and Seismic Properties in the Yuka Terrane, North Qaidam Ultrahigh-Pressure Metamorphic Belt, NW China. *J. Geophys. Res. Solid Earth* **2019**, *124*, 1–24. [CrossRef]
98. Ha, Y.; Jung, H.; Raymond, L.A. Deformation fabrics of glaucophane schists and implications for seismic anisotropy: The importance of lattice preferred orientation of phengite. *Int. Geol. Rev.* **2019**, *61*, 720–737. [CrossRef]
99. Puelles, P.; Beranoaguirre, A.; Ábalos, B.; Gil Ibarguchi, J.I.; García de Madinabeitia, S.; Rodríguez, J.; Fernández-Armas, S. Eclogite inclusions from subducted metaigneous continental crust (Malpica-Tui Allochthonous Complex, NW Spain): Petrofabric, geochronology, and calculated seismic properties. *Tectonics* **2017**, *36*, 1376–1406. [CrossRef]
100. Lee, J.; Jung, H.; Klemd, R.; Tarling, M.S.; Konopelko, D. Lattice preferred orientation of talc and implications for seismic anisotropy in subduction zones. *Earth Planet. Sci. Lett.* **2020**, *537*, 116178. [CrossRef]
101. Kim, D.; Jung, H.; Lee, J. Strain-induced fabric transition of chlorite and implications for seismic anisotropy in subduction zones. *Minerals* **2020**, *10*, 503. [CrossRef]
102. Passchier, C.W.; Trouw, R.A.J. *Microtectonics*, 2nd ed.; Springer: New York City, NY, USA, 2005; ISBN 978-3-540-64003-5.
103. Mares, V.M.; Kronenberg, A.K. Experimental deformation of muscovite. *J. Struct. Geol.* **1993**, *15*, 1061–1075. [CrossRef]
104. Okudaira, T.; Takeshita, T.; Hara, I.; Ando, J. A new estimate of the conditions for transition from basal $\langle a \rangle$ to prism [c] slip in naturally deformed quartz. *Tectonophysics* **1995**, *250*, 31–46. [CrossRef]
105. Bons, P.D.; Den Brok, B. Crystallographic preferred orientation development by dissolution-precipitation creep. *J. Struct. Geol.* **2000**, *22*, 1713–1722. [CrossRef]
106. Stallard, A.; Shelley, D. Quartz c-axes parallel to stretching directions in very low-grade metamorphic rocks. *Tectonophysics* **1995**, *249*, 31–40. [CrossRef]
107. Tullis, J.; Yund, R.A. Transition from cataclastic flow to dislocation creep of feldspar: Mechanisms and microstructures. *Geology* **1987**, *15*, 606–609. [CrossRef]
108. Fitz Gerald, J.D.; Stünitz, H. Deformation of granitoids at low metamorphic grade. I: Reactions and grain size reduction. *Tectonophysics* **1993**, *221*, 299–324. [CrossRef]
109. Wang, K.; Jiang, C.; Yang, Y.; Schulte-Pelkum, V.; Liu, Q. Crustal Deformation in Southern California Constrained by Radial Anisotropy From Ambient Noise Adjoint Tomography. *Geophys. Res. Lett.* **2020**, *47*, 1–12. [CrossRef]

110. Han, C.; Xu, M.; Huang, Z.; Wang, L.; Xu, M.; Mi, N.; Yu, D.; Gou, T.; Wang, H.; Hao, S.; et al. Layered crustal anisotropy and deformation in the SE Tibetan plateau revealed by Markov-Chain-Monte-Carlo inversion of receiver functions. *Phys. Earth Planet. Inter.* **2020**, *306*, 106522. [CrossRef]
111. Cao, Y.; Jung, H. Seismic properties of subducting oceanic crust: Constraints from natural lawsonite-bearing blueschist and eclogite in Sivrihisar Massif, Turkey. *Phys. Earth Planet. Inter.* **2016**, *250*, 12–30. [CrossRef]
112. Meltzer, A.; Christensen, N. Nanga Parbat crustal anisotropy: Implications for interpretation of crustal velocity structure and shear-wave splitting. *Geophys. Res. Lett.* **2001**, *28*, 2129–2132. [CrossRef]
113. Jung, H. Seismic anisotropy produced by serpentine in mantle wedge. *Earth Planet. Sci. Lett.* **2011**, *307*, 535–543. [CrossRef]

Article

Lattice-Preferred Orientation and Seismic Anisotropy of Minerals in Retrograded Eclogites from Xitieshan, Northwestern China, and Implications for Seismic Reflectance of Rocks in the Subduction Zone

Jaeseok Lee and Haemyeong Jung *

Tectonophysics Laboratory, School of Earth and Environmental Sciences, Seoul National University, Seoul 08826, Korea; shoo3680@snu.ac.kr
* Correspondence: hjung@snu.ac.kr; Tel.: +82-2-880-6733

Abstract: Various rock phases, including those in subducting slabs, impact seismic anisotropy in subduction zones. The seismic velocity and anisotropy of rocks are strongly affected by the lattice-preferred orientation (LPO) of minerals; this was measured in retrograded eclogites from Xitieshan, northwest China, to understand the seismic velocity, anisotropy, and seismic reflectance of the upper part of the subducting slab. For omphacite, an S-type LPO was observed in three samples. For amphibole, the <001> axes were aligned subparallel to the lineation, and the (010) poles were aligned subnormal to foliation. The LPOs of amphibole and omphacite were similar in most samples. The misorientation angle between amphibole and neighboring omphacite was small, and a lack of intracrystalline deformation features was observed in the amphibole. This indicates that the LPO of amphibole was formed by the topotactic growth of amphibole during retrogression of eclogites. The P-wave anisotropy of amphibole in retrograded eclogites was large (approximately 3.7–7.3%). The seismic properties of retrograded eclogites and amphibole were similar, indicating that the seismic properties of retrograded eclogites are strongly affected by the amphibole LPO. The contact boundary between serpentinized peridotites and retrograded eclogites showed a high reflection coefficient, indicating that a reflected seismic wave can be easily detected at this boundary.

Keywords: retrograded eclogite; amphibole; topotactic growth; reflection coefficient; omphacite; subduction zone; lattice-preferred orientation; Xitieshan eclogite; seismic anisotropy

Citation: Lee, J.; Jung, H. Lattice-Preferred Orientation and Seismic Anisotropy of Minerals in Retrograded Eclogites from Xitieshan, Northwestern China, and Implications for Seismic Reflectance of Rocks in the Subduction Zone. *Minerals* **2021**, *11*, 380. https://doi.org/10.3390/min11040380

Academic Editor: Tatsuki Tsujimori

Received: 17 February 2021
Accepted: 31 March 2021
Published: 2 April 2021

Publisher's Note: MDPI stays neutral with regard to jurisdictional claims in published maps and institutional affiliations.

1. Introduction

Subduction zones are known to have varying seismic velocity structures and anisotropies caused by subducting slabs [1–4]. Seismic velocity and anisotropy are strongly affected by the lattice-preferred orientation (LPO) of elastically anisotropic minerals [5–24]. When a slab is subducted under high-pressure and high-temperature conditions, the major rock phase of the upper subducting slab transforms into eclogite, which consists mainly of omphacite and garnet [8,25–28]. Garnet has a relatively high seismic velocity, and omphacite has a strong seismic anisotropy [5,6,23,29]. Eclogites that are formed at the deep part of the subducting slab are known to have a faster seismic velocity—similar to that of the upper mantle ($V_P \sim 8.0$ km/s)—than the surrounding rocks, and they have a strong seismic anisotropy, thus affecting various seismic velocities and anisotropies in subduction zones [30,31].

In addition, when eclogite appears near the surface due to slab break-off in the subduction zone, the eclogite is retrograded to form amphibole under low-pressure and low-temperature conditions [22,32]. Amphibole exists in subducting slabs under lower-pressure and lower-temperature conditions than eclogite stability conditions, and in the middle to lower crust [33], whereas the LPO development of amphibole is known to produce a strong seismic anisotropy [18,34–36]. Therefore, deformed retrograded eclogites may represent seismic velocity and anisotropy in relatively shallow subducting slabs [5,6,20,31,37–49]. Measuring the LPOs of omphacite, amphibole, and garnet in retrograded eclogites can therefore be useful in understanding the seismic velocity and anisotropy of subducting slabs in subduction zones.

In addition, owing to the various seismic velocity structures of subduction zones, seismic waves tend to be reflected at the boundary of rock phases with varying seismic velocities. In particular, the upper parts of subducting slabs have a fast seismic velocity [6], which can cause a difference in seismic velocity from surrounding rocks and may reflect seismic waves due to this difference in seismic velocity [6,50–53]. In previous studies, the reflection coefficient (R_C) was suggested to quantify the degree of seismic wave reflection [6,50,51]. Calculating the R_C of the boundaries between retrograded eclogites and surrounding rocks, using their P-wave velocities, and comparing the degree of retrogression of eclogites and R_C of these boundaries may explain the degree of reflection of seismic waves at these boundaries.

Garnet is a seismically nearly isotropic mineral with a cubic shape that develops little LPO [54]. The LPOs of other minerals in retrograded eclogites—amphibole and omphacite—have been studied in both naturally and experimentally deformed rocks. Four types of LPOs of amphibole have been reported: type I, type II, type III, and type IV [18,36]. A previous experimental study on amphibolite, which was conducted at a pressure of 1 GPa and temperatures of 480–700 °C [18], reported that the type I amphibole LPO was formed under relatively low-stress and low-temperature conditions and is characterized as the <100> axes aligned subnormal to foliation and <001> axes aligned subparallel to lineation. The type II amphibole LPO was formed under relatively high-stress and moderate-temperature deformation conditions and is characterized as <100> axes aligned subnormal to foliation and (010) poles aligned subparallel to lineation. The type III amphibole LPO was formed under relatively low-stress and high-temperature deformation conditions and is characterized as <100> axes aligned subnormal to foliation, with both <001> axes and (010) poles forming a girdle shape subparallel to foliation [18]. A recent experimental study reported that the type IV amphibole LPO can be formed owing to high strain [36] and is characterized as <001> axes aligned subparallel to lineation, with both <100> axes and (110) poles forming a girdle shape subnormal to the shear direction [36]. Previously, the LPOs of amphibole, similar to the type IV amphibole LPO, have been reported in naturally deformed rocks, which were observed in metabasites from southwestern Spain and northwest Scotland, retrogressed Limo harzburgites from northwestern Spain, and amphibolites from Spain [44,55–58]. On the other hand, an amphibole LPO has also been reported in which the (010) poles are aligned subnormal to foliation and the <001> axes are aligned subparallel to lineation [44,59,60].

Three types of omphacite LPOs have been reported: L-, S-, and LS-type omphacite LPOs [61]. The S-type omphacite LPO is characterized as the <001> axis of omphacite forming a girdle parallel to foliation and the (010) pole of omphacite aligned subnormal to foliation. The L-type omphacite LPO is characterized as the <001> axis of omphacite aligned subparallel to lineation and the (010) pole of omphacite forming a girdle subnormal to lineation. The LS-type omphacite LPO is characterized as the <001> axis of omphacite aligned subparallel to lineation and the (010) pole of omphacite aligned subnormal to foliation, indicating a (010) <001> slip system [20,61].

The seismic reflection coefficient (R_C) was previously calculated only for the boundary between fresh eclogite and neighboring rocks [6,50]. In this study, the omphacite and amphibole LPOs in the retrograded eclogites from Xitieshan, northwest China, which represent various degrees of retrogression (35–90%), were measured. The Xitieshan eclogites are useful for studying the relationship between the degree of retrogression and seismic velocity, anisotropy, and reflectance. The seismic velocity and anisotropy of retrograded eclogites were calculated using the LPOs of minerals, and the effect of the amphibole LPO on the seismic velocity and anisotropy of retrograded eclogites was also investigated. R_C was calculated using the P-wave velocity, and R_C was examined to understand the seismic wave reflectance of the boundaries between the retrograded eclogites and neighboring rocks. The results of the present study provide a valuable opportunity to understand the seismic velocity, anisotropy, and reflectance of retrograded eclogites and to identify the boundary between the subducting slab and mantle wedge in various subduction zones.

2. Geological Settings

Xitieshan is located in the central part of the North Qaidam ultra-high-pressure (UHP) metamorphic belt, which is located at the northeastern margin of the Tibetan Plateau in northwestern China (Figures 1 and 2). The North Qaidam UHP metamorphic belt is a typical Alpine-type continental collision/subduction zone, containing the characteristic continental rock association of granitic and pelitic gneiss with intercalated eclogite and peridotite [33,34,62–64]. From the southeastern part, four individual UHP terranes were observed: the Dulan eclogite-bearing terrane, the Xitieshan eclogite-bearing terrane, the Lulingshan garnet peridotite-bearing terrane, and the Yuka eclogite-bearing terrane. Coesite and diamonds were found as inclusions in zircon and garnet in all four terranes, proving that these four terranes had been under UHP metamorphic conditions, similar to other continental UHP metamorphic terranes in China [64].

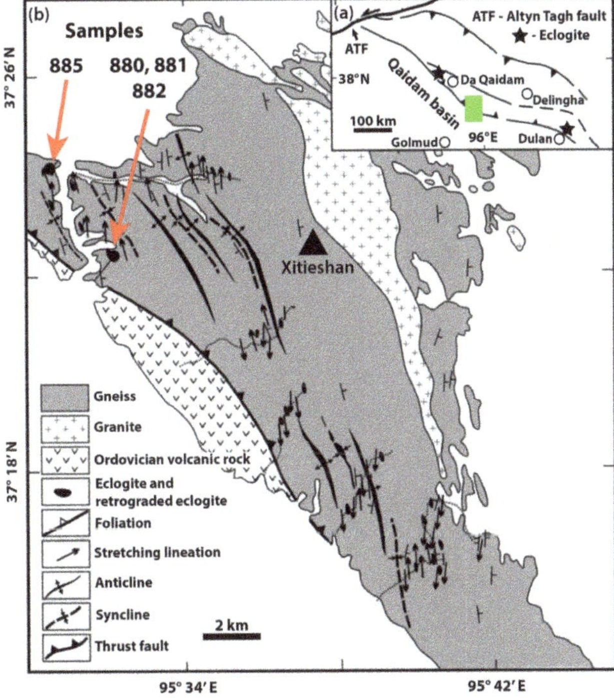

Figure 1. (**a**) Schematic geological map showing the North Qaidam ultra-high-pressure (UHP) metamorphic belt, northwestern China. (**b**) Study area and geological map of Xitieshan area. Green rectangle area in (**a**) was magnified (modified after Zhang et al. [65]).

Figure 2. Photographs of the Xitieshan outcrop. (**a**) Large view of the Xitieshan Mountain containing retrograded eclogites and (**b**) magnified view of the rectangle in (**a**) showing a body of retrograded eclogite.

The eclogites in the Xitieshan terrane can be divided into two groups: bimineralic eclogite and phengite-bearing eclogite. Bimineralic eclogite is the major component of the Xitishan terrane, whereas phengite-bearing eclogite concentrations are minor and can only be seen in the Huangyanggou area. Most eclogites in the Xitishan terrane were retrograded into garnet amphibolite or amphibolite. The original eclogite can be found in the center of the large bodies.

The Xitieshan eclogites experienced two stages of retrogression. According to previous studies [66,67], the peak pressure and temperature conditions of metamorphism of the Xitieshan eclogite were P = 2.71–3.17 GPa and T = 751–791 °C. In the first stage (isothermal decompression), the proportion of omphacite and garnet in the sample decreased, while the proportion of amphibole and plagioclase increased at a temperature of T ~ 750 °C when pressure decreased from P ~ 2.3 GPa to P ~ 1.5 GPa. In the second stage (symplectite formation), symplectite of diopside, plagioclase, and edenite appeared under the condition of P = 8–12 kbar and T = 660–720 °C [67].

In addition to the Xitieshan area, geochronology data reported that there were two major different ages measured using zircon: 750–800 Ma and 877 ± 8 Ma were calculated in the cores of the zircon, and 440–460 Ma and 433 ± 3 Ma were calculated at the rim of the zircon. Based on the ages calculated in the core of the zircon, some of the Xitieshan terrane eclogites were produced in Neoproterozoic protoliths much before the Paleozoic subduction or collision event occurred in the North Qaidam UHP metamorphism. Eclogite metamorphism occurred in the age calculated at the rim of zircon [67–71].

3. Methods

3.1. Measurement of Lattice-Preferred Orientation (LPO) of Minerals

Foliation of a rock was defined by compositional layering of amphibole and garnet, and lineation was determined by the shape-preferred orientation of minerals on foliation, using the projection function method [72]. Thin sections were made in the XZ plane to measure the LPO, the X axis was set parallel to lineation, and the Z axis was set normal to foliation.

The LPOs of the samples were measured using electron back-scattered diffraction (EBSD) using a JEOL-JSM-7100 field emission scanning electron microscope (FE-SEM) with AZTEC and HKL Channel 5 software (5.12.74.0) installed at the School of Earth and Environmental Sciences (SEES), Seoul National University (SNU). Detailed settings of the EBSD mapping are provided in Table 1. An accelerating voltage of 20 kV and a working distance of 20 mm with a tilting angle of 70° were the main settings of the FE-SEM with EBSD. The samples were observed using automatic mapping mode, with a step size of 23 to 35 μm, depending on the grain size of most of the samples. To eliminate the oversampling of large grains, pole figures were constructed from one point per grain [73]. The omphacite LPOs, which were in direct contact with amphibole single crystals, were measured using a smaller step size (1 μm). The misorientation index (M-index) [74] was calculated to estimate the fabric strength of the sample using the uncorrelated grain pairs obtained from the orientation data. The M-index ranged from 0 (random) to 1 (single grain). The misorientation angle was measured using HKL Channel 5 software (Tango).

Table 1. Modal composition, aspect ratio, settings of electron back-scattered diffraction (EBSD) mapping, and estimated equilibrium temperature of samples.

Sample	Modal Composition (%)						Aspect Ratio of Minerals			EBSD Mapping		Temperature [a] (°C)
	Amp	Omp	Plag	Grt	Qtz	Others	XY	XZ	YZ	Step Size (μm)	Hit Rate (%)	
885	33.0	14.0	8.8	33.3	10.1	Mi 1.0	1.129	1.317	1.199	25	91.2	727 ± 40
881	51.9	19.6	10.7	16.2	1.4	Il 0.2	1.222	1.424	1.154	25	95.4	746 ± 50
880	53.4	13.2	25.2	2.0	5.6	Il 0.7	1.273	1.340	1.176	35	96.1	759 ± 50
882 *	91.5	1.3 *	3.3	3.6	0.1	Il 0.2	1.138	1.489	1.403	23	84.4	780 ± 30

Amp: amphibole; Omp: omphacite; Plag: plagioclase; Grt: garnet; Qtz: quartz; Mi: mica; Il: ilmenite. EBSD: electron back-scattered diffraction. [a] Equilibrium temperature using geothermometry (Ravna and Terry [75]; Ravna [76]). * Sample 882 is the most retrograded eclogite, and the composition of omphacite was changed to diopside. X: lineation, Z: normal to foliation, and Y: normal to X and Z.

3.2. Calculation of Seismic Velocity and Anisotropy

Seismic velocity and anisotropy were calculated using the orientation data of the minerals and the single-crystal elastic constants of amphibole [77] and omphacite [29] in sample 885, 880, and 881; diopside [78] in sample 882; garnet [54]; and plagioclase [79] using the Fortran program [80]. The whole-rock seismic velocity and anisotropy were calculated by combining the calculated results of individual minerals, including amphibole, garnet, and omphacite in samples 885, 880, and 881; diopside in sample 882; and all minor minerals, quartz [81], ilmenite [82], and mica [83], with weighted values for the number of grains.

AV_P indicates the anisotropy of the P-wave velocity in all propagation directions, calculated using Equation (1). AV_S indicates the anisotropy of split fast and slow S-wave velocity in the same propagation direction, calculated using Equation (2) [80].

$$AV_P = (V_{Pmax} - V_{Pmin})/[(V_{Pmax} + V_{Pmin})/2], \qquad (1)$$

$$AV_S = (V_{Sfast} - V_{Sslow})/[(V_{Sfast} + V_{Sslow})/2] \qquad (2)$$

3.3. Calculation of Reflection Coefficient

The reflection coefficient (R_C) was calculated for a seismic wave propagating perpendicular to the boundary between two rock types or layers. This concept is simple and allows the calculation of R_C as follows:

$$R_C = [V_P(//Z)_{1\rho 1} - V_P(//Z)_{2\rho 2}]/[V_P(//Z)_{1\rho 1} + V_P(//Z)_{2\rho 2}], \tag{3}$$

where $V_P(//Z)_{1\rho 1}$ and $V_P(//Z)_{2\rho 2}$ indicate the P-wave velocity propagating normal to the foliation in rocks/layers 1 and 2, respectively [6,50,51].

Reflection coefficients were calculated assuming a boundary between the studied samples and rocks expected to be stable near existing eclogite zones, such as amphibole peridotite [84], chlorite peridotite [85], lawsonite blueschist [8,86], epidote blueschist [87], garnet peridotite [87–89], peridotite complexes (lherzolite, harzburgite, dunite) [90–92], lawsonite eclogite [17], olivine complexes [93], spinel peridotite [94], serpentinite [95,96], serpentinized peridotite [92,97], felsic and mafic blueschist [98], dunite [99], and glaucophane schist (containing epidote or phengite) [100].

3.4. Micro-Raman Spectroscopy

A micro-Raman spectrometer located at the Tectonophysics Laboratory in SEES, SNU, was used to identify minerals clearly. This spectrometer was equipped with a 532 nm laser (10 mW power) and an optical microscope with a 50× objective lens. It had a resolution of 0.01 cm^{-1} over the wavenumber range of 50 to 3550 cm^{-1}, with a beam size of 0.67 μm. The spectrum was overlapped 32 times to reduce noise.

3.5. Analysis of Chemical Composition and Temperature Estimation

The chemical composition of the minerals in the samples was measured using a JEOL JXA-8530F Plus electron hyper probe installed at the Center for Research facility at Gyeongsang National University and at the National Center for Inter-University Research Facilities at Seoul National University, South Korea, using an acceleration voltage of 15 kV, a beam current of 10 nA, and a beam size of 5 μm in diameter. The amphibole phase was classified using the Excel spreadsheet provided by Locock [101] based on the classification of Hawthorne et al. [102].

The equilibrium temperature of the samples was estimated by thermometry of garnet–clinopyroxene Fe–Mg exchange [75,76], and the measured oxide mass percentages of garnet, clinopyroxene, amphibole, and minor minerals were used to calculate the equilibrium temperature.

4. Results

4.1. Sample Description and Microstructures

All samples were identified as retrograded eclogites, and the degree of retrogression of the samples varied depending on the sample. The samples had a high proportion of amphibole and plagioclase, and half of the clinopyroxene grains exhibited a symplectite structure of omphacite, diopside, and plagioclase, indicating that our samples were retrograded. The modal composition of rocks and the aspect ratio of minerals are shown in Table 1, and photomicrographs of the samples are shown in Figure 3. Sample 885 was a relatively fresh eclogite, whereas sample 882 was the most retrograded eclogite. Samples 881 and 880 showed an intermediate degree of retrogression, between that of samples 885 and 882. The phase of the minerals in sample 885 (the relatively fresh eclogite) was identified using micro-Raman spectroscopy. Clinopyroxene was identified as omphacite, with Raman peaks at 142, 338, 674, and 1016 cm^{-1}, whereas garnet in the same sample showed Raman peaks at 356 and 911 cm^{-1}. An unknown mineral found in the relatively fresh eclogite showing Raman peaks at 175, 300, and 1099 cm^{-1} was identified as zircon (Figure 4).

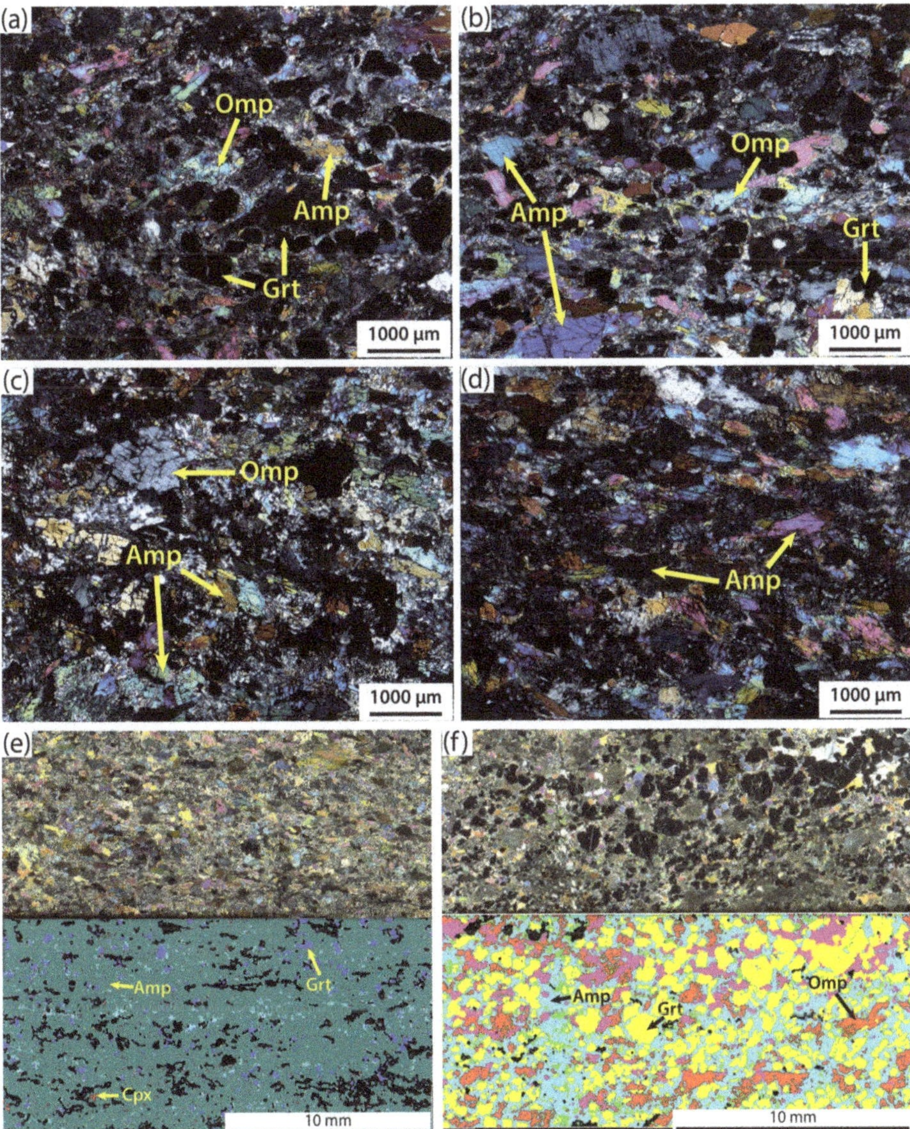

Figure 3. Photomicrograph of a thin section in the XZ plane of samples. (**a**) Sample 885, showing relatively fresh eclogite; (**b**) sample 881, showing retrograded eclogite illustrating a lack of intracrystalline deformation in amphibole; (**c**) sample 880, showing retrograded eclogite; (**d**) sample 882, the most severely retrograded eclogite showing elongated amphiboles and a well-developed foliation; (**e**) thin-section image of sample 882 in cross-polarized light (top) and EBSD phase map (bottom) (color codes: greenish blue, amphibole; blue, garnet; light blue, plagioclase; red, clinopyroxene; black, non-indexed points); and (**f**) thin-section image of sample 885 in cross-polarized light (top) and EBSD phase map (bottom) (color codes: light blue, amphibole; yellow, garnet; red, omphacite; purple, quartz; green, plagioclase; black, non-indexed points). Omp: omphacite; Amp: amphibole; Grt: garnet; Cpx: clinopyroxene.

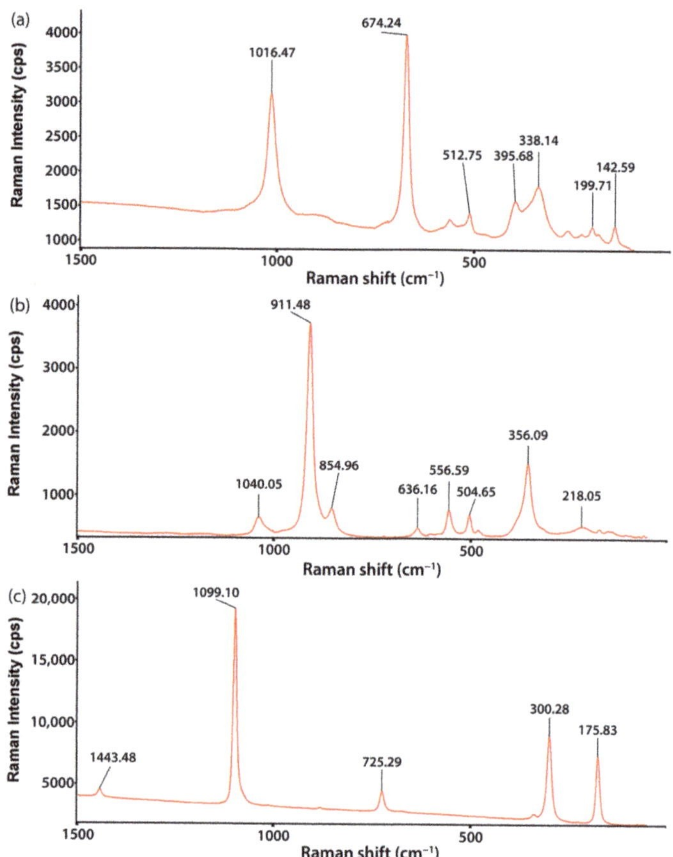

Figure 4. Micro-Raman spectra of (**a**) omphacite, (**b**) garnet, and (**c**) zircon in sample 885.

Amphibole and omphacite in the samples showed well-developed foliation (Figure 3b–f), but garnet showed less elongated grains than other minerals, indicating that garnet grains are less affected by deformation (Figure 3a,b,f). All mineral grains, except garnet, showed curved or angular-shaped grain boundaries, whereas garnet grains showed straight grain boundaries. Most of the amphibole grains showed cleavage in grains, with elongated grain shapes (Figure 3b,d). However, amphibole grains showed a lack of intracrystalline deformation features, such as undulose extinction or subgrains (Figure 3b,d).

4.2. Chemical Composition and Equilibrium Temperature

The chemical compositions of the minerals in the samples are listed in Table 2. Muscovite was found in two samples, 882 and 885, whereas the other samples, 880 and 881, did not contain muscovite. By analyzing the amphibole phase by chemical composition, amphibole in all the samples was found to be magnesiohornblende.

Table 2. Representative chemical composition (oxide mass, %) of minerals in samples.

Element	Sample 885								Sample 880				
	Amp	Omp	Grt	Mus	Ru	Ilme	Plg	Qtz	Amp	Omp	Grt	Plg	Qtz
SiO_2	46.08	54.65	39.99	46.23	0.04	0	56.63	99.16	45.09	54.59	40.47	44.68	100.79
TiO_2	0.46	0.16	0.06	0	99.7	54.94	0.02	0	0.42	0.16	0.02	0	0.01
Al_2O_3	12.44	7.02	22.24	37.04	0	0.03	27.83	0.01	12.29	7.05	22.73	35.85	0
Cr_2O_3	0.15	0.02	0.13	0.05	0.41	0.05	0	0.03	0.15	0	0	0	0.01
FeO	10.5	5.32	17.29	0.44	0.21	44.12	0.22	0.2	11.58	5.28	16.55	0.05	0
MgO	13.51	10.93	9.08	0.3	0	0.1	0	0	12.69	10.97	10.66	0.02	0
CaO	11.74	18.38	10.69	0.03	0.13	0	9.64	0	12.27	18.34	9.67	19.23	0
Na_2O	1.96	3.73	0	0.17	0.01	0	6.11	0.01	1.84	3.82	0	0.78	0
K_2O	0.01	0	0	10.85	0	0	0	0	0.02	0	0	0	0
MnO	0.08	0.06	0.28	0	0.02	1.44	0.02	0	0.17	0.08	0.38	0.02	0.01
NiO	0.07	0	0	0.03	0	0.03	0	0.03	0.05	0	0	0.02	0
Total	97	100.27	99.76	95.14	100.52	100.72	100.45	99.43	96.55	100.29	100.48	100.65	100.82

Element	Sample 881							Sample 882				
	Amp	Omp	Grt	Ru	Ilme	Plg	Qtz	Amp	Cpx	Grt	Mus	Ilme
SiO_2	49.29	53.96	40.56	0	0.04	43.55	99.32	48.27	51.48	40.41	48.99	0
TiO_2	0.24	0.18	0	99.35	54.31	0	0.01	0.3	0.14	0.01	0.02	54.59
Al_2O_3	9.77	7.87	23.2	0.08	0.01	36.65	0	10.15	5.29	23.16	30.24	0.05
Cr_2O_3	0.13	0.13	0.23	0.5	0.06	0	0.02	0.07	0.73	0.12	0.47	0.03
FeO	6.92	2.53	16.51	0.52	43.03	0.05	0.02	8.68	4.98	16.57	1.33	40.98
MgO	16.89	12.3	10.49	0.01	1.83	0	0.02	15.56	12.59	11.35	2.49	0.13
CaO	12.1	19.94	9.55	0.09	0	19.75	0.01	11.86	23.62	8.23	0.01	0.06
Na_2O	1.49	2.97	0	0.08	0.04	0.33	0	1.27	0.72	0.02	0.07	0
K_2O	0	0	0.01	0.01	0	0.01	0	0.01	0	0	11.32	0
MnO	0.11	0.03	0.3	0.03	0.59	0	0	0.1	0.16	0.32	0.02	4.96
NiO	0.05	0.13	0	0.02	0	0	0	0.03	0.04	0.03	0.02	0
Total	96.98	100.04	100.85	100.7	99.91	100.34	99.4	96.31	99.74	100.21	94.98	100.79

Amp: amphibole; Omp: omphacite; Cpx: clinopyroxene; Grt: garnet; Mus: muscovite; Ru: rutile; Ilme: ilmenite; Plg: plagioclase; Qtz: quartz.

Based on the chemical composition of the analyzed minerals and the calculation of the temperature conditions of the samples using geothermometry [75,76], the samples were found to be equilibrated at a temperature of ~750 °C. The equilibrium temperatures of samples are presented in Table 1.

4.3. Lattice-Preferred Orientations of Minerals

The amphibole LPOs are shown in Figure 5, and the strengths of the LPOs of the samples are described in Table 3. For amphibole in retrograded eclogites with relatively little retrogression (samples 880, 881, and 885), the <001> axes of amphibole were strongly aligned subparallel to lineation (L), with a weak girdle distribution subparallel to foliation (S). The amphibole (010) poles were strongly aligned subnormal to foliation, and the amphibole (110) poles formed a girdle distribution subnormal to lineation. The most retrograded eclogite (sample 882) showed that the amphibole <001> axes were strongly aligned subparallel to lineation, with a girdle distribution subparallel to foliation (S), but both amphibole (010) and (110) poles were weakly aligned subnormal to foliation (Figure 5).

The omphacite LPOs are shown in the lower part of Figure 5, and the strengths of the LPOs of the samples are listed in Table 3. Omphacites in samples 880, 881, and 885 showed that the <001> axes were strongly aligned subparallel to lineation, with a girdle distribution subparallel to foliation, and the (010) poles were strongly aligned subnormal to foliation. This omphacite alignment is known as the S-type omphacite LPO [61]. The most retrograded sample 882 contained little clinopyroxene, and the number of clinopyroxene grains in the pole figure was too small (<150 grains) to say LPO [74]. The pole figure did not provide a reliable indication.

The garnet and plagioclase LPOs are shown in Figure 6. They showed a weak LPO, which was considerably dispersed compared to the amphibole and omphacite LPO. The garnet LPO was almost random, independent of the degree of retrogression. In the case of the plagioclase LPO, two samples (885 and 880) showed (010) poles aligned subparallel to lineation.

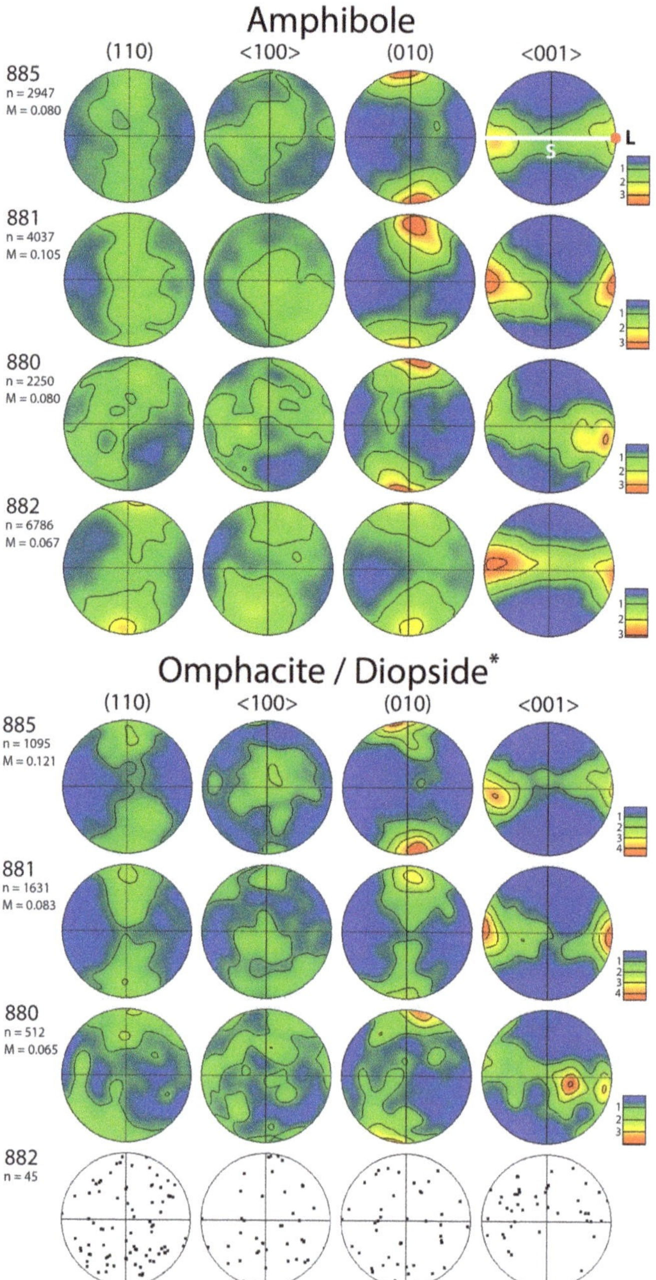

Figure 5. Amphibole and omphacite pole figures (PFs). PFs are presented in the lower hemisphere using an equal area projection. S indicates foliation, and L indicates lineation. Fabric strength of the mineral is shown as the M-index (M). A half scatter width of 20° was used. The color coding (different for each sample) refers to the density of data points, and contours are shown in multiples of uniform distribution. n: number of grains. * Diopside for sample 882.

Table 3. Fabric strength of samples.

Sample	Fabric Strength [a]			
	Amp	**Omp**	**Grt**	**Plag**
885	0.08	0.121	0.018	0.049
881	0.105	0.083	0.016	0.043
880	0.08	0.065	0.045	0.048
882	0.067	-	0.022	0.06

Amp: amphibole; Omp: omphacite; Grt: garnet; Plag: plagioclase. [a] Fabric strength was calculated as the M-index (Skemer et al. [74]).

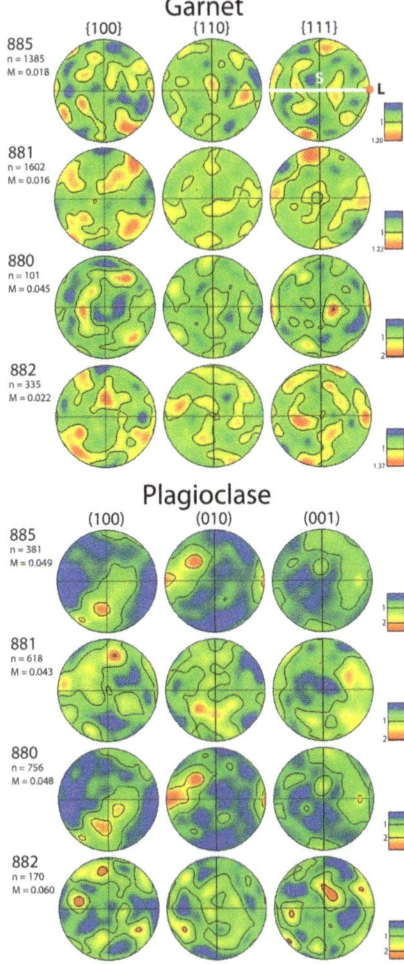

Figure 6. Pole figures (PFs) of garnet and plagioclase. PFs are presented in the lower hemisphere using an equal area projection. S indicates foliation, and L indicates lineation. Fabric strength of the mineral is shown as the M-index (M). A half scatter width of 20° was used. The color coding refers (different for each sample) to the density of data points (n), and contours are shown in multiples of uniform distribution. n: number of grains.

4.4. Misorientation between Amphibole and Omphacite

The misorientation angles between omphacite and amphibole in direct contact with omphacite were measured and are shown in Figure 7. The misorientation angles were significantly low—lower than 3° in both boundaries between omphacite grains, and amphibole in direct contact with omphacite (Figure 7a–c, <0.5°) and the boundary between omphacite and amphibole, which were trapped and surrounded by omphacite (Figure 7d–f, <2°). The symplectites in Figure 7 consist of omphacite, amphibole, diopside, and plagioclase, which were confirmed by measuring the chemical compositions of minerals using the electron probe micro analyzer installed at the National Center for Inter-University Research Facilities at Seoul National University, South Korea.

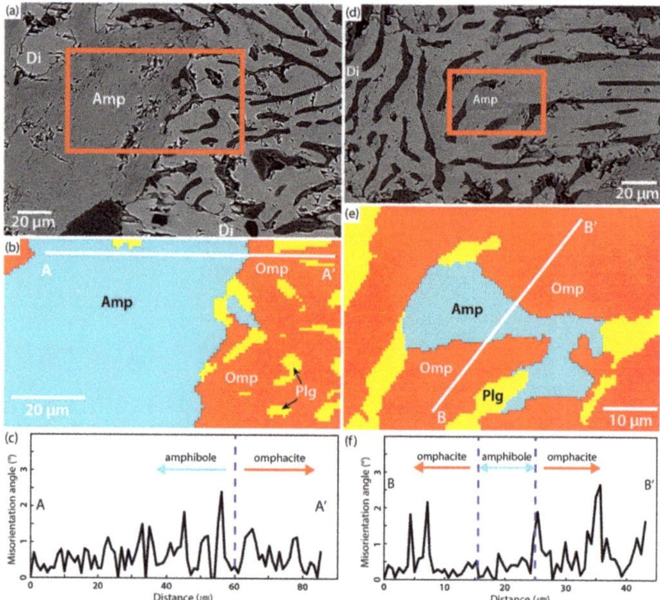

Figure 7. (**a**) and (**d**) Back-scattered electron image of amphibole and directly contacting omphacite symplectite in sample 885; (**b**) and (**e**) EBSD phase map of the area located in the red rectangle in (**a**) and (**d**); and (**c**) and (**f**) misorientation angle distribution spectra along A–A' in (**b**) and B–B' in (**e**). Vertical dashed lines indicate the phase boundary between amphibole and omphacite. Omp: omphacite; Amp: amphibole; Di: diopside; Plg: plagioclase.

4.5. Seismic Velocity and Anisotropy

The seismic velocities and anisotropies of the samples are summarized in Table 4. The seismic velocity and anisotropy of the samples are shown in the stereonet for the single-mineral amphibole, omphacite, and garnet and for the whole rock consisting of all minerals, including minor minerals such as plagioclase and ilmenite (Figures 8–10). In the case of amphibole, P-wave velocity (V_P) was in the range of 6.59 to 7.09 km/s, and P-wave anisotropy (AV_P) was in the range of 3.7% to 7.3%, whereas the maximum S-wave anisotropy (max. AV_S) was in the range of 2.72% to 3.61%. In the case of omphacite, the V_P of omphacite was in the range of 8.31 to 8.54 km/s, which is faster than that of amphibole. However, the AV_P of omphacite was in the range of 2.1% to 2.6% and the maximum AV_S was in the range of 1.23% to 1.56%, which is smaller than that of amphibole. In the case of garnet, V_P was the fastest among the minerals, ranging from 9.31 to 9.34 km/s. However, garnet AV_P and maximum AV_S were the smallest compared to other minerals, ranging from 0.1% to 0.3% and 0.16% to 0.55%, respectively (Table 4, Figure 9).

Table 4. Seismic velocity and anisotropy of minerals in samples.

Sample	Amphibole				Omphacite/Diopside *			
	V_P (km/s)	V_S (km/s)	AV_P (%)	Max. AV_S (%)	V_P (km/s)	V_S (km/s)	AV_P (%)	Max. AV_S (%)
885	6.71–6.97	3.66–3.79	3.7	3.28	8.31–8.53	4.85–4.92	2.6	1.41
881	6.67–7.02	3.64–3.79	5.2	3.61	8.36–8.54	4.84–4.92	2.1	1.56
880	6.71–6.97	3.60–3.77	3.8	3.21	8.36–8.53	4.84–4.91	2.1	1.23
882	6.59–7.09	3.65–3.76	7.3	2.72	7.99–8.13	4.66–4.75	1.7	1.68
Sample	Garnet				Whole Rock			
	V_P (km/s)	V_S (km/s)	AV_P (%)	Max. AV_S (%)	V_P (km/s)	V_S (km/s)	AV_P (%)	Max. AV_S (%)
885	9.32–9.33	5.49–5.50	0.1	0.16	7.57–7.66	4.36–4.44	1.2	1.57
881	9.31–9.33	5.48–5.50	0.2	0.31	7.36–7.55	4.12–4.21	2.6	1.87
880	9.31–9.34	5.47–5.51	0.3	0.55	6.90–7.05	3.84–3.90	2.1	1.40
882	9.31–9.33	5.48–5.51	0.2	0.35	6.70–7.16	3.73–3.83	6.6	2.46

V_P: P-wave velocity; V_S: S-wave velocity; AV_P: anisotropy of P-wave velocity; max. AV_S: maximum anisotropy of S-wave velocity.
* Diopside for sample 882.

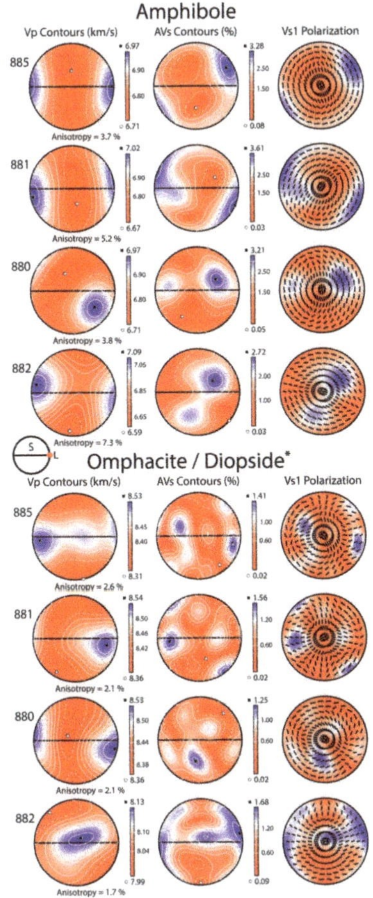

Figure 8. Seismic velocity and anisotropy of amphibole and omphacite. P-wave velocity (V_P) and seismic anisotropy of S-wave (AV_S) are shown. V_S1 polarization indicates the polarization direction of the fast shear wave (V_S1). S and L represent foliation and lineation, respectively. * Diopside for sample 882.

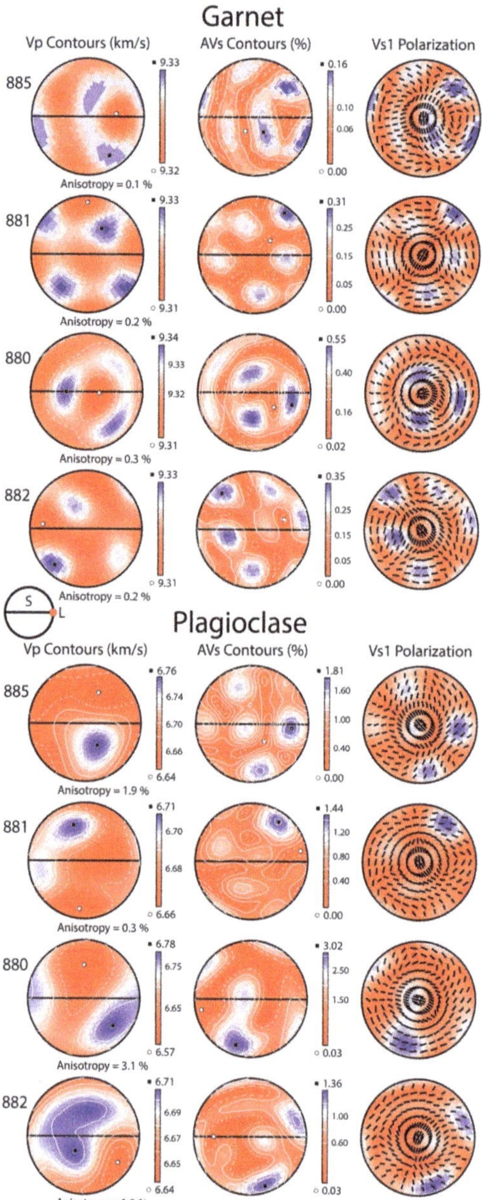

Figure 9. Seismic velocity and anisotropy of garnet and plagioclase. P-wave velocity (V_P) and seismic anisotropy of S-wave (AV_S) are shown. V_S1 polarization indicates the polarization direction of the fast shear wave (V_S1). S and L represent foliation and lineation, respectively.

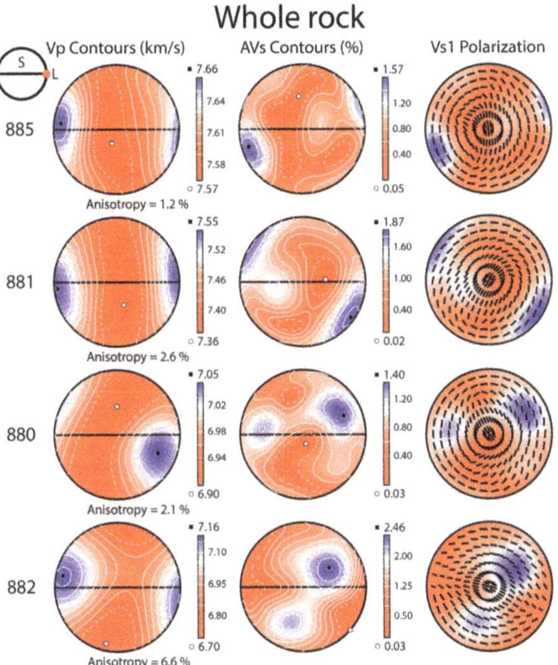

Figure 10. Seismic velocity and anisotropy of whole-rock P-wave velocity (V_P) and seismic anisotropy of S-wave (AV_S) are shown. V_S1 polarization indicates the polarization direction of the fast shear wave (V_S1). S and L represent foliation and lineation, respectively.

The seismic velocity and anisotropy of whole rock showed a value between the seismic velocity and anisotropy of omphacite and amphibole. The V_P of whole rock was in the range of 6.70 to 7.66 km/s, AV_P was in the range of 1.2% to 6.6%, and the maximum AV_S of whole rock was in the range of 1.40% to 2.46%. The seismic velocity and anisotropy profile of whole rock (Figure 10, color-coded and contour lines in pole figures) were much more similar to those of amphibole than those of other minerals (Figures 8 and 9).

4.6. Reflection Coefficient of Samples

The reflection coefficient (R_C) of the boundaries between the studied retrograded eclogites and other rocks in subduction zones is shown in Table 5. The R_C of the boundary between serpentinites and the most retrograded eclogite (sample 882) was in the range of 0.024 to 0.147. The R_C of the boundary between blueschists and the most retrograded eclogite was in the range of 0.011 to 0.05. In addition, the R_C of the boundary between dunite and the most retrograded eclogite was in the range of 0.060 to 0.092. Finally, the R_C of the boundary between peridotites and the most retrograded eclogite was in the range of 0.067 (for amphibole peridotite) to 0.129 (for garnet peridotite). Furthermore, the maximum R_C ($R_C = 0.204$) was observed at the contact boundary between the serpentinized peridotite [97] and relatively fresh eclogite (sample 885), and the minimum R_C ($R_C = 0.002$) was observed at the contact boundary between dunite [91] and relatively fresh eclogite.

Table 5. Reflection coefficient (R_C) of the contact boundary between the studied eclogites and other rocks.

Other Rocks	Sample 882 *	Sample 880	Sample 881	Sample 885 +
LT serpentinized peridotite (OHM-B) [a]	0.147	0.161	0.192	0.204
HT serpentinized peridotite (OHM-C) [a]	0.081	0.095	0.126	0.139
Serpentinite [b]	0.049	0.064	0.096	0.108
Glaucophane schist (Ph) [c]	0.041	0.056	0.087	0.100
LT serpentinized peridotite (HPS-J) [a]	0.041	0.055	0.087	0.099
Felsic blueschist [d]	0.026	0.040	0.072	0.084
Epidote blueschist (Gln:Ep = 94:6) [e]	0.011	0.004	0.004	0.048
Lawsonite blueschist (Gln:Lws = 91:9) [e]	0.019	0.004	0.028	0.040
Mafic blusechist [d]	0.019	0.004	0.027	0.039
Epidote blueschist (Gln:Ep = 26:74) [e]	0.021	0.007	0.025	0.037
HT serpentinized peridotite (HPS-I) [a]	0.024	0.009	0.023	0.035
Harzburgite + Melt [f]	0.024	0.009	0.023	0.035
Glaucophane schist (Ep) [c]	0.033	0.019	0.013	0.025
Lawsonite blueschist (Gln: Lws = 75:25) [e]	0.035	0.020	0.012	0.024
Chlorite peridotite [g]	0.044	0.029	0.003	0.015
Lawsonite blueschist [h]	0.050	0.035	0.003	0.009
Harzburgite + Dunite [f]	0.051	0.036	0.005	0.008
Harzburgite [f]	0.053	0.038	0.007	0.006
Dunite [f]	0.060	0.046	0.014	0.002
100% olivine [i]	0.060	0.046	0.014	0.002
Ol 70% + OPX 20% + CPX 10% [i]	0.067	0.052	0.021	0.008
Amphibole peridotite [j]	0.067	0.052	0.021	0.008
Ol 50% + OPX 30% + CPX 20% [i]	0.073	0.059	0.027	0.015
Dunite (NJ437-1) [i]	0.080	0.065	0.003	0.021
Lherzolite [k]	0.080	0.065	0.033	0.021
Dunite [l]	0.080	0.065	0.033	0.021
Lherzolite [m]	0.083	0.068	0.037	0.024
Spinel peridotite [n]	0.089	0.074	0.043	0.030
Lherzolite (NJ437-8) [i]	0.092	0.077	0.046	0.033
Dunite [m]	0.092	0.077	0.046	0.033
Lawsonite eclogite [o]	0.104	0.090	0.058	0.046
Garnet peridotite [p]	0.105	0.090	0.059	0.046
Garnet peridotite (150-4) [q]	0.110	0.096	0.064	0.052
Garnet peridotite [r]	0.112	0.098	0.066	0.054
Garnet peridotite (160-9) [q]	0.129	0.115	0.083	0.071

* Sample 882 is the most retrograded eclogite. + Sample 885 is the relatively fresh eclogite. [a] Watanabe et al. [97]; [b] Jung [95]; [c] Ha et al. [100]; [d] Cao et al. [98]; [e] Kim et al. [86]; [f] Tommasi et al. [91]; [g] Kim and Jung [85]; [h] Cao and Jung [8]; [i] Michibayashi et al. [93]; [j] Kang and Jung [84]; [k] Tommasi et al. [92]; [l] Cao et al. [99]; [m] Tommasi et al. [90]; [n] Jung et al. [94]; [o] Kim et al. [17]; [p] Skemer et al. [87]; [q] Xu et al. [89]; [r] Wang et al. [88]. LT: low temperature; HT: high temperature; Ph: phengite; Gln: glaucophane; Ep: epidote; Lws: lawsonite; Ol: olivine; OPX: orthopyroxene; CPX: clinopyroxene.

5. Discussion

5.1. Lattice-Preferred Orientations of Minerals

Omphacite in retrograded eclogites (samples 880, 881, and 885) from Xitieshan, northwestern China, showed that the <001> axes are strongly aligned subparallel to lineation, with a girdle distribution subparallel to foliation, and the (010) poles are strongly aligned subnormal to foliation. This is called S-type omphacite LPO [61]. The observed S-type omphacite LPOs were consistent with those reported in previous studies [8,22]. The S-type omphacite LPO has been reported in eclogites from the eastern Alps [103] and Cabo Ortegal eclogites from northwestern Spain [104].

Amphibole in retrograded eclogites from Xitieshan, northwest China, showed that the amphibole (010) poles were aligned subnormally to foliation, and the maximum concentration of amphibole <001> axes were aligned subparallel to lineation, which is referred to as type V amphibole LPO. Previously, four other types of amphibole LPOs have been reported in both naturally and experimentally deformed amphibole [12,18,36]. The type V amphibole LPO in this study is different from the amphibole LPOs (types I, II, III, and IV) reported previously by high-pressure experimental studies. Similar amphibole LPOs to type V LPO have been reported only in naturally deformed amphibole from the Malpica Tui allochthonous complex, northwestern Spain, and Cabo Ortegal, Spain [44,59], and in retrograded eclogites from the Sanbagawa metamorphic belt, central Shikoku, southwestern Japan [60]. To verify that there was no error in the reference frame (X: lineation, Z: normal to foliation) of samples in our study, the aspect ratios of amphibole and omphacite in the XZ, YZ, and XY planes were determined and compared (Table 1). Grains on the XZ plane were more elongated compared to those on the other planes. This result indicates that the reference frame is correct.

The amphibole LPO (type V) in the studied samples (samples 880, 881, and 885) has only been reported in retrograded eclogites, and the amphibole LPO was similar to that of omphacite in the same samples (Figure 5). The lack of intracrystalline deformation in amphibole indicates that the samples were not deformed after the retrogression of eclogite (Figure 3b,d). In addition, comparison of the misorientation angle between amphibole and the contacting omphacite (Figure 7) showed small angles in both the contact boundary (<2°) and the grain itself (<2.5°), suggesting that type V amphibole LPOs are strongly affected by omphacite LPOs [105,106]. Our results indicate that the amphibole LPO is formed due to the topotactic growth of amphibole during the retrogression of eclogite. Previous studies have also reported that amphibole LPOs are similar to those of omphacite in the same sample, and they suggested that amphibole is formed due to the topotactic growth during retrogression of eclogite [44,59,60].

5.2. Seismic Velocity, Anisotropy, and Reflection Coefficient (R_C)

The amphibole P-wave velocity (V_P) was slow (in the range of 6.59 to 7.09 km/s), and amphibole P-wave anisotropy (AV_P) was large (in the range of 3.7% to 7.3%) (Table 4). The amphibole S-wave velocity (V_S) was also slow (in the range of 3.64 to 3.79 km/s) and amphibole S-wave anisotropy (max. AV_S) large (in the range of 2.72% to 3.61%). When the omphacite seismic velocity and anisotropy were compared with those of amphibole, the seismic velocity of omphacite was much faster than that of amphibole in both P- and S-waves, but the seismic anisotropy of omphacite was much smaller than that of amphibole in both P- and S-waves (Table 4). This is consistent with previous studies on the seismic velocity and anisotropy of these minerals in eclogites [32,107] and the seismic velocity and anisotropy analysis of mineral mixtures in eclogite [22].

The garnet P-wave velocity was the fastest in the studied samples, whereas the amphibole P-wave velocity was the slowest. The P-wave and S-wave anisotropies of garnet were the slowest in the samples, similar to previous studies [22]. The trends in seismic velocity and anisotropy of whole retrograded eclogites (color coded in pole figures in Figure 10) were similar to those of amphibole (Figure 8). This result indicates that the seismic properties of retrograded eclogites are strongly affected by the seismic properties of amphibole due to the high fraction of amphibole (33–91.5%) in retrograded eclogites.

When the seismic velocity and anisotropy of relatively fresh eclogite (sample 885) were compared with those of the most retrograded eclogite (sample 882), the P-wave velocity decreased with increasing retrogression (Figures 8–10 and Table 4), but the P- and S-wave anisotropies increased with increasing retrogression (Figures 8–10 and Table 4). This result indicates that when eclogite increases to the eclogite instability zone and gets retrograded, the seismic velocity of retrograded eclogite decreases, but the seismic anisotropy of retrograded eclogite increases.

The R_C of the studied samples was in the range of 0.002 to 0.204 (Table 5). The contact boundary between relatively fresh eclogite and serpentinized peridotites showed a high R_C (up to 0.204), which was high enough to reflect the seismic wave. However, at the contact boundary between the studied eclogites and the dunites and blueschists, the R_C was small (0.002–0.05), independent of the degree of retrogression of eclogite (Table 5). A small R_C indicates that the seismic wave reflects little at this boundary, and the seismic wave reflected by this boundary is difficult to detect [6].

In Figure 11, the R_C between various rocks and the studied eclogites is compared with the P-wave velocity. In the case of serpentinized peridotites, which have low V_P ($V_P < 6.8$ km/s), the R_C was high for the contact boundary with relatively fresh eclogite (sample 885), indicating that the boundary between serpentinized peridotites and the relatively fresh eclogites reflects a large seismic wave, resulting in this boundary being easily detected compared to the other contact boundaries of rocks. In contrast, in the case of dry mantle rocks (spinel peridotites and garnet peridotites), which have a high V_P ($V_P > 7.5$ km/s), the R_C was high for the contact boundary with the most retrograded eclogite (sample 882). This result indicates that the contact boundary between dry mantle rocks and the most retrograded eclogite reflects more seismic waves, resulting in the boundary being easily detected compared to the other contact boundaries of rocks, similar to the contact boundary between serpentinized peridotites and relatively fresh eclogite. However, in the case of blueschists, which have an intermediate V_P (6.8 km/s $< V_P <$ 7.5 km/s), the R_C was low for the contact boundary with the studied eclogites, whether eclogites were retrograded or not, indicating that the boundaries between blueschists and eclogites reflected little seismic waves, independent of the degree of retrogression of eclogite, indicating that these boundaries may be difficult to detect.

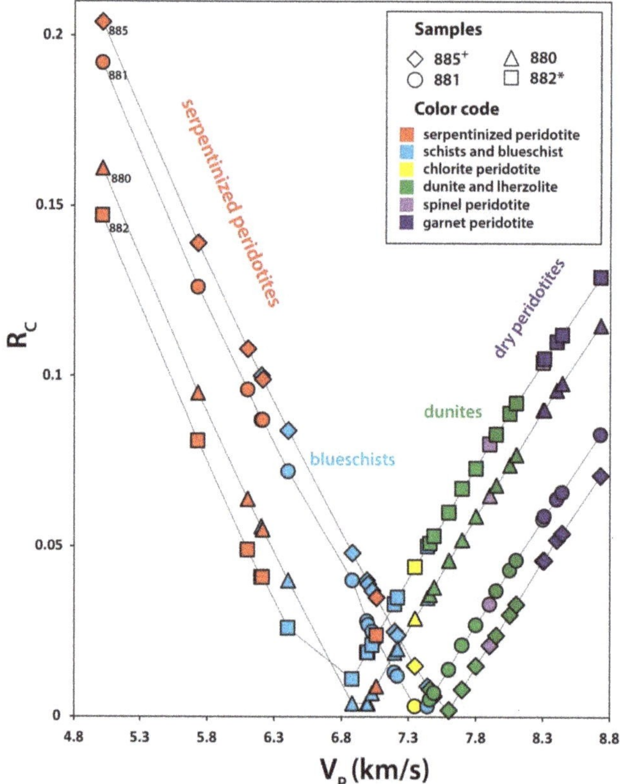

+Sample 885 is relatively fresh eclogite. *Sample 882 is the most retrograded eclogite.

Figure 11. Relationship between P-wave velocity (V_P) of a surrounding rock and reflection coefficient (R_C) of the contact boundary between the studied retrograded eclogite and the surrounding rock (Table 5).

6. Conclusions

The LPOs of minerals in retrograded eclogites from Xitieshan, northwestern China, were studied to understand the seismic anisotropies and seismic reflection coefficients of the rocks that are in contact with other surrounding rocks. The amphibole LPOs in the retrograded eclogites were mostly similar to omphacite LPOs. The low misorientation angle between amphibole in direct contact with omphacite suggests that the amphibole LPO is formed due to topotactic growth of amphibole during the retrogression process of eclogites. There was a lack of intracrystalline deformation features in amphibole, such as undulose extinction and subgrains, indicating that there might have been little deformation after the retrogression of eclogites.

The seismic properties of retrograded eclogites were more severely affected by amphibole than by other minerals due to the high fraction of amphibole (33–91.5%) in retrograded eclogites. During the retrogression process of eclogites, the seismic velocity slowed and the seismic anisotropy increased. R_C, which was calculated using the V_P of minerals, showed that the contact boundaries between relatively fresh eclogite and hydrated mantle rocks, such as serpentinized peridotites, showed high R_C, indicating that these contact boundaries can be easily detected using the reflected seismic wave, but the contact boundaries between dunites or blueschists and eclogites showed low R_C, independent of the degree of retrogression. This indicates that these contact boundaries are expected to be difficult to detect using the reflected seismic wave.

Author Contributions: Conceptualization, H.J.; methodology, J.L. and H.J.; software, J.L.; validation, J.L. and H.J.; formal analysis, J.L.; investigation, J.L.; resources, J.L. and H.J.; data curation, J.L.; writing—original draft preparation, J.L.; writing—review and editing, J.L. and H.J.; visualization, J.L.; supervision, H.J.; project administration, H.J.; funding acquisition, H.J. All authors have read and agreed to the published version of the manuscript.

Funding: This research was supported by a grant to H.J. from the Korea Meteorological Administration Research Development Program (KMI2019-00110).

Institutional Review Board Statement: Not applicable.

Informed Consent Statement: Not applicable.

Data Availability Statement: Not applicable.

Acknowledgments: The authors are grateful to the anonymous reviewers for constructive comments and corrections.

Conflicts of Interest: The authors declare no conflict of interest.

References

1. Savage, M.K. Seismic anisotropy and mantle deformation: What have we learned from shear wave splitting? *Rev. Geophys.* **1999**, *37*, 65–106. [CrossRef]
2. Long, M.D.; Silver, P.G. The subduction zone flow field from seismic anisotropy: A global view. *Science* **2008**, *319*, 315–318. [CrossRef] [PubMed]
3. Stern, R.J. Subduction zones. *Rev. Geophys.* **2002**, *40*, 3-1–3-38. [CrossRef]
4. Cao, Y.; Jung, H.; Song, S. Microstructures and petro-fabrics of lawsonite blueschist in the North Qilian suture zone, NW China: Implications for seismic anisotropy of subducting oceanic crust. *Tectonophysics* **2014**, *628*, 140–157. [CrossRef]
5. Ábalos, B.; Fountain, D.M.; Ibarguchi, J.I.G.; Puelles, P. Eclogite as a seismic marker in subduction channels: Seismic velocities, anisotropy, and petrofabric of Cabo Ortegal eclogite tectonites (Spain). *Geol. Soc. Am. Bull.* **2011**, *123*, 439–456. [CrossRef]
6. Bascou, J.; Barruol, G.; Vauchez, A.; Mainprice, D.; Egydio-Silva, M. EBSD-measured lattice-preferred orientations and seismic properties of eclogites. *Tectonophysics* **2001**, *342*, 61–80. [CrossRef]
7. Ben Ismail, W.; Mainprice, D. An olivine fabric database: An overview of upper mantle fabrics and seismic anisotropy. *Tectonophysics* **1998**, *296*, 145–157. [CrossRef]
8. Cao, Y.; Jung, H. Seismic properties of subducting oceanic crust: Constraints from natural lawsonite-bearing blueschist and eclogite in Sivrihisar Massif, Turkey. *Phys. Earth Planet. Inter.* **2016**, *250*, 12–30. [CrossRef]
9. Godard, G.; van Roermund, H.L.M. Deformation-induced clinopyroxene fabrics from eclogites. *J. Struct. Geol.* **1995**, *17*, 1425–1443. [CrossRef]

10. Ji, S.; Saruwatari, K.; Mainprice, D.; Wirth, R.; Xu, Z.; Xia, B. Microstructures, petrofabrics and seismic properties of ultra high-pressure eclogites from Sulu region, China: Implications for rheology of subducted continental crust and origin of mantle reflections. *Tectonophysics* **2003**, *370*, 49–76. [CrossRef]

11. Ji, S.; Wang, Q.; Xia, B. P-wave velocities of polymineralic rocks: Comparison of theory and experiment and test of elastic mixture rules. *Tectonophysics* **2003**, *366*, 165–185. [CrossRef]

12. Jung, H. Crystal preferred orientations of olivine, orthopyroxene, serpentine, chlorite, and amphibole, and implications for seismic anisotropy in subduction zones: A review. *Geosci. J.* **2017**, *21*, 985–1011. [CrossRef]

13. Jung, H.; Karato, S. Water-induced fabric transitions in olivine. *Science* **2001**, *293*, 1460–1463. [CrossRef]

14. Jung, H.; Katayama, I.; Jiang, Z.; Hiraga, I.; Karato, S. Effect of water and stress on the lattice-preferred orientation of olivine. *Tectonophysics* **2006**, *421*, 1–22. [CrossRef]

15. Karato, S.; Jung, H.; Katayama, I.; Skemer, P. Geodynamic significance of seismic anisotropy of the upper mantle: New insights from laboratory studies. *Annu. Rev. Earth Planet. Sci.* **2008**, *36*, 59–95. [CrossRef]

16. Keppler, R.; Ullemeyer, K.; Behrmann, J.H.; Stipp, M.; Kurzawski, R.M.; Lokajíček, T. Crystallographic preferred orientations of exhumed subduction channel rocks from the Eclogite Zone of the Tauern Window (Eastern Alps, Austria), and implications on rock elastic anisotropies at great depths. *Tectonophysics* **2015**, *647*, 89–104. [CrossRef]

17. Kim, D.; Wallis, S.; Endo, S.; Ree, J.-H. Seismic properties of lawsonite eclogites from the southern Motagua fault zone, Guatemala. *Tectonophysics* **2016**, *677–678*, 88–98. [CrossRef]

18. Ko, B.; Jung, H. Crystal preferred orientation of an amphibole experimentally deformed by simple shear. *Nat. Commun.* **2015**, *6*, 6586. [CrossRef]

19. Lee, J.; Jung, H.; Klemd, R.; Tarling, M.S.; Konopelko, D. Lattice preferred orientation of talc and implications for seismic anisotropy in subduction zones. *Earth Planet. Sci. Lett.* **2020**, *537*, 116178. [CrossRef]

20. Mauler, A.; Burlini, L.; Kunze, K.; Philippot, P.; Burg, J.P. P-wave anisotropy in eclogites and relationship to the omphacite crystallographic fabric. *Phys. Chem. Earth Part A Solid Earth Geod.* **2000**, *25*, 119–126. [CrossRef]

21. Nicolas, A.; Christensen, N.I. Formation of Anisotropy in Upper Mantle Peridotites—A Review. In *Composition, Structure and Dynamics of the Lithosphere-Asthenosphere System*; Fuchs, K., Froidevaux, C., Eds.; American Geophysical Union: Washington, DC, USA, 1987; Volumn 16, pp. 111–123.

22. Park, M.; Jung, H. Relationships between eclogite-facies mineral assemblages, deformation microstructures, and seismic properties in the Yuka terrane, North Qaidam ultrahigh-pressure metamorphic belt, NW China. *J. Geophys. Res. Solid Earth* **2019**, *124*, 13168–13191. [CrossRef]

23. Worthington, J.R.; Hacker, B.R.; Zandt, G. Distinguishing eclogite from peridotite: EBSD-based calculations of seismic velocities. *Geophys. J. Int.* **2013**, *193*, 489–505. [CrossRef]

24. Almqvist, B.S.; Mainprice, D. Seismic properties and anisotropy of the continental crust: Predictions based on mineral texture and rock microstructure. *Rev. Geophys.* **2017**, *55*, 367–433. [CrossRef]

25. Evans, B.W. Phase-relations of epidote-blueschists. *Lithos* **1990**, *25*, 3–23. [CrossRef]

26. Okamoto, K.; Maruyama, S. The high-pressure synthesis of lawsonite in the MORB+H2O system. *Am. Mineral.* **1999**, *84*, 362–373. [CrossRef]

27. Tsujimori, T.; Ernst, W.G. Lawsonite blueschists and lawsonite eclogites as proxies for palaeo-subduction zone processes: A review. *J. Metamorph. Geol.* **2014**, *32*, 437–454. [CrossRef]

28. Wei, C.J.; Clarke, G.L. Calculated phase equilibria for MORB compositions: A reappraisal of the metamorphic evolution of lawsonite eclogite. *J. Metamorph. Geol.* **2011**, *29*, 939–952. [CrossRef]

29. Bhagat, S.S.; Bass, J.D.; Smyth, J.R. Single-crystal elastic properties of omphacite-C2/C by Brillouin spectroscopy. *J. Geophys. Res. Solid Earth* **1992**, *97*, 6843–6848. [CrossRef]

30. Barruol, G.; Mainprice, D. A quantitative-evaluation of the contribution of crustal rocks to the shear-wave splitting of teleseismic SKS waves. *Phys. Earth Planet. Inter.* **1993**, *78*, 281–300. [CrossRef]

31. Christensen, N.I.; Mooney, W.D. Seismic velocity structure and composition of the continental-crust—A global view. *J. Geophys. Res. Solid Earth* **1995**, *100*, 9761–9788. [CrossRef]

32. Shi, F.; Wang, Y.; Xu, H.; Zhang, J. Effects of lattice preferred orientation and retrogression on seismic properties of eclogite. *J. Earth Sci.* **2010**, *21*, 569–580. [CrossRef]

33. Rudnick, R.; Gao, S. Composition of the continental crust. *Treatise Geochem.* **2003**, *3*, 659.

34. Fountain, D.M.; Salisbury, M.H. Exposed cross-sections through the continental-crust—Implications for crustal structure, petrology, and evolution. *Earth Planet. Sci. Lett.* **1981**, *56*, 263–277. [CrossRef]

35. Ji, S.; Shao, T.; Michibayashi, K.; Long, C.; Wang, Q.; Kondo, Y.; Zhao, W.; Wang, H.; Salisbury, M.H. A new calibration of seismic velocities, anisotropy, fabrics, and elastic moduli of amphibole-rich rocks. *J. Geophys. Res. Solid Earth* **2013**, *118*, 4699–4728. [CrossRef]

36. Kim, J.; Jung, H. New Crystal Preferred Orientation of Amphibole Experimentally Found in Simple Shear. *Geophys. Res. Lett.* **2019**, *46*. [CrossRef]

37. Bezacier, L.; Reynard, B.; Bass, J.D.; Wang, J.; Mainprice, D. Elasticity of glaucophane, seismic velocities and anisotropy of the subducted oceanic crust. *Tectonophysics* **2010**, *494*, 201–210. [CrossRef]

38. Cao, Y.; Song, S.G.; Niu, Y.L.; Jung, H.; Jin, Z.M. Variation of mineral composition, fabric and oxygen fugacity from massive to foliated eclogites during exhumation of subducted ocean crust in the North Qilian suture zone, NW China. *J. Metamorph. Geol.* **2011**, *29*, 699–720. [CrossRef]

39. Fountain, D.M.; Boundy, T.M.; Austrheim, H.; Rey, P. Eclogite-facies shear zones—Deep-crustal reflectors. *Tectonophysics* **1994**, *232*, 411–424. [CrossRef]

40. Fujimoto, Y.; Kono, Y.; Hirajima, T.; Kanagawa, K.; Ishikawa, M.; Arima, M. P-wave velocity and anisotropy of lawsonite and epidote blueschists: Constraints on water transportation along subducting oceanic crust. *Phys. Earth Planet. Inter.* **2010**, *183*, 219–228. [CrossRef]

41. Kern, H.; Gao, S.; Jin, Z.M.; Popp, T.; Jin, S.Y. Petrophysical studies on rocks from the Dabie ultrahigh-pressure (UHP) metamorphic belt, Central China: Implications for the composition and delamination of the lower crust. *Tectonophysics* **1999**, *301*, 191–215. [CrossRef]

42. Kern, H.; Jin, Z.M.; Gao, S.; Popp, T.; Xu, Z.Q. Physical properties of ultrahigh-pressure metamorphic rocks from the Sulu terrain, eastern central China: Implications for the seismic structure at the Donghai (CCSD) drilling site. *Tectonophysics* **2002**, *354*, 315–330. [CrossRef]

43. Kim, D.; Katayama, I.; Michibayashi, K.; Tsujimori, T. Rheological contrast between glaucophane and lawsonite in naturally deformed blueschist from Diablo Range, California. *Island Arc* **2013**, *22*, 63–73. [CrossRef]

44. Llana-Funez, S.; Brown, D. Contribution of crystallographic preferred orientation to seismic anisotropy across a surface analog of the continental Moho at Cabo Ortegal, Spain. *Geol. Soc. Am. Bull.* **2012**, *124*, 1495–1513. [CrossRef]

45. Mookherjee, M.; Bezacier, L. The low velocity layer in subduction zone: Structure and elasticity of glaucophane at high pressures. *Phys. Earth Planet. Inter.* **2012**, *208*, 50–58. [CrossRef]

46. Rudnick, R.L.; Fountain, D.M. Nature and composition of the continental-crust—A lower crustal perspective. *Rev. Geophys.* **1995**, *33*, 267–309. [CrossRef]

47. Sun, S.; Ji, S.; Wang, Q.; Xu, Z.; Salisbury, M.; Long, C. Seismic velocities and anisotropy of core samples from the Chinese Continental Scientific Drilling borehole in the Sulu UHP terrane, eastern China. *J. Geophys. Res. Solid Earth* **2012**, *117*. [CrossRef]

48. Wang, Q.; Burlini, L.; Mainprice, D.; Xu, Z. Geochemistry, petrofabrics and seismic properties of eclogites from the Chinese Continental Scientific Drilling boreholes in the Sulu UHP terrane, eastern China. *Tectonophysics* **2009**, *475*, 251–266. [CrossRef]

49. Zhang, J.; Wang, Y.; Jin, Z. CPO-induced seismic anisotropy in UHP eclogites. *Sci. China Ser. D Earth Sci.* **2008**, *51*, 11–21. [CrossRef]

50. Cooke, D.A.; Schneider, W.A. Generalized linear inversion of reflection seismic data. *Geophysics* **1983**, *48*, 665–676. [CrossRef]

51. Xu, D.; Wang, Y.; Gan, Q.; Tang, J. Frequency-dependent seismic reflection coefficient for discriminating gas reservoirs. *J. Geophys. Eng.* **2011**, *8*, 508–513. [CrossRef]

52. Laubscher, H. The problem of the moho in the Alps. *Tectonophysics* **1990**, *182*, 9–20. [CrossRef]

53. Warner, M.; Morgan, J.; Barton, P.; Morgan, P.; Price, C.; Jones, K. Seismic reflections from the mantle represent relict subduction zones within the continental lithosphere. *Geology* **1996**, *24*, 39–42. [CrossRef]

54. Bass, J.D. Elasticity of grossular and spessartite garnets by Brillouin spectroscopy. *J. Geophys. Res. Solid Earth Planets* **1989**, *94*, 7621–7628. [CrossRef]

55. Díaz Aspiroz, M.; Lloyd, G.E.; Fernández, C. Development of lattice preferred orientation in clinoamphiboles deformed under low-pressure metamorphic conditions. A SEM/EBSD study of metabasites from the Aracena metamorphic belt (SW Spain). *J. Struct. Geol.* **2007**, *29*, 629–645. [CrossRef]

56. Getsinger, A.J.; Hirth, G. Amphibole fabric formation during diffusion creep and the rheology of shear zones. *Geology* **2014**, *42*, 535–538. [CrossRef]

57. Puelles, P.; Gil Ibarguchi, J.I.; Beranoaguirre, A.; Ábalos, B. Mantle wedge deformation recorded by high-temperature peridotite fabric superposition and hydrous retrogression (Limo massif, Cabo Ortegal, NW Spain). *Int. J. Earth Sci.* **2012**, *101*, 1835–1853. [CrossRef]

58. Tatham, D.; Lloyd, G.; Butler, R.; Casey, M. Amphibole and lower crustal seismic properties. *Earth Planet. Sci. Lett.* **2008**, *267*, 118–128. [CrossRef]

59. Puelles, P.; Beranoaguirre, A.; Ábalos, B.; Gil Ibarguchi, J.I.; García de Madinabeitia, S.; Rodríguez, J.; Fernández-Armas, S. Eclogite inclusions from subducted metaigneous continental crust (Malpica-Tui Allochthonous Complex, NW Spain): Petrofabric, geochronology, and calculated seismic properties. *Tectonics* **2017**, *36*, 1376–1406. [CrossRef]

60. Rehman, H.U.; Mainprice, D.; Barou, F.; Yamamoto, H.; Okamoto, K. EBSD-measured crystal preferred orientation of eclogites from the Sanbagawa metamorphic belt, central Shikoku, SW Japan. *Eur. J. Mineral.* **2016**, *28*, 1155–1168. [CrossRef]

61. Zhang, J.F.; Green, H.W.; Bozhilov, K.N. Rheology of omphacite at high temperature and pressure and significance of its lattice preferred orientations. *Earth Planet. Sci. Lett.* **2006**, *246*, 432–443. [CrossRef]

62. Chen, N.S.; Xia, X.P.; Li, X.Y.; Sun, M.; Xu, P.; Liu, X.M.; Wang, X.Y.; Wang, Q.Y. Timing of magmatism of the gneissic-granite plutons along North Qaidam margin and implications for Precambrian crustal accretions: Zircon U-Pb dating and Hf isotope evidences. *Acta Petrol. Sin.* **2007**, *23*, 501–512.

63. Fu, J.; Liang, X.; Zhou, Y.; Wang, C.; Jiang, Y.; Zhong, Y. Geochemistry, zircon U-Pb geochronology and Hf isotopes of granitic rocks in the Xitieshan area, North Qaidam, Northwest China: Implications for Neoproterozoic geodynamic evolutions of North Qaidam. *Precambrian Res.* **2015**, *264*, 11–29. [CrossRef]

64. Zhao, Z.X.; Wei, J.H.; Fu, L.B.; Liang, S.N.; Zhao, S.Q. The Early Paleozoic Xitieshan syn-collisional granite in the North Qaidam ultrahigh-pressure metamorphic belt, NW China: Petrogenesis and implications for continental crust growth. *Lithos* **2017**, *278*, 140–152. [CrossRef]

65. Zhang, J.; Yang, J.; Meng, F.; Wan, Y.; Li, H.; Wu, C. U–Pb isotopic studies of eclogites and their host gneisses in the Xitieshan area of the North Qaidam mountains, western China: New evidence for an early Paleozoic HP–UHP metamorphic belt. *J. Asian Earth Sci.* **2006**, *28*, 143–150.

66. Zhang, C.; Zhang, L.; Bader, T.; Song, S.; Lou, Y. Geochemistry and trace element behaviors of eclogite during its exhumation in the Xitieshan terrane, North Qaidam UHP belt, NW China. *J. Asian Earth Sci.* **2013**, *63*, 81–97. [CrossRef]

67. Zhang, C.; Zhang, L.; van Roermund, H.; Song, S.; Zhang, G. Petrology and SHRIMP U-Pb dating of Xitieshan eclogite, North Qaidam UHP metamorphic belt, NW China. *J. Asian Earth Sci.* **2011**, *42*, 752–767. [CrossRef]

68. Liu, X.; Wu, Y.; Gao, S.; Liu, Q.; Wang, H.; Qin, Z.; Li, Q.; Li, X.-H.; Gong, H. First record and timing of UHP metamorphism from zircon in the Xitieshan terrane: Implications for the evolution of the entire North Qaidam metamorphic belt. *Am. Mineral.* **2012**, *97*, 1083–1093. [CrossRef]

69. Song, S.; Zhang, C.; Li, X.; Zhang, L. HP/UHP metamorphic time of eclogite in the Xitieshan terrane, North Qaidam UHPM belt, NW China. *Acta Petrol. Sin.* **2011**, *27*, 1191–1197.

70. Zhang, C.; Bader, T.; Zhang, L.F.; van Roermund, H. The multi-stage tectonic evolution of the Xitieshan terrane, North Qaidam orogen, western China: From Grenville-age orogeny to early-Paleozoic ultrahigh-pressure metamorphism. *Gondwana Res.* **2017**, *41*, 290–300. [CrossRef]

71. Zhang, J.X.; Yang, J.S.; Mattinson, C.G.; Xu, Z.Q.; Meng, F.C.; Shi, R.D. Two contrasting eclogite cooling histories, North Qaidam HP/UHP terrane, western China: Petrological and isotopic constraints. *Lithos* **2005**, *84*, 51–76. [CrossRef]

72. Panozzo, R. Two-dimensional strain from the orientation of lines in a plane. *J. Struct. Geol.* **1984**, *6*, 215–221. [CrossRef]

73. Park, M.; Jung, H. Analysis of electron backscattered diffraction (EBSD) mapping of geological materials: Precautions for reliably collecting and interpreting data on petro-fabric and seismic anisotropy. *Geosci. J.* **2020**, *24*, 679–687. [CrossRef]

74. Skemer, P.; Katayama, B.; Jiang, Z.T.; Karato, S. The misorientation index: Development of a new method for calculating the strength of lattice-preferred orientation. *Tectonophysics* **2005**, *411*, 157–167. [CrossRef]

75. Krogh Ravna, E.J.; Terry, M.P. Geothermobarometry of UHP and HP eclogites and schists–an evaluation of equilibria among garnet–clinopyroxene–kyanite–phengite–coesite/quartz. *J. Metamorph. Geol.* **2004**, *22*, 579–592. [CrossRef]

76. Ravna, K. The garnet–clinopyroxene Fe^{2+}–Mg geothermometer: An updated calibration. *J. Metamorph. Geol.* **2000**, *18*, 211–219. [CrossRef]

77. Aleksandrov, K.S.; Ryzhova, T.V. The elastic properties of rock forming minerals, pyroxenes and amphiboles. *Bull. Acad. Sci. USSR Geophys. Ser.* **1961**, *871*, 1339–1344.

78. Isaak, D.G.; Ohno, I.; Lee, P.C. The elastic constants of monoclinic single-crystal chrome-diopside to 1,300 K. *Phys. Chem. Miner.* **2006**, *32*, 691–699. [CrossRef]

79. Aleksandrov, K.S. Velocities of elastic waves in minerals at atmospheric pressure and increasing precision of elastic constants by means of EVM. *Izv. Acad. Sci. USSR Geol. Ser.* **1974**, *10*, 15–24.

80. Mainprice, D. A Frotran program to calculate seismic anisotropy from the lattice preferred orientation of minerals. *Comput. Geosci.* **1990**, *16*, 385–393. [CrossRef]

81. McSkimin, H.J.; Andreatch, P., Jr.; Thurston, R.N.L. Elastic moduli of quartz versus hydrostatic pressure at 25 and -195.8 C. *J. Appl. Phys.* **1965**, *36*, 1624–1632. [CrossRef]

82. Weidner, D.J.; Ito, E. Elasticity of $MgSiO_3$ in the ilmenite phase. *Phys. Earth Planet. Inter.* **1985**, *40*, 65–70. [CrossRef]

83. Vaughan, M.T.; Guggenheim, S. Elasticity of muscovite and its relationship to crystal structure. *J. Geophys. Res. Solid Earth* **1986**, *91*, 4657–4664. [CrossRef]

84. Kang, H.; Jung, H. Lattice-preferred orientation of amphibole, chlorite, and olivine found in hydrated mantle peridotites from Bjørkedalen, southwestern Norway, and implications for seismic anisotropy. *Tectonophysics* **2019**, *750*, 137–152. [CrossRef]

85. Kim, D.; Jung, H. Deformation microstructures of olivine and chlorite in chlorite peridotites from Almklovdalen in the Western Gneiss Region, SW Norway and implications for seismic anisotropy. *Int. Geol. Rev.* **2015**. [CrossRef]

86. Kim, D.; Katayama, I.; Michibayashi, K.; Tsujimori, T. Deformation fabrics of natural blueschists and implications for seismic anisotropy in subducting oceanic crust. *Phys. Earth Planet. Inter.* **2013**, *222*, 8–21. [CrossRef]

87. Skemer, P.; Katayama, I.; Karato, S.I. Deformation fabrics of the Cima di Gagnone peridotite massif, Central Alps, Switzerland: Evidence of deformation at low temperatures in the presence of water. *Contrib. Mineral. Petrol.* **2006**, *152*, 43–51. [CrossRef]

88. Wang, Y.F.; Zhang, J.F.; Shi, F. The origin and geophysical implications of a weak C-type olivine fabric in the Xugou ultrahigh pressure garnet peridotite. *Earth Planet. Sci. Lett.* **2013**, *376*, 63–73. [CrossRef]

89. Xu, Z.Q.; Wang, Q.; Ji, S.C.; Chen, J.; Zeng, L.S.; Yang, J.S.; Chen, F.Y.; Liang, F.H.; Wenk, H.R. Petrofabrics and seismic properties of garnet peridotite from the UHP Sulu terrane (China): Implications for olivine deformation mechanism in a cold and dry subducting continental slab. *Tectonophysics* **2006**, *421*, 111–127. [CrossRef]

90. Tommasi, A.; Godard, M.; Coromina, G.; Dautria, J.-M.; Barsczus, H. Seismic anisotropy and compositionally induced velocity anomalies in the lithosphere above mantle plumes: A petrological and microstructural study of mantle xenoliths from French Polynesia. *Earth Planet. Sci. Lett.* **2004**, *227*, 539–556. [CrossRef]

91. Tommasi, A.; Vauchez, A.; Godard, M.; Belley, F. Deformation and melt transport in a highly depleted peridotite massif from the Canadian Cordillera: Implications to seismic anisotropy above subduction zones. *Earth Planet. Sci. Lett.* **2006**, *252*, 245–259. [CrossRef]

92. Tommasi, A.; Vauchez, A.; Ionov, D.A. Deformation, static recrystallization, and reactive melt transport in shallow subcontinental mantle xenoliths (Tok Cenozoic volcanic field, SE Siberia). *Earth Planet. Sci. Lett.* **2008**, *272*, 65–77. [CrossRef]

93. Michibayashi, K.; Kusafuka, Y.; Satsukawa, T.; Nasir, S.J. Seismic properties of peridotite xenoliths as a clue to imaging the lithospheric mantle beneath NE Tasmania, Australia. *Tectonophysics* **2012**, *522*, 218–223. [CrossRef]

94. Jung, S.; Jung, H.; Austrheim, H. Characterization of olivine fabrics and mylonite in the presence of fluid and implications for seismic anisotropy and shear localization. *Earth Planets Space* **2014**, *66*. [CrossRef]

95. Jung, H. Seismic anisotropy produced by serpentine in mantle wedge. *Earth Planet. Sci. Lett.* **2011**, *307*, 535–543. [CrossRef]

96. Katayama, I.; Hirauchi, H.; Michibayashi, K.; Ando, J. Trench-parallel anisotropy produced by serpentine deformation in the hydrated mantle wedge. *Nature* **2009**, *461*, 1114–1117. [CrossRef]

97. Watanabe, T.; Kasami, H.; Ohshima, S. Compressional and shear wave velocities of serpentinized peridotites up to 200 MPa. *Earth, Planets Space* **2007**, *59*, 233–244. [CrossRef]

98. Cao, Y.; Jung, H.; Song, S.G. Petro-fabrics and seismic properties of blueschist and eclogite in the North Qilian suture zone, NW China: Implications for the low-velocity upper layer in subducting slab, trench-parallel seismic anisotropy, and eclogite detectability in the subduction zone. *J. Geophys. Res. Solid Earth* **2013**, *118*, 3037–3058. [CrossRef]

99. Cao, Y.; Jung, H.; Song, S. Seismic anisotropies of the Songshugou peridotites (Qinling orogen, central China) and their seismic implications. *Tectonophysics* **2018**, *722*, 432–446. [CrossRef]

100. Ha, Y.; Jung, H.; Raymond, L.A. Deformation fabrics of glaucophane schists and implications for seismic anisotropy: The importance of lattice preferred orientation of phengite. *Int. Geol. Rev.* **2018**, 1–18. [CrossRef]

101. Locock, A.J. An Excel spreadsheet to classify chemical analyses of amphiboles following the IMA 2012 recommendations. *Comput. Geosci.* **2014**, *62*, 1–11. [CrossRef]

102. Hawthorne, F.C.; Oberti, R.; Harlow, G.E.; Maresch, W.V.; Martin, R.F.; Schumacher, J.C.; Welch, M.D. Nomenclature of the amphibole supergroup. *Am. Mineral.* **2012**, *97*, 2031–2048. [CrossRef]

103. Neufeld, K.; Ring, U.; Heidelbach, F.; Dietrich, S.; Neuser, R. Omphacite textures in eclogites of the Tauern Window: Implications for the exhumation of the Eclogite Zone, Eastern Alps. *J. Struct. Geol.* **2008**, *30*, 976–992. [CrossRef]

104. Ábalos, B. Omphacite fabric variation in the Cabo Ortegal eclogite (NW Spain): Relationships with strain symmetry during high-pressure deformation. *J. Struct. Geol.* **1997**, *19*, 621–637. [CrossRef]

105. Boudier, F.; Baronnet, A.; Mainprice, D. Serpentine mineral replacements of natural olivine and their seismic implications: Oceanic lizardite versus subduction-related antigorite. *J. Petrol.* **2010**, *51*, 495–512. [CrossRef]

106. Cao, S.; Liu, J.; Leiss, B. Orientation-related deformation mechanisms of naturally deformed amphibole in amphibolite mylonites from the Diancang Shan, SW Yunnan, China. *J. Struct. Geol.* **2010**, *32*, 606–622. [CrossRef]

107. Cao, Y.; Jung, H.; Ma, J. Seismic Properties of a Unique Olivine-Rich Eclogite in the Western Gneiss Region, Norway. *Minerals* **2020**, *10*, 774. [CrossRef]

Article

Twin Induced Reduction of Seismic Anisotropy in Lawsonite Blueschist

Seungsoon Choi [1], Olivier Fabbri [2], Gültekin Topuz [3], Aral I. Okay [3] and Haemyeong Jung [1,*]

[1] Tectonophysics Laboratory, School of Earth and Environmental Sciences, Seoul National University, Seoul 08826, Korea; seungshum@snu.ac.kr
[2] UMR CNRS 6249, Université de Franche-Comté, 25030 Besançon, France; olivier.fabbri@univ-fcomte.fr
[3] Avrasya Yer Bilimleri Enstitüsü, İstanbul Teknik Üniversitesi, 34469 İstanbul, Turkey; topuzg@itu.edu.tr (G.T.); okay@itu.edu.tr (A.I.O.)
* Correspondence: hjung@snu.ac.kr; Tel.: +82-2-880-6733

Abstract: Lawsonite is an important mineral for understanding seismic anisotropy in subducting oceanic crust due to its large elastic anisotropy and prevalence in cold subduction zones. However, there is insufficient knowledge of how lawsonite twinning affects seismic anisotropy, despite previous studies demonstrating the presence of twins in lawsonite. This study investigated the effect of lawsonite twinning on the crystal preferred orientation (CPO), CPO strength, and seismic anisotropy using lawsonite blueschists from Alpine Corsica (France) and the Sivrihisar Massif (Turkey). The CPOs of the minerals are measured with an electron backscatter diffraction instrument attached to a scanning electron microscope. The electron backscatter diffraction analyses of lawsonite reveal that the {110} twin in lawsonite is developed, the [001] axes are strongly aligned subnormal to the foliation, and both the [100] and [010] axes are aligned subparallel to the foliation. It is concluded that the existence of twins in lawsonite could induce substantial seismic anisotropy reduction, particularly for the maximum S-wave anisotropy in lawsonite and whole rocks by up to 3.67% and 1.46%, respectively. Lawsonite twinning needs to be considered when determining seismic anisotropy in the subducting oceanic crust in cold subduction zones.

Keywords: lawsonite; twin; blueschist; crystal preferred orientation; seismic anisotropy

Citation: Choi, S.; Fabbri, O.; Topuz, G.; Okay, A.I.; Jung, H. Twin Induced Reduction of Seismic Anisotropy in Lawsonite Blueschist. *Minerals* **2021**, *11*, 399. https://doi.org/10.3390/min11040399

Academic Editor: Paris Xypolias

Received: 10 February 2021
Accepted: 7 April 2021
Published: 10 April 2021

Publisher's Note: MDPI stays neutral with regard to jurisdictional claims in published maps and institutional affiliations.

1. Introduction

Seismic anisotropy, which is useful for studying tectonic fabric in the Earth, has been widely observed in subduction zones [1–4], including subducting oceanic slabs [5,6]. It can be caused by the crystal preferred orientation (CPO) of elastically anisotropic minerals [7,8]. Since lawsonite is an elastically anisotropic mineral in blueschist and eclogite facies rocks at the top of the subducting slab in cold subduction zones, the development of the CPO of lawsonite could cause substantial seismic anisotropy. The CPO of lawsonite has been used to interpret various seismic anisotropies observed in cold subduction zones from naturally and experimentally deformed lawsonite-bearing rocks such as blueschist and eclogite [9–14].

Several studies performed in the last decade have shown the existence of twins in lawsonite from the Sivrihisar Massif (Turkey) [10,15,16], South Motagua (Guatemala) [16,17], Southern New England Orogen (Australia) [18], Alpine Corsica (France) [19], and experimentally deformed lawsonites [14]. However, there is an insufficient understanding of how lawsonite twinning affects seismic anisotropy. In this study, the effect of lawsonite twinning on the CPO, CPO strength, and seismic anisotropy of lawsonite was investigated utilizing natural lawsonite blueschists collected from Alpine Corsica (France) and the Sivrihisar Massif in the Tavşanlı Zone (Turkey). The main focus of this study is the direct effect of twins in lawsonite on the seismic anisotropy. We compare the CPO and seismic anisotropy of lawsonite in four natural samples with those of the modelled samples where

177

the twins are not considered. The results show that the existence of lawsonite twins could induce reductions in the CPO strength of lawsonite and whole rock, as well as the resultant seismic anisotropy, which is caused by the CPO of elastically anisotropic minerals in cold subduction zones.

2. Geological Overview

Alpine Corsica (France) is part of the Alpine orogenic belt [20,21] and mainly consists of Schistes Lustrés that were deposited on the seafloor of a branch of the Tethys Ocean and underwent high-pressure, low-temperature (HP-LT) metamorphism in a subduction zone [22–25]. The estimated metamorphic pressure-temperature conditions for the blueschist and eclogite facies conditions are P = 1.0 and 2.4 GPa, and T = 420 and 520 °C, respectively [16,22–24,26]. The blueschist and eclogite metamorphisms are dated at ~37 and ~34 Ma utilizing the Lu-Hf in lawsonite and garnet, respectively [16,22,27]. A portion of the Schistes Lustrés in Alpine Corsica, Monte Pinatelle (study area) also experienced HP-LT metamorphism and primarily consists of lawsonite blueschist and eclogite [22,23,28].

The Tavşanlı Zone in Turkey is an Alpine high-pressure belt and was formed during the Mesozoic convergence between Eurasia and the Anatolide-Tauride Block [29–31]. The timing of HP-LT metamorphism is estimated by Rb-Sr phengite and Lu-Hf garnet as ca. 80 Ma and 86–92 Ma, respectively [32,33]. As a part of the Tavşanlı Zone, the Sivrihisar Massif mainly consists of high-pressure metamorphic rocks [34], and the Halilbağı region (study area) in the Sivrihisar Massif includes several types of lawsonite blueschist and eclogite. The estimated peak pressure and temperature conditions are 1.2–2.5 GPa and 350–650 °C [10,35].

In this study, lawsonite blueschist samples were collected from Alpine Corsica in Monte Pinatelle, France, and the Halilbağı area in the Tavşanlı Zone, Turkey (Figure 1a,b). Both regions are considered to be structurally coherent terrains [16] and experience progressive lawsonite blueschist metamorphism and rapid exhumation [22,35,36]. The Monte Pinatelle outcrop shows a well-foliated blueschist body (Figure 1c) in contact with weakly foliated metagabbro. The Halilbağı lawsonite blueschists and eclogites exhibit a steeply dipping foliation and a mineral compositional layering (Figure 1d).

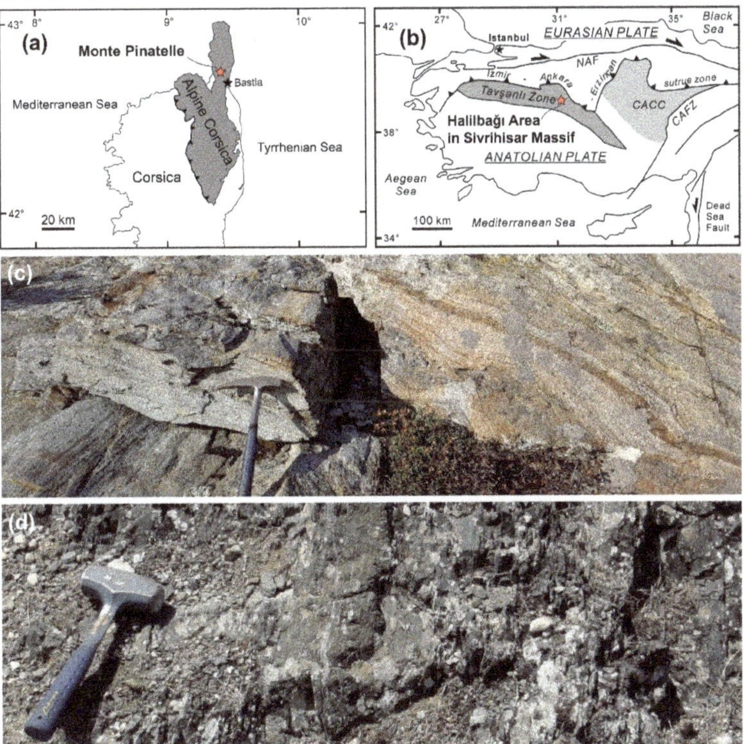

Figure 1. (**a**) Simplified tectonic map of Alpine Corsica (France) with sample locations indicated by red star (samples 3034 and 3033), modified from Danišík et al. [37]. (**b**) Simplified tectonic map of western and central Turkey with sample locations indicated by red star (samples 2023 and 2021), modified from Cao and Jung [10] and Davis and Whitney [34]. CACC: Central Anatolian crystalline complex; CAFZ: Central Anatolian fault zone; and NAF: North Anatolian fault. (**c**) Outcrop of Monte Pinatelle in Alpine Corsica. The blueschist shows a well-developed foliation. (**d**) Outcrop of the Halilbağı area in the Sivrihisar Massif, in the Tavşanlı Zone. The blueschist exhibits compositional layering and foliation. Hammer provided for scale in subfigures (**c**,**d**).

3. Methods

Foliation is determined by parallel alignment of glaucophane, lawsonite, and white mica, and compositional layering of minerals. The lineation of the samples is defined by the shape preferred orientations of the lawsonite and glaucophane [38], and thin sections were made in the x–z plane (x: parallel to the lineation; z: normal to the foliation). The thin sections were polished with diamond paste and colloidal silica (0.06 μm) and coated with carbon to avoid charging in the scanning electron microscope. Electron backscatter diffraction (EBSD) analysis, utilized to measure the CPOs of lawsonite and glaucophane, was performed in the x–z plane of the samples utilizing Aztec software (Version 4.3, Oxford Instruments, Abingdon, UK) with a symmetry detector attached to a field emission scanning electron microscope (JSM 7100F, JEOL, Tokyo, Japan) at the School of Earth and Environmental Sciences at Seoul National University, Korea. The system was operated at a 20 kV acceleration voltage and a 25 mm working distance. The lawsonite and glaucophane EBSD patterns were automatically indexed by mapping the sample with step sizes ranging from ~1.2–3.0 μm, which were at least 10 times smaller than the average grain size of the samples. To prevent misinterpretation of the original data, the raw EBSD data was post-cleaned utilizing Aztec software (Oxford Instruments) in three steps: (1) wild spikes with a pixel size were removed, (2) zero solution pixels that neighbor six pixels with solutions were eliminated, and (3) Step 1 was repeated.

The procedure to remove twins in lawsonite for plotting the CPOs is described as follows: Twin boundaries having a [001] rotation axis with a 67° misorientation angle and a 5° deviation were automatically disregarded as grain boundaries in EBSD orientation maps using Channel 5 software (Version 5.12.74.0). Then, the crystallographic orientation representing each grain was determined by the mean orientation of the lawsonite grains. This procedure was reiterated consistently for all grains in all samples. The lawsonite CPO with and without twins and glaucophane CPO were plotted in a pole figure as one point per grain to avoid introducing a bias by the large grains [39]. The misorientation index [40] was calculated to determine the CPO strength of the minerals.

The seismic velocity and anisotropy of the P- and S-waves of lawsonite were calculated utilizing the CPO, crystal density, and elastic constants of lawsonite [41], with a modified crystallographic reference [9]. They were calculated for glaucophane [42] utilizing a FORTRAN software program [43]. The seismic velocity and anisotropy of lawsonite and whole rocks were calculated for samples with and without twins. To calculate the seismic velocity and anisotropy in the absence of twins in lawsonite, narrow twin areas in lawsonite grains were removed manually in EBSD orientation maps using HKL Channel 5 Tango software. The choice of narrow twin removal in lawsonite was consistent for all samples. All point-per-grain crystallographic orientation data of lawsonite and glaucophane were used for accurate seismic properties with and without twins [39]. The seismic properties of the whole rocks were calculated utilizing the normalized volume fraction of the major minerals as lawsonite and glaucophane (Table 1). The normalized volume proportions of lawsonite and glaucophane were based on the mineral fractions from the large area EBSD mapping. Minor minerals such as garnet, omphacite, titanite, quartz, phengite, and epidote were ignored in the calculations.

Table 1. Modal composition, twin frequency, and CPO strength of samples.

Sample	Location	Major Minerals [1]		Grain Size (μm) [2]	Twin Frequency [3]	Fraction of Twinned Area (%) [3]	CPO Strength [4]	
		Lws (%)	Gln (%)				M_1	M_2
3034	Corsica	37	63	104	High	31.4	0.15	0.17
3033	Corsica	45	55	29	Middle	15.5	0.23	0.24
2023	Sivrihisar	43	57	106	Low	12.9	0.12	0.09
2021	Sivrihisar	53	47	38	Very low	0.6	0.078	0.080

[1] Normalized volume proportions of major minerals (lawsonite and glaucophane) are based on the mineral fractions by large area EBSD mapping. [2] Grain size is determined as the average length of the maximum Feret diameters [44] of the lawsonite grains. [3] Twin frequency is determined by the number of twins per length in lawsonite. The fraction of the twinned area (area of the twinned domain) is determined by large-area EBSD mapping. [4] M1 and M2 represent the misorientation index of lawsonite with and without twins, respectively. Lws: lawsonite; Gln: glaucophane; CPO: crystal preferred orientation.

4. Results

4.1. Microstructures

Four representative lawsonite blueschist samples collected from Alpine Corsica at Monte Pinatelle and the Halilbağı area in the Tavşanlı Zone were studied. The major minerals in the blueschists were lawsonite and glaucophane, and the minor minerals were garnet, omphacite, titanite, quartz, phengite, and epidote. Blueschists from Alpine Corsica showed higher phengite and omphacite content than those from the Halilbağı area. All samples showed high volume proportions of lawsonite (Table 1) in the 37–53% range.

The lawsonite crystals were primarily euhedral or subhedral, and the glaucophane crystals exhibited needle shapes (Figures 2 and 3, respectively). The samples were characterized by well-developed non-folded foliation and mineral stretching lineation (Figure 2a–d,g,h), except for sample 2023 (Figure 2e,f). Sample 3034 from Monte Pinatelle contained coarse-grained lawsonite, whereas sample 3033 contained fine-grained lawsonite (Table 1). Sample 2023 from the Halilbağı area contained coarse-grained lawsonite, whereas sample 2021 contained fine-grained lawsonite. Samples 3034 and 3033 showed high and medium frequencies of twins in the lawsonite, respectively (Figure 3a,b), and samples 2023 and 2021 exhibited low and very low frequencies of twins, respectively (Figure 3c,d).

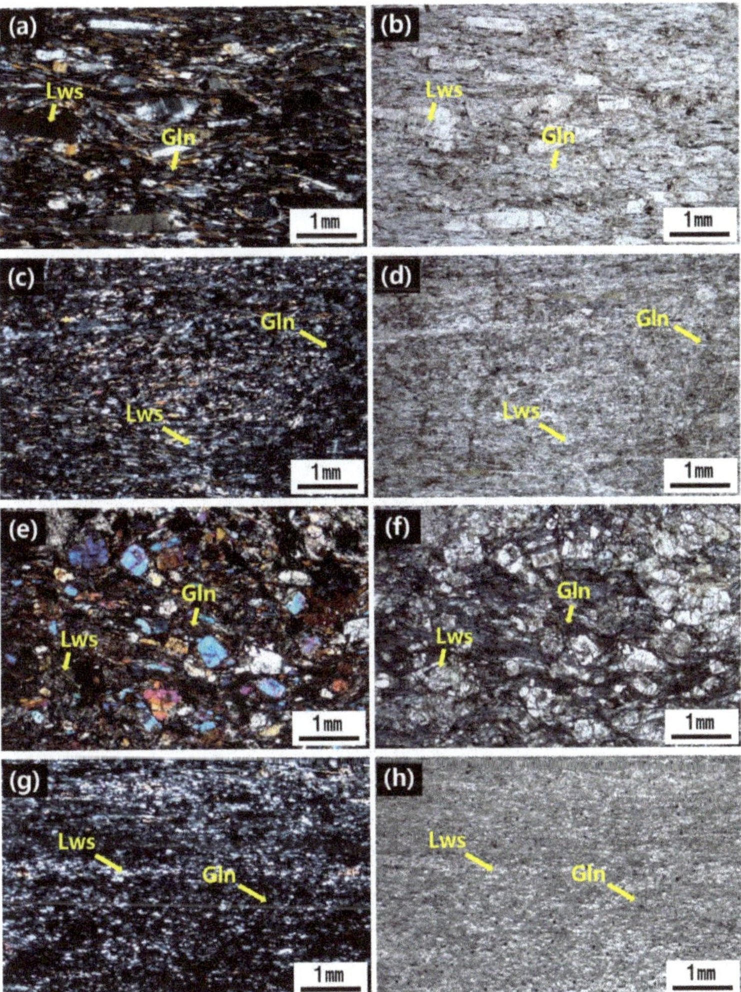

Figure 2. Photomicrographs of lawsonite blueschist samples from (**a**–**d**) Alpine Corsica and (**e**–**h**) the Sivrihisar Massif (**a**,**c**,**e**,**g**) are in cross-polarized light, and (**b**,**d**,**f**,**h**) are in plane-polarized light). Photomicrographs are from thin sections prepared in the x–z plane. (**a**,**b**) Sample 3034, characterized by coarse-grained lawsonite. (**c**,**d**) Sample 3033, characterized by fine-grained lawsonite. (**e**,**f**) Sample 2023, characterized by coarse-grained lawsonite. (**g**,**h**) Sample 2021, characterized by fine-grained lawsonite. According to the consistent arrangement of needle-shaped glaucophanes parallel to the lineation and short direction of lawsonite normal to the foliation, samples are considered to have been deformed in a non-coaxial mode. Lws: lawsonite; Gln: glaucophane.

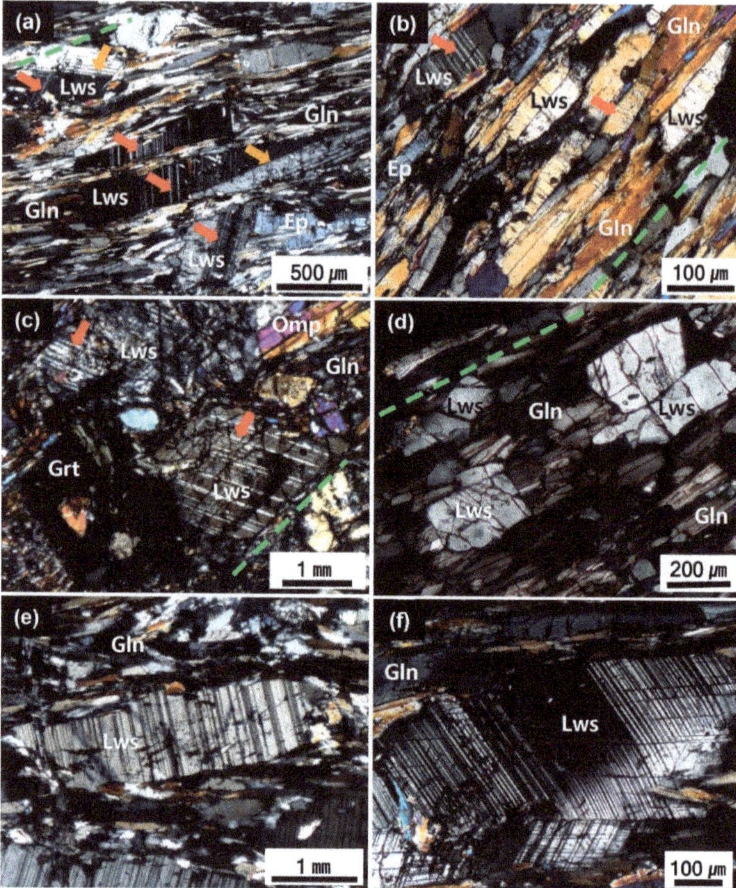

Figure 3. Enlarged photomicrographs of lawsonite blueschist samples. All photos are from thin sections in the x–z plane in cross-polarized light. (**a**) High lawsonite twin frequency (sample 3034); (**b**) medium lawsonite twin frequency (sample 3033); (**c**) low lawsonite twin frequency (sample 2023); (**d**) very low lawsonite twin frequency and some fracturing in lawsonite (sample 2021); (**e**) representative polysynthetic twin in lawsonite (sample 3034); and (**f**) representative deformation twin with curved textures in lawsonite (sample 3034). Some crossed twins can be observed as the second generation of twins, which are horizontal in (**f**). The green dashed lines indicate the foliations of the samples, and red and orange arrows denote the polysynthetic and deformation twins, respectively. Lws: lawsonite; Gln: glaucophane; Grt: garnet; Omp: omphacite; Ep: epidote.

Examples of polysynthetic and deformation twins are shown in Figure 3e,f. In lawsonite, polysynthetic twinning showed repeated contacts on {110} planes, whereas deformation twinning exhibited curved textures and crossing behaviors on {110} planes. Polysynthetic twins in lawsonites were also shown in the EBSD phase and grain maps in Figure 4. The {110} twin was detected in lawsonite, where twin boundaries were shown as red lines in Figure 4b and indicated by white arrows in the enlarged EBSD grain map in Figure 4c. The EBSD phase map (Figure 4b) showed {110} twin planes with a 67° rotation of lawsonite, indicating that the red boundaries are not lattice bending or subgrains but twin boundaries in lawsonite.

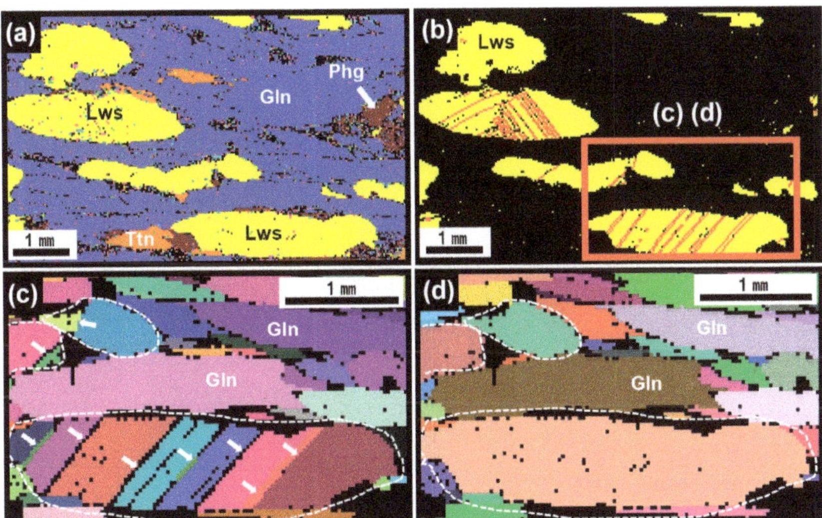

Figure 4. (**a**) Electron backscatter diffraction (EBSD) phase and grain maps of sample 3034. (**b**) EBSD phase map of sample 3034 showing the twin boundaries (red lines) in the lawsonite. Only lawsonite grains are shown (yellow color). Two sets of polysynthetic twins, which are almost perpendicular to each other, are shown in lawsonite. (**c**) EBSD grain map magnified from the EBSD phase map in Figure 4b. White arrows indicate the twin boundaries in lawsonite. (**d**) EBSD grain map without twins magnified from the EBSD phase map in Figure 4b. The twin boundaries in the lawsonite were eliminated using Aztec Software. The white dashed lines denote the true grain boundary of lawsonite. Lws: lawsonite; Gln: glaucophane; Phg: phengite; Ttn: titanite.

4.2. CPOs of Lawsonite and Glaucophane

4.2.1. CPOs and CPO Strength of Lawsonite

The CPOs of lawsonite with and without twins in the samples are presented in Figure 5. The deformed samples with and without twins showed that the maxima of the [001] axes of lawsonite are aligned subnormal to the foliation, and the [100] and [010] axes are aligned subparallel to the foliation (Figure 5a–c,e–g), except for sample 2021 (Figure 5d,h). Sample 2021 showed a weak lawsonite CPO, where the [010] axes of lawsonite were strongly aligned subparallel to the lineation, and the [100] axes were aligned as a weak girdle subnormal to the lineation. There were some differences in the lawsonite CPO strength depending on the fraction (size) of the twinned area (the area of the twinned domain) in the sample (Table 1 and Figure 5). The CPO strength of lawsonite in the samples with twins (twin fraction > 15%) was relatively low (M = 0.15–0.23) compared to that in the samples without twins (M = 0.17–0.24) (Figure 5). Sample 2021, which was characterized by a very low twin fraction (0.6%), showed the almost same CPO strength of lawsonite with twins (M = 0.078) and lawsonite without twins (M = 0.080).

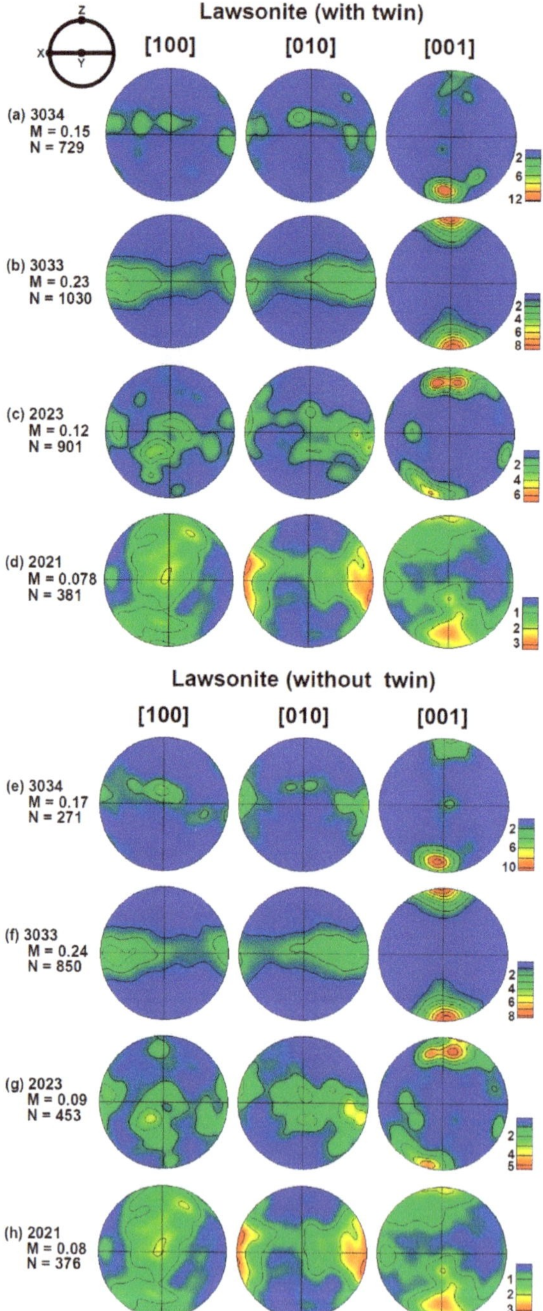

Figure 5. (**a–h**) Pole figures of lawsonite depending on the existence of twins are presented with a lower hemisphere equal-area projection. The x and z directions correspond to the lineation and direction normal to the foliation, respectively. A 20° half-scatter width and a 5° cluster size are used. Contours represent the CPO strength, and multiples of uniform distribution (m.u.d.) are shown as color bars. The higher the number of the bar, the higher is the CPO strength. M: misorientation index; N: number of grains.

4.2.2. CPO of Glaucophane

The glaucophane CPOs are presented in Figure 6. The maxima of the glaucophane [001] axes were aligned subparallel to the lineation, and the maxima of the (110) poles and [100] axes were aligned subnormal to the foliation (Figure 6a,b,d). The (010) poles showed a high concentration subnormal to the lineation near the center of the pole figure, except for sample 2023, which showed weak girdle distributions of the (110) and (010) poles subnormal to the lineation (Figure 6c).

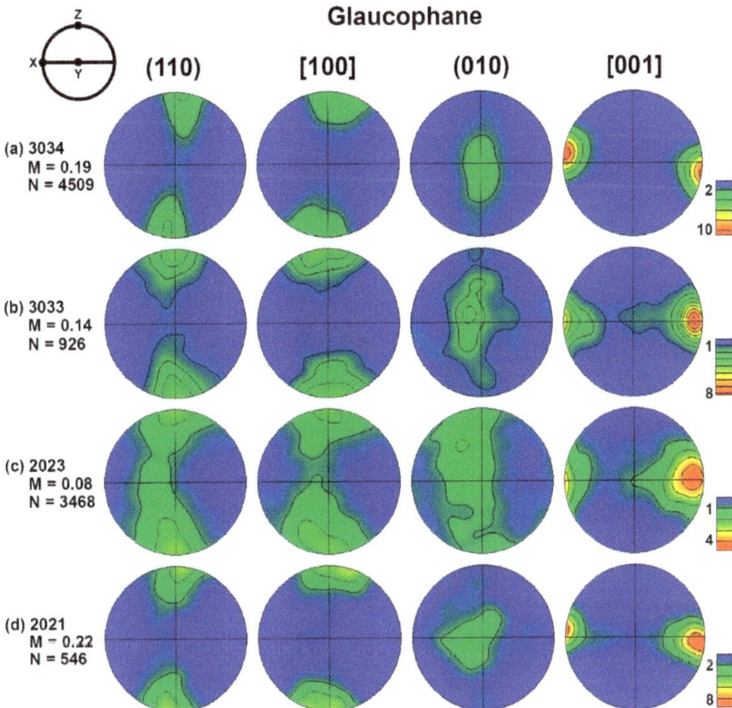

Figure 6. (a–d) Pole figures of glaucophane. The x and z directions correspond to the lineation and direction normal to the foliation, respectively. A 20° half-scatter width and a 5° cluster size are used. Contours represent the CPO strength, and multiples of uniform distribution (m.u.d.) are shown as color bars. The higher the number of the bar, the higher is the CPO strength. M: misorientation index; N: number of grains.

4.3. Seismic Velocity and Anisotropy of Minerals and Whole Rocks

The seismic velocities and anisotropies of lawsonite aggregates with and without twins are presented in Figure 7. For the lawsonite aggregate with twins, the P-wave seismic anisotropy (AVp) ranged from 5.1% to 12.6%, and the maximum S-wave seismic anisotropy (max AVs) ranged from 10.15% to 20.87% (Figure 7a–d). For the lawsonite aggregate without twins, AVp and max AVs ranged from 5.1% to 13.0% and from 10.12% to 22.64%, respectively (Figure 7e–h). Therefore, AVp and max AVs are lower in lawsonite aggregates with twins, except for sample 2021, which shows a very low frequency of lawsonite twinning (Figure 7d,h). The direction of the P-wave velocity is subnormal to the foliation regardless of the existence of twins.

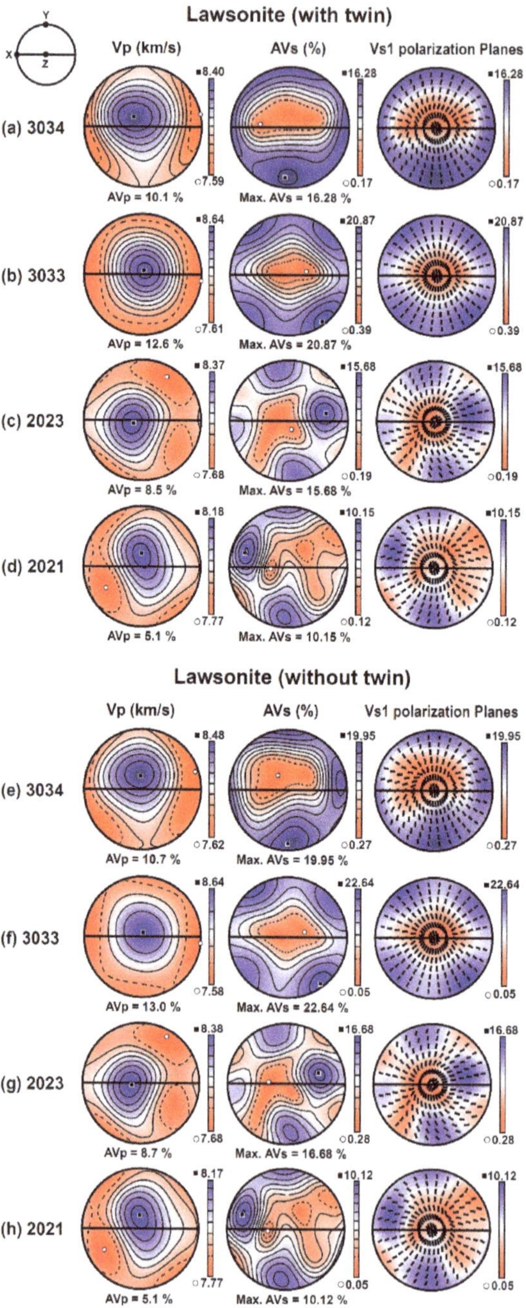

Figure 7. (a–h) Seismic properties of lawsonite with and without twins. The poles are presented with a lower hemisphere equal-area projection. The x and z directions correspond to the lineation and direction normal to the foliation, respectively. Vp: P-wave velocity; AVp: anisotropy of P-wave velocity; AVs: anisotropy of S-wave velocity; max AVs: maximum anisotropy of S-wave velocity; Vs1: fast S-wave velocity.

The seismic anisotropy of the glaucophane aggregates is presented in Figure 8. The AVp and max AVs of glaucophane range from 13.0% to 25.1% and from 6.33% to 14.48%, respectively, and the direction of the P-wave velocity is aligned subparallel to the lineation (Figure 8a–d).

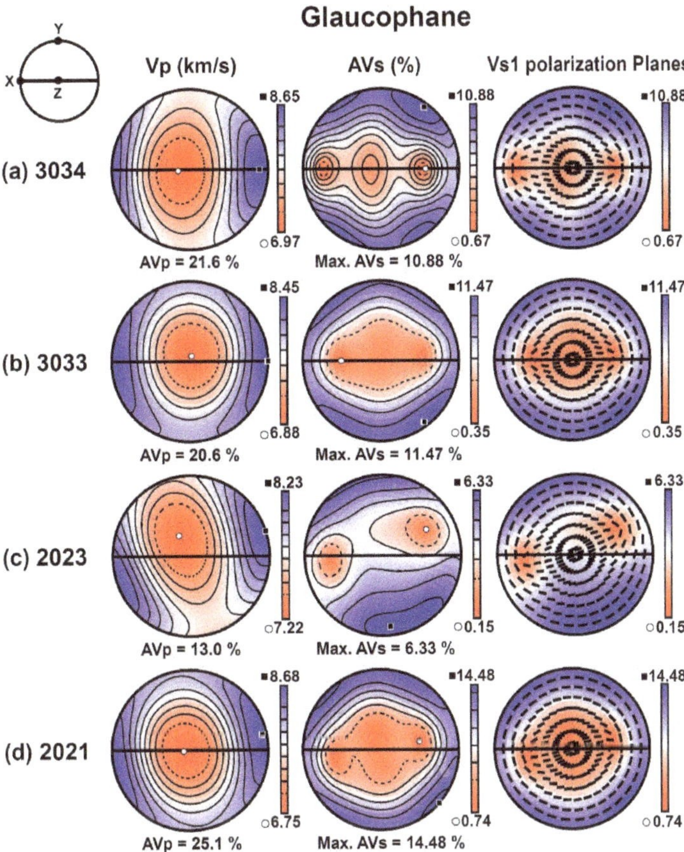

Figure 8. (a–d) Seismic properties of glaucophane. The poles are presented with a lower hemisphere equal-area projection. The x and z directions correspond to the lineation and direction normal to the foliation, respectively. Vp: P-wave velocity; AVp: anisotropy of P-wave velocity; AVs: anisotropy of S-wave velocity; max AVs: maximum anisotropy of S-wave velocity; Vs1: fast S-wave velocity.

The seismic anisotropies of the whole rocks are presented in Figure 9. For whole rocks with lawsonite twins, AVp and max AVs vary from 4.3% to 9.5% and from 4.54% to 7.63%, respectively (Figure 9a–d). For whole rocks without lawsonite twins, AVp and max AVs ranged from 4.3% to 9.5% and from 6.00% to 8.13%, respectively (Figure 9e–h). Generally, AVp and max AVs increased in the samples without lawsonite twins, and the direction of the P-wave velocity was similar to that in glaucophane regardless of the existence of twins.

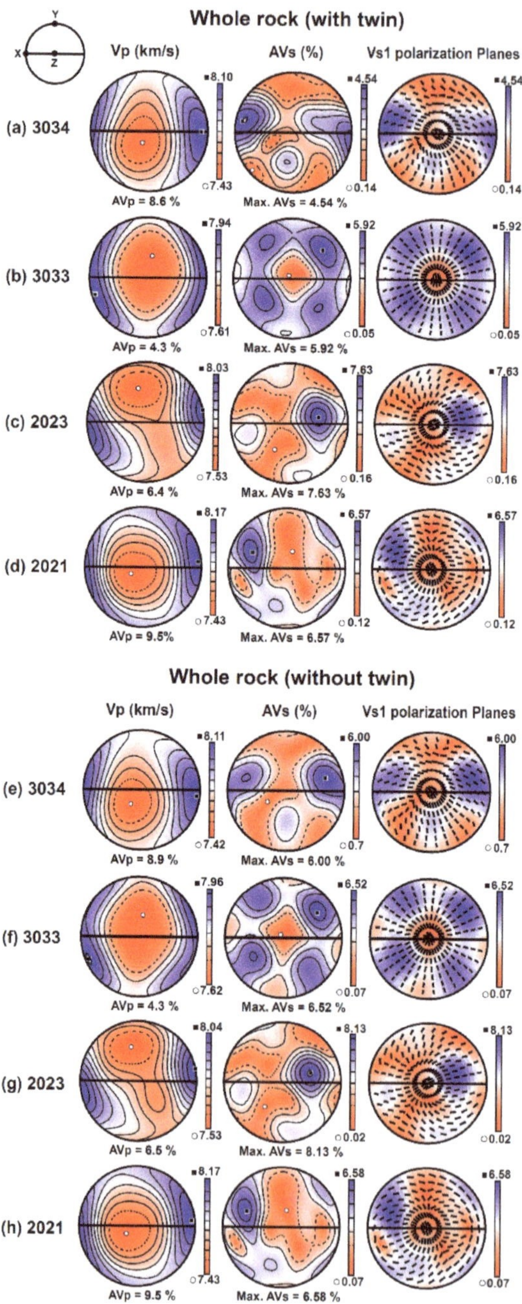

Figure 9. (**a–h**) Seismic properties of whole rocks with and without twins. The poles are presented with a lower hemisphere equal-area projection. The x and z directions correspond to the lineation and direction normal to the foliation, respectively. Vp: P-wave velocity; AVp: anisotropy of P-wave velocity; AVs: anisotropy of S-wave velocity; max AVs: maximum anisotropy of S-wave velocity; Vs1: fast S-wave velocity.

5. Discussion

5.1. CPO of Lawsonite and Effect of Twins on the CPO Strength of Lawsonite

All the CPOs of lawsonite in deformed blueschists from Alpine Corsica and the Sivrihisar Massif are characterized by the maxima of the [001] axes aligned subnormal to the foliation regardless of twin existence and by [100] and [010] axes aligned subparallel to the foliation (Figure 5). These features are consistent with those of previous studies on blueschists and eclogites from the North Qilian suture zone in China [9], the Sivrihisar Massif in Turkey [10], the Diablo Range in California, USA [11], the southern Motagua fault zone in Guatemala [12], and the Kurosegawa Belt in Japan [45]. However, this study reveals that the CPO strength of lawsonite is reduced when considering the existence of twins in lawsonite. The CPO strength of lawsonite with twins is lower (M = 0.078–0.23) than that of lawsonite without twins (M = 0.080–0.24), except in sample 2023 (Figure 5). Sample 2023 shows a higher CPO strength of lawsonite with twins than without twins (M = 0.12 and 0.09, respectively). This may be related to the scattered distribution of the [001] axes that are partly aligned subparallel to the lineation.

Figure 10 illustrates the influence of lawsonite twins on the CPO of lawsonite in a single grain. The {110} twins in the grain are easily identified in the inverse pole figure orientation map (Figure 10a), and the misorientation angles of the twin boundaries are 67° (Figure 10b). This angle is large enough to diffuse the [100] and [010] axes of lawsonite, but not the [001] axis, which is the rotation axis of the {110} twin (Figure 10c). The existence of lawsonite twinning thus produces a weak CPO on the [100] and [010] axes (Figure 10c,d), but a strong alignment of the [001] axes (Figure 5a,b,e,f). Therefore, lawsonite twinning could be one of the mechanisms that weakens the CPO strength of lawsonite. However, we do not know the exact origin of twins in lawsonite. There is a lack of information on the pressure, temperature, and differential stress conditions in which lawsonite twins occur.

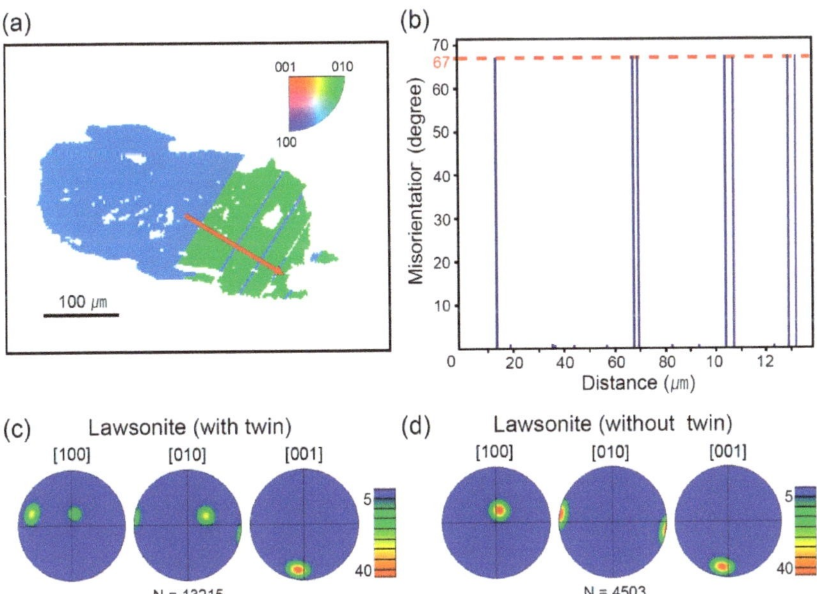

Figure 10. (a) Inverse pole figure orientation map parallel to the lineation direction of a single lawsonite grain; (b) misorientation line profile of the red line in Figure 10a; pole figures of the single lawsonite grain in Figure 10a (c) with twins and (d) without twins. Pole figures were produced using all points per grain. A 20° half-scatter width and a 5° cluster size were utilized. N: number of data points.

5.2. Effect of Lawsonite Twinning on Seismic Anisotropy in Subducting Oceanic Crust

The seismic anisotropy of lawsonite reduces the AVp and AVs by up to 0.60% and 3.67%, respectively, when lawsonite twinning is taken into account (Table 2). Notably, S-wave anisotropy is significantly greater than P-wave anisotropy, possibly due to the seismic anisotropy of the lawsonite single crystal [41] which shows a much larger max AVs value (65%) than the AVp value (24%) [9]. The effect of lawsonite twinning on the seismic anisotropy of whole-rock blueschist is also presented in Table 2. The seismic anisotropy of the whole rock also reduces the AVp and AVs by up to 0.30% and 1.46%, respectively, when lawsonite twinning is considered. Lawsonite twinning has only a minor effect on the P-wave anisotropy, but a relatively larger effect on the S-wave anisotropy of lawsonite and whole rock. The seismic anisotropy reduces with increasing area of twinned domain in lawsonite (Tables 1 and 2). For example, the difference of the max AVs of lawsonite with and without twins is very small for sample 2021, which has a very small area of twinned domain; however, the difference of max AVs of lawsonite is 3.67% for sample 3034, which has a large area of twinned domain.

Table 2. Seismic anisotropy of lawsonite and whole-rock.

Sample	Twin	Lawsonite		Whole Rock	
		AVp (%)	**Max AVs (%)**	**AVp (%)**	**Max AVs (%)**
3034	with twin	10.10	16.28	8.60	4.54
	without twin	10.70	19.95	8.90	6.00
		-0.60% [1]	-3.67%	-0.30%	-1.46%
3033	with twin	12.60	20.87	4.30	5.92
	without twin	13.00	22.64	4.30	6.52
		-0.40%	-1.77%	0%	-0.60%
2023	with twin	8.50	15.68	6.40	7.63
	without twin	8.70	16.68	6.50	8.13
		-0.20%	-1.00%	-0.10%	-0.50%
2021	with twin	5.10	10.15	9.50	6.57
	without twin	5.10	10.12	9.50	6.58
		0%	0.30%	0%	-0.01%

[1] Change of seismic anisotropy of lawsonite and whole rock with twin compared to that of lawsonite and whole rock without twin. Anisotropy of P-wave (AVp) = $200 \times (Vp_{max} - Vp_{min})/(Vp_{max} + Vp_{min})$, and anisotropy of S-wave (AVs) = $200 \times (Vs1 - Vs2)/(Vs1 + Vs2)$. Max AVs: maximum anisotropy of S-wave velocity.

In Figure 10, we manually removed the blue part (narrow area) of the twin in the lawsonite grain. This is because most of the twins in lawsonite in the studied samples are observed as narrow areas (for example, see Figure 3). The choice of twin removal may have an influence on the resulting CPO and seismic anisotropy of lawsonite. According to our study, the seismic anisotropy reduces with increasing area of the twinned domain in lawsonite. If wide areas of twins are removed in lawsonite, the area of the twinned domain in the sample would become larger, resulting in a large effect of twins in lawsonite on the reduction of seismic anisotropy. Thus, the choice of narrow twin removal in this study has produced the minimum effect of twins on the reduction of seismic anisotropy.

Based on these findings, the existence of twins in lawsonite can induce the reduction of seismic anisotropy in blueschist, particularly S-wave anisotropy, which is attributed to the ability of the {110} twin to weaken the lawsonite CPO strength (i.e., weak alignment of the [100] and [010] axes) (Figure 5). The CPO strength, along with other factors, including rock type and interactions between the CPOs of different minerals, can lead to a change in the seismic anisotropy in lawsonite blueschist [10]. This study clearly shows that lawsonite twinning can be an important factor in determining the seismic anisotropy of lawsonite-bearing rocks if the area of the twinned domain in lawsonite is large.

Lawsonite twinning has also been reported in natural lawsonite eclogites [10,18]. As eclogite mainly consists of anhydrous minerals such as garnet and omphacite, it has a weaker seismic anisotropy than blueschist [46,47]. Therefore, the influence of lawsonite on the seismic anisotropy of eclogite is significant [12,13]. Lawsonite twinning can also produce seismic anisotropy reduction in lawsonite eclogite, particularly S-wave anisotropy. As lawsonite is prevalent in the subducting oceanic crust of cold subduction zones, lawsonite twinning can be important in determining seismic anisotropy.

6. Conclusions

The effect of lawsonite twinning on the CPO and seismic anisotropy of lawsonite and lawsonite blueschist was studied using lawsonite blueschists collected from Alpine Corsica and the Sivrihisar Massif. The results indicate that polysynthetic and deformation twins of lawsonite exist on the {110} plane with the rotation of the [001] symmetry axis by 67°. The CPO of lawsonite with twins is weakened due to the scattered distribution of the [100] and [010] axes. Lawsonite twinning induces a substantial reduction in the maximum S-wave anisotropy (max AVs) by up to 3.67% and 1.46% for lawsonite and whole rock, respectively, and a minor reduction of the P-wave anisotropy (AVp) by up to 0.6% and 0.3% for lawsonite and whole rock, respectively. Considering the distribution of lawsonite in blueschist and eclogite in cold subduction zones, lawsonite twinning needs to be taken into account in determining seismic anisotropy.

Author Contributions: Conceptualization, S.C and H.J.; methodology, S.C.; software, S.C.; validation, S.C., O.F., G.T., A.I.O., and H.J.; formal analysis, S.C.; investigation, S.C.; resources, O.F., G.T., A.I.O., and H.J.; data curation, S.C.; writing—original draft preparation, S.C.; writing—review and editing, O.F., G.T., A.I.O., and H.J.; visualization, S.C.; supervision, H.J.; project administration, H.J.; funding acquisition, H.J. All authors have read and agreed to the published version of the manuscript.

Funding: This research was funded by the Mid-Career Research Program through the National Research Foundation of Korea (NRF: 2020R1A2C2003765) and the Korea Meteorological Administration Research Development Program (KMI2019-00110) to H.J. G.T. and A.I.O. gratefully acknowledge financial support from Turkish Academy of Sciences for field work.

Data Availability Statement: Not applicable.

Conflicts of Interest: The authors declare no conflict of interest.

References

1. Savage, M. Seismic anisotropy and mantle deformation: What have we learned from shear wave splitting? *Rev. Geophys.* **1999**, *37*, 65–106. [CrossRef]
2. Long, M.D.; Silver, P.G. The subduction zone flow field from seismic anisotropy: A global view. *Science* **2008**, *319*, 315–318. [CrossRef]
3. Long, M.D. Constraints on Subduction Geodynamics from Seismic Anisotropy. *Rev. Geophys.* **2013**, *51*, 76–112. [CrossRef]
4. Zhao, D.; Yu, S.; Liu, X. Seismic anisotropy tomography: New insight into subduction dynamics. *Gondwana Res.* **2016**, *33*, 24–43. [CrossRef]
5. Healy, D.; Reddy, S.M.; Timms, N.E.; Gray, E.M.; Brovarone, A.V. Trench-parallel fast axes of seismic anisotropy due to fluid-filled cracks in subducting slabs. *Earth Planet. Sci. Lett.* **2009**, *283*, 75–86. [CrossRef]
6. Wang, J.; Zhao, D. Mapping P-wave anisotropy of the Honshu arc from Japan Trench to the back-arc. *J. Asian Earth Sci.* **2010**, *39*, 396–407. [CrossRef]
7. Jung, H. Crystal preferred orientations of olivine, orthopyroxene, serpentine, chlorite, and amphibole, and implications for seismic anisotropy in subduction zones: A review. *Geosci. J.* **2017**, *21*, 985–1011. [CrossRef]
8. Almqvist, B.S.; Mainprice, D. Seismic properties and anisotropy of the continental crust: Predictions based on mineral texture and rock microstructure. *Rev. Geophys.* **2017**, *55*, 367–433. [CrossRef]
9. Cao, Y.; Jung, H.; Song, S. Microstructures and petro-fabrics of lawsonite blueschist in the North Qilian suture zone, NW China: Implications for seismic anisotropy of subducting oceanic crust. *Tectonophysics* **2014**, *628*, 140–157. [CrossRef]
10. Cao, Y.; Jung, H. Seismic properties of subducting oceanic crust: Constraints from natural lawsonite-bearing blueschist and eclogite in Sivrihisar Massif, Turkey. *Phys. Earth Planet. Inter.* **2016**, *250*, 12–30. [CrossRef]
11. Kim, D.; Katayama, I.; Michibayashi, K.; Tsujimori, T. Deformation fabrics of natural blueschists and implications for seismic anisotropy in subducting oceanic crust. *Phys. Earth Planet. Inter.* **2013**, *222*, 8–21. [CrossRef]

12. Kim, D.; Wallis, S.; Endo, S.; Ree, J.-H. Seismic properties of lawsonite eclogites from the southern Motagua fault zone, Guatemala. *Tectonophysics* **2016**, *677–678*, 88–98. [CrossRef]

13. Park, M.; Jung, H. Relationships Between Eclogite-Facies Mineral Assemblages, Deformation Microstructures, and Seismic Properties in the Yuka Terrane, North Qaidam Ultrahigh-Pressure Metamorphic Belt, NW China. *J. Geophys. Res. Solid Earth* **2019**, *124*, 13168–13191. [CrossRef]

14. Iizuka-Oku, R.; Soustelle, V.; Miyajima, N.; Walte, N.P.; Frost, D.J.; Yagi, T. Experimentally deformed lawsonite at high pressure and high temperature: Implication for low velocity layers in subduction zones. *Phys. Earth Planet. Inter.* **2019**, *295*, 106282. [CrossRef]

15. Fornash, K.F.; Whitney, D.L.; Seaton, N.C.A. Lawsonite composition and zoning as an archive of metamorphic processes in subduction zones. *Geosphere* **2018**, *15*, 24–46. [CrossRef]

16. Whitney, D.L.; Fornash, K.F.; Kang, P.; Ghent, E.D.; Martin, L.; Okay, A.I.; Vitale Brovarone, A. Lawsonite composition and zoning as tracers of subduction processes: A global review. *Lithos* **2020**, *370–371*, 105636. [CrossRef]

17. Tsujimori, T.; Ernst, W.G. Lawsonite blueschists and lawsonite eclogites as proxies for palaeo-subduction zone processes: A review. *J. Metamorph. Geol.* **2014**, *32*, 437–454. [CrossRef]

18. Tamblyn, R.; Hand, M.; Morrissey, L.; Zack, T.; Phillips, G.; Och, D. Resubduction of lawsonite eclogite within a serpentinite-filled subduction channel. *Contrib. Mineral. Petrol.* **2020**, *175*. [CrossRef]

19. Vitale-Brovarone, A.; Groppo, C.; HetÉNyi, G.; Compagnoni, R.; Malavieille, J. Coexistence of lawsonite-bearing eclogite and blueschist: Phase equilibria modelling of Alpine Corsica metabasalts and petrological evolution of subducting slabs. *J. Metamorph. Geol.* **2011**, *29*, 583–600. [CrossRef]

20. Fabbri, O.; Magott, R.; Fournier, M.; Etienne, L. Pseudotachylyte in the Monte Maggiore ophiolitic unit (Alpine Corsica): a possible lateral extension of the Cima di Gratera intermediate-depth Wadati-Benioff paleo-seismic zone. *Earth Sci. Bull.* **2018**, *189*, 18. [CrossRef]

21. Vitale-Brovarone, A.; Beltrando, M.; Malavieille, J.; Giuntoli, F.; Tondella, E.; Groppo, C.; Beyssac, O.; Compagnoni, R. Inherited Ocean–Continent Transition zones in deeply subducted terranes: Insights from Alpine Corsica. *Lithos* **2011**, *124*, 273–290. [CrossRef]

22. Vitale-Brovarone, A.; Picatto, M.; Beyssac, O.; Lagabrielle, Y.; Castelli, D. The blueschist–eclogite transition in the Alpine chain: P–T paths and the role of slow-spreading extensional structures in the evolution of HP–LT mountain belts. *Tectonophysics* **2014**, *615–616*, 96–121. [CrossRef]

23. Ravna, E.J.K.; Andersen, T.B.; Jolivet, L.; de Capitani, C. Cold subduction and the formation of lawsonite eclogite — Constraints from prograde evolution of eclogitized pillow lava from Corsica. *J. Metamorph. Geol.* **2010**, *28*, 381–395. [CrossRef]

24. Caron, J.-M.; Péquignot, G. The transition between blueschists and lawsonite-bearing eclogites based on observations from Corsican metabasalts. *Lithos* **1986**, *19*, 205–218. [CrossRef]

25. Brunet, C.; Monié, P.; Jolivet, L.; Cadet, J.-P. Migration of compression and extension in the Tyrrhenian Sea, insights from 40Ar/39Ar ages on micas along a transect from Corsica to Tuscany. *Tectonophysics* **2000**, *321*, 127–155. [CrossRef]

26. Vitale-Brovarone, A.; Beyssac, O.; Malavieille, J.; Molli, G.; Beltrando, M.; Compagnoni, R. Stacking and metamorphism of continuous segments of subducted lithosphere in a high-pressure wedge: The example of Alpine Corsica (France). *Earth-Sci. Rev.* **2013**, *116*, 35–56. [CrossRef]

27. Vitale-Brovarone, A.; Herwartz, D. Timing of HP metamorphism in the Schistes Lustrés of Alpine Corsica: New Lu–Hf garnet and lawsonite ages. *Lithos* **2013**, *172*, 175–191. [CrossRef]

28. Austrheim, H.; Andersen, T.B. Pseudotachylytes from Corsica: Fossil earthquakes from a subduction complex. *Terra Nova* **2004**, *16*, 193–197. [CrossRef]

29. Plunder, A.; Agard, P.; Chopin, C.; Okay, A.I. Geodynamics of the Tavşanlı zone, western Turkey: Insights into subduction/obduction processes. *Tectonophysics* **2013**, *608*, 884–903. [CrossRef]

30. Topuz, G.; Okay, A.; Altherr, R.; Meyer, H.P.; Nasdala, L. Partial high-pressure aragonitization of micritic limestones in an accretionary complex, Tavşanlı Zone, NW Turkey. *J. Metamorph. Geol.* **2006**, *24*, 603–613. [CrossRef]

31. Okay, A.I.; Harris, N.B.; Kelley, S.P. Exhumation of blueschists along a Tethyan suture in northwest Turkey. *Tectonophysics* **1998**, *285*, 275–299. [CrossRef]

32. Pourteau, A.; Scherer, E.E.; Schorn, S.; Bast, R.; Schmidt, A.; Ebert, L. Thermal evolution of an ancient subduction interface revealed by Lu–Hf garnet geochronology, Halilbağı Complex (Anatolia). *Geosci. Front.* **2019**, *10*, 127–148. [CrossRef]

33. Sherlock, S.; Kelley, S.; Inger, S.; Harris, N.; Okay, A. 40 Ar-39 Ar and Rb-Sr geochronology of high-pressure metamorphism and exhumation history of the Tavsanli Zone, NW Turkey. *Contrib. Mineral. Petrol.* **1999**, *137*, 46–58.

34. Davis, P.B.; Whitney, D.L. Petrogenesis and structural petrology of high-pressure metabasalt pods, Sivrihisar, Turkey. *Contrib. Mineral. Petrol.* **2008**, *156*, 217–241. [CrossRef]

35. Davis, P.B.; Whitney, D.L. Petrogenesis of lawsonite and epidote eclogite and blueschist, Sivrihisar Massif, Turkey. *J. Metamorph. Geol.* **2006**, *24*, 823–849. [CrossRef]

36. Malavieille, J.; Chemenda, A.; Larroque, C. Evolutionary model for Alpine Corsica: Mechanism for ophiolite emplacement and exhumation of high-pressure rocks. *Terra Nova-Oxf.* **1998**, *10*, 317–322. [CrossRef]

37. Danišík, M.; Kuhlemann, J.; Dunkl, I.; Evans, N.J.; Székely, B.; Frisch, W. Survival of ancient landforms in a collisional setting as revealed by combined fission track and (U-Th)/He thermochronometry: A case study from Corsica (France). *J. Geol.* **2012**, *120*, 155–173. [CrossRef]
38. Panozzo, R. Two-dimensional strain from the orientation of lines in a plane. *J. Struct. Geol.* **1984**, *6*, 215–221. [CrossRef]
39. Park, M.; Jung, H. Analysis of electron backscattered diffraction (EBSD) mapping of geological materials: Precautions for reliably collecting and interpreting data on petro-fabric and seismic anisotropy. *Geosci. J.* **2020**, *24*, 679–687. [CrossRef]
40. Skemer, P.; Katayama, I.; Jiang, Z.; Karato, S.-i. The misorientation index: Development of a new method for calculating the strength of lattice-preferred orientation. *Tectonophysics* **2005**, *411*, 157–167. [CrossRef]
41. Sinogeikin, S.V.; Schilling, F.R.; Bass, J.D. Single crystal elasticity of lawsonite. *Am. Mineral.* **2000**, *85*, 1834–1837. [CrossRef]
42. Bezacier, L.; Reynard, B.; Bass, J.D.; Wang, J.; Mainprice, D. Elasticity of glaucophane, seismic velocities and anisotropy of the subducted oceanic crust. *Tectonophysics* **2010**, *494*, 201–210. [CrossRef]
43. Mainprice, D. A FORTRAN program to calculate seismic anisotropy from the lattice preferred orientation of minerals. *Comput. Geosci.* **1990**, *16*, 385–393. [CrossRef]
44. Walton, W. Feret's statistical diameter as a measure of particle size. *Nature* **1948**, *162*, 329–330. [CrossRef]
45. Fujimoto, Y.; Kono, Y.; Hirajima, T.; Kanagawa, K.; Ishikawa, M.; Arima, M. P-wave velocity and anisotropy of lawsonite and epidote blueschists: Constraints on water transportation along subducting oceanic crust. *Phys. Earth Planet. Inter.* **2010**, *183*, 219–228. [CrossRef]
46. Keppler, R.; Behrmann, J.H.; Stipp, M. Textures of eclogites and blueschists from Syros island, Greece: Inferences for elastic anisotropy of subducted oceanic crust. *J. Geophys. Res. Solid Earth* **2017**, *122*, 5306–5324. [CrossRef]
47. Abdullah, S.; Misra, S.; Sarvesha, R.; Ghosh, B. Resurfacing of deeply buried oceanic crust in Naga Hills Ophiolite, North-East India: Petrofabric, microstructure and seismic properties. *J. Struct. Geol.* **2020**, *139*. [CrossRef]

MDPI

St. Alban-Anlage 66

4052 Basel

Switzerland

Tel. +41 61 683 77 34

Fax +41 61 302 89 18

www.mdpi.com

Minerals Editorial Office

E-mail: minerals@mdpi.com

www.mdpi.com/journal/minerals